# Nanomaterials

## A Guide to Fabrication and Applications

T0225579

# Devices, Circuits, and Systems

Series Editor
*Krzysztof Iniewski*
Emerging Technologies CMOS Inc.
Vancouver, British Columbia, Canada

PUBLISHED TITLES:

# PUBLISHED TITLES:

## FORTHCOMING TITLES:

## FORTHCOMING TITLES:

**Power Management Integrated Circuits and Technologies**
*Mona M. Hella and Patrick Mercier*

**Radio Frequency Integrated Circuit Design**
*Sebastian Magierowski*

**Semiconductor Devices in Harsh Conditions**
*Kirsten Weide-Zaage and Malgorzata Chrzanowska-Jeske*

**Smart eHealth and eCare Technologies Handbook**
*Sari Merilampi, Lars T. Berger, and Andrew Sirkka*

**Structural Health Monitoring of Composite Structures Using Fiber Optic Methods**
*Ginu Rajan and Gangadhara Prusty*

**Tunable RF Components and Circuits: Applications in Mobile Handsets**
*Jeffrey L. Hilbert*

**Wireless Medical Systems and Algorithms: Design and Applications**
*Pietro Salvo and Miguel Hernandez-Silveira*

# Nanomaterials

## A Guide to Fabrication and Applications

Edited by
## Sivashankar Krishnamoorthy
Luxembourg Institute of Science and Technology (LIST),
Luxembourg

## Krzysztof Iniewski MANAGING EDITOR
Emerging Technologies CMOS Inc.
Vancouver, British Columbia, Canada

CRC Press
Taylor & Francis Group
Boca Raton London New York

CRC Press is an imprint of the
Taylor & Francis Group, an **informa** business

CRC Press
Taylor & Francis Group
6000 Broken Sound Parkway NW, Suite 300
Boca Raton, FL 33487-2742

First issued in paperback 2020

ISBN-13: 978-1-4665-9125-7 (hbk)
ISBN-13: 978-0-367-73751-1 (pbk)

**Visit the Taylor & Francis Web site at**
**http://www.taylorandfrancis.com**

**and the CRC Press Web site at**
**http://www.crcpress.com**

# Contents

# Foreword

Nanomaterials are being incorporated into products all around us and are having an incredible impact on durability, strength, functionality, and other material properties. There are a vast number of nanomaterials presently available, and new formulations and chemistries are being announced daily. The range of options and available materials creates a difficulty for the researcher who wishes to improve his product and doesn't understand the properties and potentialities of the various nanomaterials.

This book was compiled with a view to helping product developers and material scientists to understand the various types of nanomaterials and their applications.

In Chapter 1, Kushagra et al. provide an overview and discussion of the scale of nanomaterials and nanomachines, relating specifically to integrated circuits and microelectromechanical systems.

Nanomaterials can offer extremely high ratios of surface area to volume, and this has an impact on both the physics and the chemistry of reactions. Chapters 2 through 4 offer insight into the interactions of different nanomaterials with chemical reactions, with biological processes, and the effects on the environment.

The strength of nanotubes, 2D materials such as graphene, and nanocoatings is transforming material science. Chapters 5 and 6 relate to the mechanical properties of nanomaterials and the potential treatments that can be used to improve their performance.

Although electronic devices have followed Moore's law for decades, basic physical limitations may be bringing predictable performance improvements to an end. Chapters 7 and 8 describe some of the most recent accomplishments in the use of nanomaterials to create new forms of electronic devices. Some of these techniques may usher in a new era of high-speed computation or storage.

Chapters 9 and 10 discuss the optical properties of certain nanomaterials and their real-world applications in improving lasers and optical absorbers.

Finally, Chapter 11 describes an application in the area of energy storage and ways in which nanomaterials from waste products may be used to improve capacitors.

We hope that the readers find this book informative and exciting and that it serves to lead them to the discovery of new applications of nanomaterials.

**Gordon Harling**
*Glen Hart Inc.*

# Acknowledgments

We thank Jessica Vakili and Nora Konopka at Taylor & Francis Group for their perseverance and professionalism.

# Editors

**Sivashankar Krishnamoorthy** leads efforts in nano-enabled medicine and cosmetics domains within materials research and technology department at Luxembourg Institute of Science and Technology (www.list.lu). He has spent his entire career to date in technology development environments, successfully integrating fundamental, cross-disciplinary, and translational aspects of fabrication, processing and investigation of nanostructured materials and interfaces. He brings over 10 years of transnational experience to advantage in driving innovation at the interface of nanotechnology, biology, and medicine, with active engagement in several professional activities including member of organizing committee of international conferences, journal editorial boards, peer-reviewing of journal articles, and supervision of researchers at different levels. He can be reached at shivchem@gmail.com.

**Krzysztof (Kris) Iniewski** is managing R&D at Redlen Technologies, Inc., a start-up company in Vancouver, Canada. Redlen's revolutionary production process for advanced semiconductor materials has enabled a new generation of more accurate, all-digital, radiation-based imaging solutions. Kris is also a founder of Emerging Technologies CMOS Inc. (www.etcmos.com), and organization of high-tech events covering communications, microsystems, optoelectronics, and sensors. In his career, Dr. Iniewski held numerous faculty and management positions at University of Toronto, University of Alberta, SFU, and PMC-Sierra, Inc. He has published over 100 research papers in international journals and conferences. He holds 18 international patents, granted in the USA, Canada, France, Germany, and Japan. He is a frequent invited speaker and has consulted for multiple organizations internationally. He has written and edited several books for CRC Press, Cambridge University Press, IEEE Press, Wiley, McGraw-Hill, Artech House, and Springer. His personal goal is to contribute to healthy living and sustainability through innovative engineering solutions. In his leisurly time, Kris can be found hiking, sailing, skiing, or biking in beautiful British Columbia. He can be reached at kris.iniewski@gmail.com.

# Contributors

**Hernán L. Calvo**
Instituto de Física Enrique Gaviola–
   CONICET, FaMAF
Universidad Nacional de Córdoba
Córdoba, Argentina

**Sébastien Cambier**
Environmental Research and
   Innovation
Luxembourg Institute of Science and
   Technology
Belvaux, Luxembourg

**Chih-Hung Chang**
School of Chemical, Biological and
   Environmental Engineering
Oregon State University
Corvallis, Oregon

**Chang-Ho Choi**
School of Chemical, Biological and
   Environmental Engineering
Oregon State University
Corvallis, Oregon

**Patrick Choquet**
Luxembourg Institute of Science and
   Technology
Belvaux, Luxembourg

**Liam Damewood**
Department of Physics
University of California, Davis
Davis, California

**Marie Paule Delplancke**
Universite Libre de Bruxelles
Brussels, Belgium

**Yan Duan**
Department of Mechanical Engineering
Washington State University
Vancouver, Washington

**Philip X.-L. Feng**
Electrical Engineering & Computer
   Science
Case Western Reserve University
Cleveland, Ohio

**Teresa Fernandes**
School of Life Sciences
Heriot Watt University
Edinburgh, United Kingdom

**Luis E.F. Foa Torres**
Instituto de Física Enrique Gaviola–
   CONICET, FaMAF
Universidad Nacional de Córdoba
Córdoba, Argentina

**Ching-Yao Fong**
Department of Physics
University of California, Davis
Davis, California

**Anastasia Georgantzopoulou**
Environmental Research and
   Innovation
Luxembourg Institute of Science and
   Technology
Belvaux, Luxembourg

**Raju Kumar Gupta**
Department of Chemical Engineering
   and DST Unit on Nanosciences
Indian Institute of Technology
Kanpur, India

**Arno C. Gutleb**
Environmental Research and Innovation
Luxembourg Institute of Science and
    Technology
Belvaux, Luxembourg

**Seung-Yeol Han**
School of Chemical, Biological and
    Environmental Engineering
Oregon State University
Corvallis, Oregon

**Jason Juhala**
Department of Electrical Engineering
School of Engineering and Computer
    Science
Washington State University
Vancouver, Washington

**Anupama B. Kaul**
Department of Metallurgical and
    Materials Engineering and
    Biomedical Engineering
and
Department of Electrical and
    Computer Engineering
University of Texas at El Paso
El Paso, Texas

**Vishal Khetan**
School of Mechanical Engineering
University of Leeds
Leeds, United Kingdom

**Ki-Joong Kim**
School of Chemical, Biological and
    Environmental Engineering
Oregon State University
Corvallis, Oregon

**Thomas A.J. Kuhlbusch**
Air Quality & Sustainable
    Nanotechnology
Institute of Energy and Environmental
    Technology e.V. (IUTA)
Duisburg, Germany

**Arindam Kushagra**
Department of Electrical Engineering
IIT Bombay
Mumbai, India

**Andrew J. Lee**
Institute of Microwaves and Photonics
School of Electronic and Electrical
    Engineering
University of Leeds
Leeds, United Kingdom

**Jaesung Lee**
Electrical Engineering & Computer
    Science
Case Western Reserve University
Cleveland, Ohio

**Iseult Lynch**
School of Geography, Earth and
    Environmental Sciences
University of Birmingham
Birmingham, United Kingdom

**Ailbhe Macken**
Ecotoxicology and Risk Assessment
Norwegian Institute for Water Research
Oslo, Norway

**Kahina Mehennaoui**
Environmental Research and Innovation
Luxembourg Institute of Science and
    Technology
Belvaux, Luxembourg

**Ruth Moeller**
Environmental Research and Innovation
Luxembourg Institute of Science and
    Technology
Belvaux, Luxembourg

**Carmen Nickel**
Air Quality & Sustainable
    Nanotechnology
Institute of Energy and Environmental
    Technology e.V. (IUTA)
Duisburg, Germany

**Horacio M. Pastawski**
Instituto de Física Enrique Gaviola–
  CONICET, FaMAF
Universidad Nacional de Córdoba
Córdoba, Argentina

**W. Peijnenburg**
Institute of Environmental Sciences
Leiden University
Leiden
and
Center for Safety of Substances and
  Products
National Institute of Public Health and
  the Environment
Bilthoven, the Netherlands

**V. Ramgopal Rao**
Department of Electrical Engineering
IIT Bombay
Mumbai, India

**Stephan Roche**
CIN2 (ICN–CSIC), Catalan Institute of
  Nanotechnology
Universidad Autónoma de Barcelona
and
Theoretical and Computational
  Nanoscience Group
Institució Catalana de Recerca i Estudis
  Avançats
Barcelona, Spain

**Sampath Satti**
Department of Electrical Engineering
IIT Bombay
Mumbai, India

**Tomasso Serchi**
Environmental Research and Innovation
Luxembourg Institute of Science and
  Technology
Belvaux, Luxembourg

**Santiranjan Shannigrahi**
Institute of Materials Research and
  Engineering
Agency for Science Technology and
  Research
Singapore

**Mohit Sharma**
Institute of Materials Research and
  Engineering
Agency for Science Technology and
  Research
Singapore

**Michael Shaughnessy**
RTBiQ, Inc.
San Francisco, California

**Ankit Tyagi**
Department of Chemical
  Engineering
Indian Institute of Technology
Kanpur, India

**Nathalie Valle**
Luxembourg Institute of Science and
  Technology
Belvaux, Luxembourg

**Christoph Walti**
University of Leeds
Leeds, United Kingdom

**Wei Xue**
Department of Mechanical
  Engineering
Henry M. Rowan College of
  Engineering
Rowan University
Glassboro, New Jersey

# 1 Top Down Meets Bottom Up for Nanoscale CMOS and MEMS

*Arindam Kushagra, Sampath Satti, and V. Ramgopal Rao*

## CONTENTS

The drive toward smaller feature sizes in complementary metal-oxide semiconductor technology enables increased processing power and transistor density, while reducing the cost per transistor. Considerable effort has been devoted to the top-down creation of nanoscale devices, but feature sizes are ultimately limited by the capability of lithography and processing complexity. This is especially true as the dimensions of the channel are in the atomic scale. True integration of top-down and bottom-up processes provides an opportunity to leverage the paradigm of bottom-up fabrication which is complementary to positional alignment and system complexity afforded by top-down fabrication and assembly. This chapter is broadly divided into two separate sections discussing about the role of self-assembled monolayers (SAMs) in fabricating electronic devices and the role of nanowires grown on microelectromechanical system devices, especially zinc oxide (ZnO) nanowires on cantilevers, and using them for various sensing purposes.

## 1.1 SAMs OF PORPHYRINS

SAMs provide a convenient, flexible, and simple system with which one can tailor the interfacial properties of metals, metal oxides, and semiconductors. They are assemblies of organic surfactant molecules formed by the adsorption of molecular constituents from solution or the gas phase onto the surface of solids or liquids. The adsorbate molecules organize into crystalline or semicrystalline structures on the substrate. The molecules or ligands that form SAMs consist of three parts: a headgroup that possesses a high affinity for the substrate; a tail group whose chemical nature now affects interfacial surface properties of the material such as hydrophobicity, surface potential, and roughness; and a structural group between the head and tail groups. There are a number of headgroups that bind to specific metals, metal oxides, and semiconductors. The most extensively studied class of SAMs is derived from the adsorption of alkanethiols on gold, silver, copper, palladium, platinum, and mercury. The high affinity of thiols for the surfaces of noble and coinage metals makes it possible to generate well-defined organic surfaces with useful and highly alterable chemical functionalities displayed at the exposed interface.

Porphyrins are nitrogen-containing compounds derived from the tetrapyrrole porphin molecule. The basic structure of the porphyrin macrocycle consists of four pyrrolic subunits linked by four methine bridges (Figure 1.1).

Porphyrins bind metals to form complexes, usually with a charge of 2+ or 3+, which reside in the central cavity formed by the loss of two protons. These metalloporphyrins play an important role in biology. Fe(II) porphyrin complex is the functional part of hemoglobin protein, responsible for oxygen transport and storage in living tissues. The Mg-complexed porphyrin present in chlorophyll plays a major role in the photosynthesis process in plants.

In this chapter, we aim to illustrate the synergistic effect of combining the top-down and bottom-up approaches to keep up with scaling demands as well as open

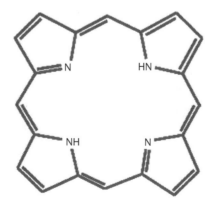

FIGURE 1.1    Simplest member of the porphyrin family, porphin.

up new possibilities for emerging nanoelectronic device platforms. We utilize SAMs of porphyrins as a means unto the end goal of keeping up with scaling laws, which are dictated by top-down processes. This has been illustrated through the following application cases:

1. Porphyrin SAM as a copper diffusion barrier for advanced CMOS technologies
2. Variable interface dipoles of metallated porphyrin SAMs for metal gate work function tuning in advanced CMOS technologies
3. Fabrication of unipolar graphene field-effect transistors (FETs) by modifying source and drain electrode interfaces with zinc Porphyrin

## 1.2 COPPER INTERCONNECT AND LOW-*k* DIELECTRICS

Interconnects distribute important global signals across an integrated circuit, as well as locally connect different transistors. With CMOS process technologies scaling to smaller dimensions, copper is the material of choice for interconnect metallization. Copper possesses one of the highest electrical conductivities among metals as well as high electromigration resistance. These properties enable a higher processing speed by virtue of a lower interconnect resistive-capacitive delay (Figure 1.2) [1,2].

The dielectric constant of the interlayer dielectrics (ILDs) is equally important to reduce signal delay and cross talk. Low-*k* dielectric materials such as silsesquioxane

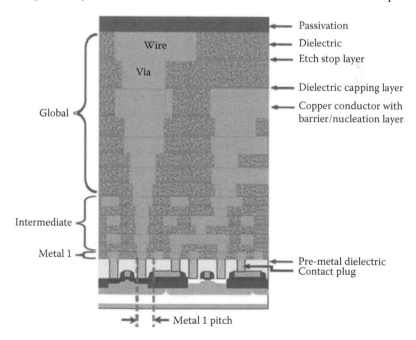

**FIGURE 1.2** Schematic of a typical interconnect–dielectric setup. (Data from The Semiconductor Industry Association, *International Technology Roadmap for Semiconductors*. Austin, TX: International SEMATECH, 2009.)

and silane-based compounds have been developed with $k$ values as low as 1.5 [4,5]. With ongoing scaling in CMOS technology node, however, major reliability concerns arise due to copper diffusion through ultrathin ILDs as well as thermal stability. Diffusion of the metal into the dielectric can lead to high leakage currents, lowering performance [6]. In such a scenario, it is desirable to have a diffusion barrier to prevent copper diffusion.

### 1.2.1 CHALLENGES TO TRADITIONAL TOP-DOWN APPROACHES

Traditionally, sputtered ternary nitride alloys such as Ta–Si–N, W–Si–N, and W–B–N have been used as barrier materials. CMOS nodes below 45 nm, however, require a diffusion barrier of 1–3 nm thick colloquially called an ultrathin barrier [7]. Barrier layers up to 10 nm have been realized by chemical vapor deposition (CVD) processes as in the case of TiSiN and WN [8]. Atomic layer deposition (ALD) processes can achieve film thicknesses of below 5 nm but are plagued by issues such as high defect densities and the presence of fast diffusion paths [9].

### 1.2.2 SAMs AS DIFFUSION BARRIER—A BOTTOM-UP SOLUTION

Mrunal et al. [9] have reported the use of a SAM of a zinc tetraphenylporphyrin (Zn-TPP-OH) molecule [10] to function as a diffusion barrier against copper diffusion. By their very definition, SAMs satisfy the thickness requirements of the CMOS node. Additionally, they can withstand back end of line (BEOL) processes, deposit in a conformal fashion, and are extremely cost effective. Krishnamurthy et al. have shown that the organosilane monolayers can inhibit Cu diffusion into $SiO_2$. The size and functional groups present in the molecule forming the monolayer have a role in the barrier properties of SAMs. SAMs with long-chain lengths screen Cu atoms from the influence of the substrate and the aromatic rings sterically hinder Cu diffusion between the molecules through the SAM layer [11]. Mikami et al. demonstrated the use of 2-(diphenylphosphino)ethyltriethoxysilane as a SAM barrier, wherein the Cu–P bond is understood to prevent copper diffusion [12]. SAMs can serve as sealants for porous low-$k$ ILDs. Caro et al. observed an improvement in the adhesion and inhibition of Cu silicide formation with $NH_2$ SAM derived from 3-aminopropyltrimethoxysilane [13].

The choice of Zn-TPP(OH) as the SAM molecule is due to the presence of several of the chemical moieties that contribute toward acting as a barrier for copper diffusion. Phenyl aromatic rings present in the molecule have been known to sterically hinder copper diffusion. The presence of zinc ion in the core of the porphyrin molecule can also prevent diffusion due to electronegativity. The pyrrole units in the molecule supplement these effects by virtue of formation of Cu–N bonds that hinder copper diffusion as well (Figure 1.3).

X-ray photoelectron spectroscopic (XPS) studies were performed on the samples with the SAM on the dielectric in a metal-insulator-semiconductor capacitor

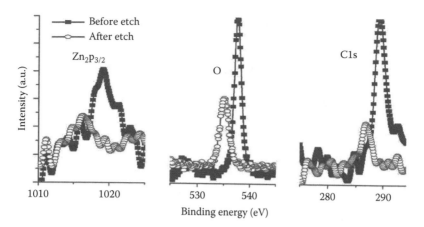

**FIGURE 1.3** $Zn_2P_{3/2}$, O, and C1s peaks of the ZnTPP(OH) monolayer on $SiO_2$ before and after etching. The peaks obtained due to the monolayer disappear or decrease in intensity as the sample is etched. (Data from Khaderbad, M.A. et al., *IEEE Trans. Electron Dev.*, 59, 1963–1969, 2012.)

(MISCAP) structure to extract information about the exposed surface of the sample, that is, the monolayer. XPS enabled an etching of the sample using a 3 keV Ar+ ion beam. The results of the chemical species before and after the etch were compared. Similar results of decrease in binding energy after etching have been described earlier in literature [15].

Bias temperature stress (BTS) conditions, which are a combination of an annealing step and electrical stress, were performed on metal-oxide-semiconductor capacitor (MOSCAP) and MISCAP structures to evaluate the diffusion of copper species into the dielectric. This diffusion was measured through electrical measurements as well as physical secondary ion mass spectrometry (SIMS) experiments.

In the case of the electrical measurements, the mobile charge created as a result of the BTS conditions was determined from the lateral shift of the capacitance–voltage curves (Figure 1.4).

$$Q_m = q\left[Cu\right]^+ = -\Delta V C_{ox}$$

where:
$Q_m$ is the mobile charge
$\Delta V$ is the voltage shift
$C_{ox}$ is the oxide capacitance

Figure 1.2 demonstrates the effect of a SAM on the *C–V* characteristics of a MISCAP structure. For MISCAPs without the SAM layer, the *C–V* curve shifts leftward on stressing for longer time intervals. This can be explained by

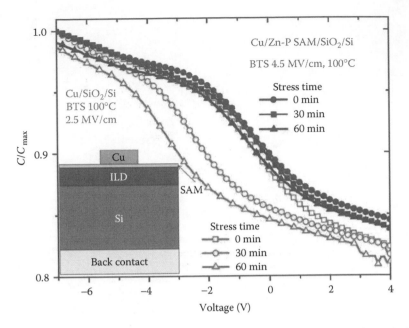

**FIGURE 1.4**  Capacitance–voltage characteristics for MISCAP structures shown in inset with (solid legends) and without (hollow legends) SAM. The temperature applied was 100°C and the field strength of 2.5 MV/cm was applied for different time intervals. (Data from Khaderbad, M.A. et al., *IEEE Trans. Electron Dev.*, 59, 1963–1969, 2012.)

formation of mobile charges in the dielectric layer, which occur due to the diffusion of copper. In the presence of the barrier layer of SAM, the copper diffusion, and therefore the shift in *C–V*, is reduced. Additionally, the presence of the SAM does not affect the *C–V* characteristics at 0 min (prior to BTS conditions), as evinced by the coincidence of the Cu/BD/Si (Figure 1.5) and Cu/SAM/SiO$_2$/Si (Figure 1.6).

SIMS was performed on the following MISCAP structures: Cu (50 nm)/SiO$_2$ (150 nm)/Si and Cu (50 nm)/SAM/SiO$_2$ (150 nm)/Si samples annealed at 400°C in a nitrogen atmosphere for 1 h. A 30 keV gold ion source was used for high mass-resolution spectroscopy. A 3 keV Cs gun was used for sputter removal of the different layers on the sample. The type as well as the number of sputtered species was measured by a time-of-flight instrument. This characterization technique enables us to extract the density of diffused copper species at different depths by alternating between sputtering the surface and sampling the species so formed. The copper/ILD interface is deduced from the copper and silicon depth profiles. A kink in the Cu profile indicates piling up of copper atoms that diffuse into ILD [37]. An abrupt gradient in the depth profile was observed in samples with SAM, which shows the effectiveness of ZnTPP(OH) monolayer as a good barrier layer to copper diffusion into ILD. Similar results were obtained for MISCAP structures with black diamond as a dielectric material.

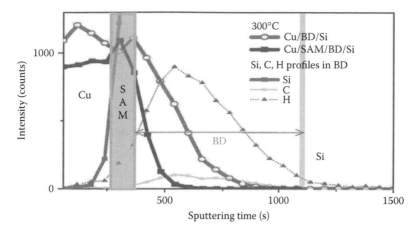

**FIGURE 1.5**  SIMS depth profile of MOSCAP devices. The number of counts of copper observed after sputtering for $t$ seconds is shown with SAM (solid legends) and without SAM (hollow legends). The drop in the number of copper counts is steeper for SAM-modified dielectric indicating its utility as a diffusion barrier. (Data from Khaderbad, M.A. et al., *IEEE Trans. Electron Dev.*, 59, 1963–1969, 2012.)

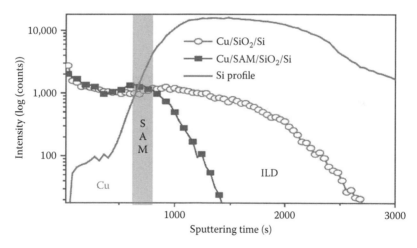

**FIGURE 1.6**  SIMS depth profile of copper obtained with and without SAM on $SiO_2$ dielectric after annealing in a nitrogen environment for 1 h. The interface between the copper and $SiO_2$ can be deduced by the increase in the count of silicon species. (Data from Khaderbad, M.A. et al., *IEEE Trans. Electron Dev.*, 59, 1963–1969, 2012.)

## 1.3  METAL GATE TECHNOLOGIES—WORK FUNCTION TUNING REQUIREMENTS

Top-down technologies have enabled aggressive scaling resulting in nanometer-level feature sizes. Traditional poly-Si gate electrodes are desired due to their ability to make a Fermi-level adjustment to the gate work function by donor or acceptor implantation. With aggressive scaling, the traditional polysilicon–$SiO_2$ gate dielectric structure shows degraded performance due to *short-channel effects*. Leakage current due to tunneling through the gate, Boron penetration causing threshold voltage fluctuations, polysilicon depletion, and high gate sheet resistance are some examples of these undesirable effects [16,17]. As a result, metal gates have been used below the 45 nm node to overcome these challenges.

The first commercial use of metal gate electrodes and high-$k$ dielectrics has been demonstrated by Intel in its 45 nm node technology [18]. The challenge was to develop metal gates that not only eliminate gate depletion and boron penetration problems, but also reduce the gate sheet resistance. Suitable metal gates are required to possess good interfacial properties with respect to the chosen high-$k$ dielectric as well as an appropriate work function, which is around 4 eV in the case of n-type metal-oxide semiconductor (NMOS) and 5 eV for p-type metal-oxide semiconductor (PMOS). The requirement of different work functions for PMOS and NMOS forms the cornerstone of the rationale behind work function tuning of metals. Use of a single midgap metal for both PMOS and NMOS would result in large threshold voltages, whereas use of unique metals for PMOS and NMOS would lead to additional processing complexity and increased cost. Hence, an analog of donor/acceptor doping of polysilicon has to be developed for metal gates.

### 1.3.1  WORK FUNCTION ENGINEERING USING SAMs—A BOTTOM-UP FACILITATION

Dipolar SAMs on various interfaces have an enormous potential to tailor the behavior of nanoelectronic devices. In a previous work done by Gu et al., dipolar aminopropyl triethoxy silane (APTES) SAM was used to tailor the work function of Ti on $SiO_2$/Si and the change was attributed to the change in the electrical potential at the Ti/SAM interface [19]. De Boer et al. demonstrated that the derivatives of alkanethiols can be used for the formation of chemisorbed SAMs of thiol molecules on coinage metals, which increases the work function of Ag ($\Phi$Ag ~ 4.4 eV) to 5.5 eV ($\Delta\Phi$ ~ 1.1 eV). The ordering of molecules in SAMs creates an effective dipole at the metal/SAM interface, which causes the change in metal work function [20,21]. Heimel et al. demonstrated that the work function modification is determined by the local ionization potential and electron affinity at the docking group side of the SAM, the interface dipole resulting from the bond formation, and the step in the vacuum potential across the SAM (resulting from the aligned molecular dipole moments) [22]. Venkataraman et al. demonstrated a spatial tuning of metal work function by utilizing surface chemical gradients of two different molecules and a controlled deposition process [23].

In Mrunal et al.'s work, tetraphenyl(OH) molecules have been used to form SAMs on $SiO_2$ and engineer the work function [24]. They have demonstrated that the dipolar properties can be tuned by incorporating various metal species in them. SAMs provide excellent thickness control, as well as the wide range of their derivatives. The dipole moment can be tuned not only by changing the central metal ion, but also by changing the groups attached to the porphyrin ring (Figure 1.7).

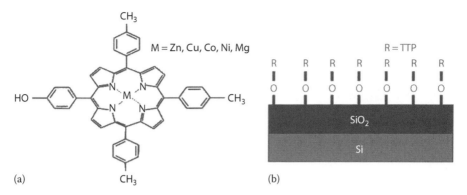

(a)    (b)

**FIGURE 1.7** (a) TPP(OH) with different central metal ions. (b) Representation of SAM formed on $SiO_2$. (Data from Zhu, Y. et al., *Adv. Mater.*, 22, 3906–3924, 2010.)

The $SiO_2$ substrate with activated hydroxyl groups was dipped in a $10^{-4}$ M solution of TPP(OH) in toluene for 5 min. For electrical characterization, Al/$SiO_2$/Si and Al/SAM/$SiO_2$/Si MOSCAP test structures were fabricated. Al gate/back contact was deposited. The hydrogen fuel cell vehicle (HFCV) technique was used for extracting the flat-band voltage ($V_{fb}$), the work function difference ($\Phi_{ms}$), and other important electrical param-eters for the test capacitors (Figures 1.8 and 1.9).

The normalized $C–V$ curves for MOSCAP structures with and without SAM are compared. As seen in Figure 1.6a, the presence of SAM results in an increase in the effective oxide thickness, decreasing the value of $C_{max}$.

The potential ($\Phi_{SAM}$) across SAM modifies the surface potential at the metal/SAM interface, thereby increasing or decreasing the $\Phi_{metal}$. Therefore, the measured metal workfunction is given by

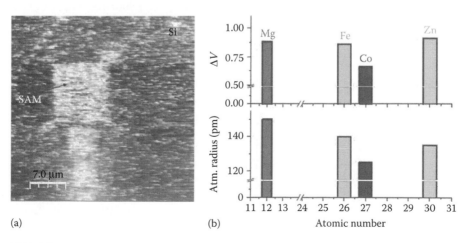

(a)    (b)

**FIGURE 1.8** (a) Surface potential plot of patterned Fe porphyrin SAM on Si obtained by Kelvin probe force microscopy. The intensity observed is directly proportional to the surface potential measured. (b) Measured surface potential for various metallated porphyrins. (Data from Zhu, Y. et al., *Adv. Mater.*, 22, 3906–3924, 2010.)

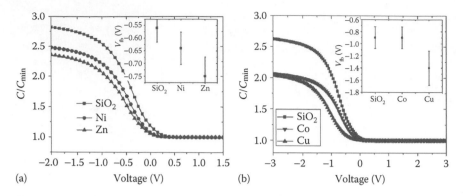

**FIGURE 1.9** Normalized ($C/C_{min}$) HFCV plots (a) $t_{ox} = 3$ nm, Al/SiO$_2$/p-Si, Al/(Zn/Ni) porphyrin-SAM/SiO$_2$/p-Si MOSCAPs. (b) $t_{ox} = 3.5$ nm, Al/SiO$_2$/p-Si,Al/(Cu,Co) porphyrin SAM/SiO$_2$/p-Si MOSCAPs. (Data from Zhu, Y. et al. *Adv. Mater.*, 22, 3906–3924, 2010.)

$$\Phi_{metal, mod} = \Phi_{metal} + \Phi_{SAM}$$

where:

$\Phi_{metal,mod}$ is the modified metal work function due to SAM

$\Phi_{metal}$ is the metal work function

$\Phi_{SAM}$ is the potential due to dipolar SAM

Due to the presence of π-conjugation in porphyrins, and the direction of dipole moment toward the Al back gate, a $V_{fb}$ shift occurs toward the left in the *C–V* curves of the fabricated MOSCAP. The magnitude of the shift is higher for porphyrins with Zn and Cu metal ions, which decreases onward to Ni and Co. This trend matches in $V_{fb}$ matches with that of the calculated dipole moment for each metallated porphyrin.

## 1.4 UNIPOLAR GRAPHENE OXIDE FETs

Graphene, a two-dimensional network of carbon atoms, possesses unique electrical and mechanical properties. It has a large specific surface area, high intrinsic mobility, high Young's modulus, and high thermal conductivity [25,26]. Reduced graphene oxide (RGO), a solution-processed form of graphene, is being considered in various sensor applications, albeit slightly defective compared to graphene [27–29]. Solution processability of RGO enables spin-coating and drop-casting deposition methods to pave the way for efficient, low-cost, and large-scale device fabrication [30–32].

The ambipolar conductance in RGO (and graphene) is undesirable, because the power consumption is higher in such logic circuits compared to unipolar logic. Wang et al. have achieved unipolar transport in graphene using nitrogen doping, whereas Nouchi et al. used cobalt electrodes in graphene FETs, resulting in asymmetric electron–hole currents in the devices [33,34]. The mechanism critical to achieving unipolar devices is asymmetry between electron and hole injection.

### 1.4.1 Modification of Electrical Transport Properties—A Bottom-Up Approach

SAMs of organic surfactants enable a flexible, versatile, and powerful method to engineer interfaces to tune the properties of electronic devices. Abe et al. have tuned the threshold voltage of pentacene organic FETs (OFETs) by carrier doping at the charge–transfer interface [20]. Asadi et al. have shown that SAMs of alkanethiols can manipulate the charge injection in OFETs [35]. Calhoun et al. have functionalized the surface of organic semiconductors with SAMs to enable their usage as sensors [36]. Fluoroalkyltrichlorosilane (FOTS) SAM was used by Lee et al. to modify graphene resulting in high-surface doping levels and increased carrier density [37]. Barrier heights in OFETs are determined by the difference between the metal electrode work function ($\varphi$) and the highest occupied molecular orbital/lowest unoccupied molecular orbital (HOMO/LUMO) level of the semiconductor, assuming a Mott–Schottky model. However, through the integration of dipoles (naturally occurring and artificial) at the metal/semiconductor interfaces, barrier heights can be modulated, significantly affecting the charge injection [38,39]. When appropriate SAMs with functional groups are grafted onto metal surfaces, an effective dipole at the metal/SAM interface is created which tunes the metal work function.

Dipolar monolayers of metalloporphyrins are excellent materials for this application, due to their diverse structural motifs and associated electrical, magnetic, optical, and chemical properties. The integration of 5-(4-hydroxyphenyl)-10,15,20-tri($p$-tolyl) zinc(II) porphyrin (Zn(II)TTPOH) SAMs at the electrode interfaces of reduced graphene oxide (RGO) transistors and study their influence on the electrical properties of the transistors is reported by Mrunal et al. [39]. A higher injection barrier for electrons has been proposed as a reason for the observed unipolar current characteristics. The injection barriers for electrons and holes have been computed through density functional theory (DFT) calculations of the ZnTPP(OH) molecule and correlate well with the work function of the porphyrin-modified electrodes obtained through Kelvin probe measurements.

Barrier heights for charge injection into the semiconductor can be expressed as (Figure 1.10)

$$\varphi_e = \varphi_M + \Delta\varphi_{dipole} - \vartheta$$

$$\varphi_h = E_g - (\varphi_M + \Delta\varphi_{dipole} - \vartheta)$$

where:
$\varphi_M$ is the metal work function
$\vartheta$ is the electron affinity
$E_g$ is the energy band gap
$\Delta\varphi_{dipole}$ is the barrier change due to interface dipole

Different peaks in cyclic voltammograms correspond to different oxidation/reduction states of the material (Figures 1.11 and 1.12). The onset of oxidation is related to the HOMO/LUMO energy corresponding to the removal/addition of electrons.

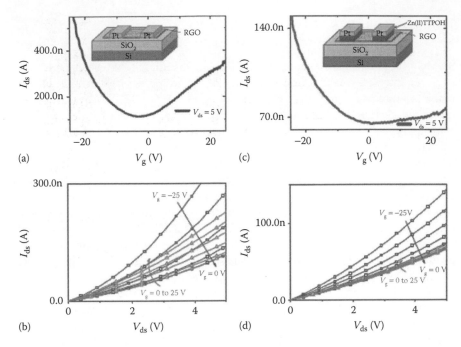

**FIGURE 1.10** Transfer ($I_d$–$V_g$) characteristics of RGO FET with and without porphyrin interlayer. Part (a) shows the ambipolar nature of the RGO FET. (c) Introduction of a porphyrin interlayer results in the inhibition of electron transport with a slight decrease in hole mobility. Part (b) shows the effect of gate voltage on the modulation of drain current. In part (d), however, the effect is not observed due to the presence of the porphyrin interlayer resulting in limited gate control, due to injection limited transport. (Data from Khaderbad, M.A. et al., *Nanotechnology*, 23, 025501.)

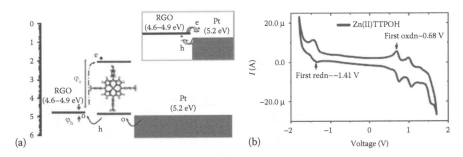

**FIGURE 1.11** (a) The energy band diagrams represent electron and hole transport at the RGO/Pt interface. In the inset, the ambipolar transport scenario is demonstrated. In the presence of the interlayer, the barrier to electron mobility is increased. (b) Extraction of HOMO/LUMO for porphyrin through cyclic voltammetry. The oxidation and reduction peaks of porphyrin are with respect to ferrocene/ferrocenium which is at 4.8 eV below vacuum. (Data from Khaderbad, M.A. et al., *Nanotechnology*, 23, 025501.)

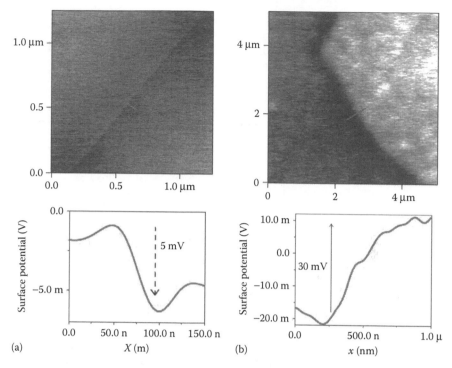

**FIGURE 1.12** Kelvin probe force microscopy was used to calculate the surface potential measurements of the ZnTPP(OH) SAM on RGO to compute the dipole effect due to the monolayer. The surface potential measurements of RGO are shown with respect to $SiO_2$ (a) with and (b) without the SAM modification, respectively. As opposed to bare RGO, ZnTPP(OH)-modified RGO showed higher surface potential. (Data from Khaderbad, M.A. et al., *Nanotechnology*, 23, 025501.)

Combining the values of HOMO and LUMO extracted through cyclic voltammetry and the dipole moment extracted through Kelvin probe force microscopy, the barriers for electron and hole injection into RGO are calculated to be 2.2 and 0.11 eV, respectively. Unequal injection barriers result in unipolar behavior.

## 1.5 INCORPORATION OF NANOWIRES IN MICROELECTROMECHANICAL SYSTEM DEVICES FOR SENSING APPLICATIONS

In this section, we discuss about the integration of bottom-up approaches with the microelectromechanical system (MEMS) platforms, especially microcantilevers. In the first case study, we explore the use of ZnO. ZnO nanowires have sparked great interest among researchers for various applications owing to their piezoelectric property. ZnO has been used as pre-concentrator material in sensing volatile organic compounds upon integration with strain-sensitive microcantilevers [41].

Its piezoelectric property has been studied showing the electrical actuation of a suspended ZnO nanowire [42]. In the second case study, we report the use of porphyrin SAMs on microcantilevers for sensing applications, with a particular example of carbon monoxide sensing using piezoresistive platform [43].

MEMS-based devices have been widely used for sensing applications. As an addition to increase selectivity and sensitivity of the devices, various modifications have been done on the sensor surface. ZnO nanowires have been encapsulated in SU-8 polymer cantilevers as shown in the adjoining schematic (Figure 1.13).

When the device thus fabricated was subjected to different deflections using an indenter, corresponding different currents were observed owing to the piezoelectric behavior of ZnO. These characteristics have been shown in Figure 1.14.

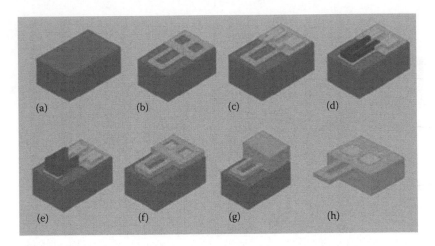

**FIGURE 1.13**  Process flow. (a) Silicon dioxide grown on a silicon wafer. (b) First encapsulation layer of SU-8. (c) Patterned Cr/Au layer for contact. (d) Pattering of ZnO seed layer. (e) Vertical growth of ZnO nanowire. (f) Bottom encapsulation layer of SU-8. (g) Pattering of SU-8 anchor layer. (h) Released device after sacrificial layer etching. (Data from Nathawat, R. et al., ZnO nanorods based ultra sensitive and selective explosive sensor, In *Proceedings of the IEEE 5th International Nanoelectronics Conference*, IEEE, pp. 40–42, 2013.)

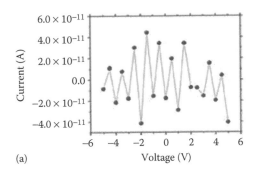

**FIGURE 1.14**  (a) Current–voltage characteristics through ZnO seed layer without nanowires.
*(Continued)*

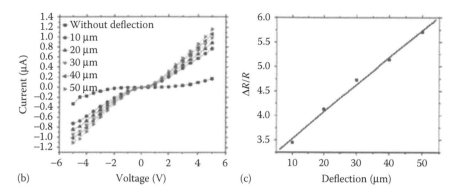

(b)

(c)

**FIGURE 1.14 (Continued)**   (b) Current–voltage plot for different values of deflection for cantilever beam with ZnO nanowires. (c) $\Delta R/R$ as a function of deflection for SU-8 cantilever with embedded ZnO nanowires. (Data from Nathawat, R. et al., ZnO nanorods based ultra sensitive and selective explosive sensor, In *Proceedings of the IEEE 5th International Nanoelectronics Conference*, IEEE, pp. 40–42, 2013.)

These ZnO nanowires embedded in the polymer microcantilevers have been shown to detect volatile organic compounds, for example, explosives [44], shown in Figure 1.15.

ZnO nanowires have been demonstrated to generate electric potential upon deflection using conventional atomic force microscopy tip [44]. Using this work as a basis, AC electrical actuation was used to determine the resonant frequency of a doubly clamped ZnO nanowire [42].

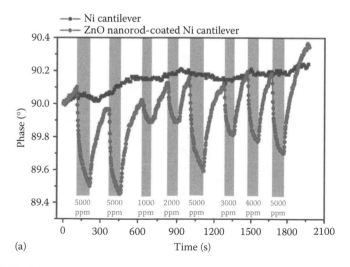

(a)

**FIGURE 1.15**   Phase versus time graphs for microcantilever devices exposed to different diethanolamine (DEA) concentrations. DEA detection of ZnO nanorod coated–uncoated Ni microcantilever (a).                                                    (*Continued*)

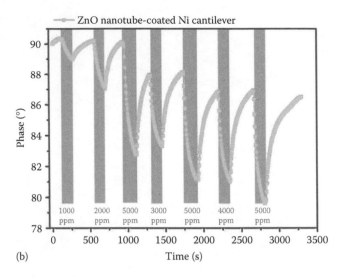

(b)

**FIGURE 1.15 (Continued)**   Phase versus time graphs for microcantilever devices exposed to different diethanolamine (DEA) concentrations. ZnO nanotube-coated microcantilever (b). (Data from Wang, Z.L. and Jinhui, S., *Science*, 312, 242–246, 2006.)

Another instance of the amalgamation of bottom-up approach with the microcantilevers that we will address is porphyrin-based SAM that is used for sensing carbon monoxide (CO). Iron(III) porphyrin *(5,10,15,20-tetra(4,5-dimethoxyphenyl)-21H,23H-porphyrin iron(III) chloride)* coated on piezoresistive SU-8/carbon black (CB) cantilevers was exposed to CO and the response in volts recorded, as shown in Figure 1.16.

Figure 1.16a shows that there is no specific signal generated when the piezoresistive microcantilever was subjected to gases when it was not coated with Fe(III) porphyrin. When the device was exposed to CO after it was coated with Fe(III) porphyrin, a distinct pattern emerges indicating the sensing of the gas (Figure 1.16b). This suggests

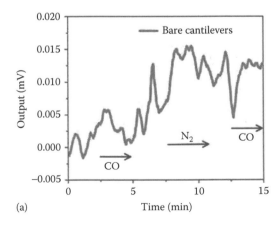

(a)

**FIGURE 1.16**   (a) Response of a bare SU-8/CB polymer composite microcantilever for consecutive cycles of CO and $N_2$.                                                                                  (*Continued*)

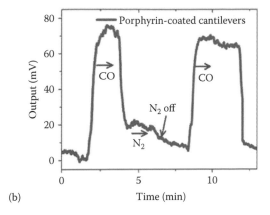

(b)

**FIGURE 1.16 (Continued)** (b) Response of a Fe(III) porphyrin-coated microcantilever for consecutive cycles of CO and $N_2$. (Data from Kilinc, N. et al., *Sensors Actuat. B Chem.*, 202, 357–364, 2014.)

that the pattern has been generated because of the iron porphyrin layer. Sensitivity and selectivity of the porphyrin SAM to CO have also been checked by exposing it to different concentrations of CO and also with different gases/a mixture of gases and CO showing the desired responses of the porphyrin molecule toward CO (Figure 1.17).

Zinc oxide nanorods have also been demonstrated to sense explosive vapors such as dinitrobenzene (DNB), trinitrotoluene (TNT), and research development explosive (RDX; IUPAC name: 1,3,5-trinitroperhydro-1,3,5-triazine) on a microcantilever platform, thus corroborating the observations as discussed earlier. In this work, the change in the resistance was observed when the microcantilever setup was exposed to periodic heating and cooling processes, as shown in Figure 1.18. Figure1.18a shows the control behavior of the setup demonstrating the baseline stability when it was not exposed to explosive chemical compounds. The graph shows a stable behavior of the setup. When the microcantilever coated with ZnO nanorods was exposed with TNT

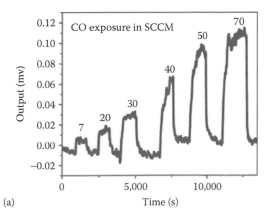

(a)

**FIGURE 1.17** (a) Response of porphyrin-functionalized SU-8/CB microcantilevers for different CO flow rates. (*Continued*)

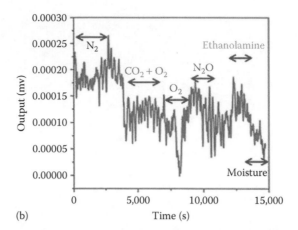

(b)                                          Time (s)

**FIGURE 1.17 (Continued)** (b) Response of porphyrin-functionalized SU-8/CB microcanti-levers for different gases. (Data from Kilinc, N. et al., *Sensors Actuat. B Chem.*, 202, 357–364, 2014.)

and the subsequent temperature variations as in the aforementioned cycles, it showed a resistance change accompanied with degrading sensor response with every cycle, depicted in Figure 1.18b.

This kind of response hints at the temperature-dependent sensing behavior of ZnO nanorods, that is, with every cycle of temperature rise and fall, the sensor response followed the same trend. This module shows a selective response only for TNT giving a positive change in resistance from the base value, whereas a negative quantum for other volatile compounds such as RDX, DNB, ethanol, and water vapor, respectively (as shown in Figure 1.19).

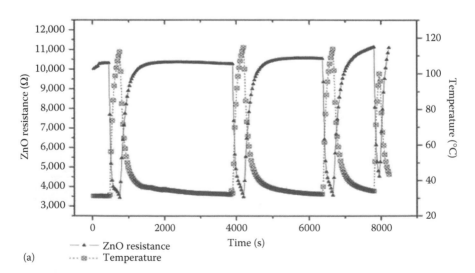

(a)

**FIGURE 1.18** (a) Temperature stability: Base resistance of sensor remains constant after applying successive heating cycles.                                                    (*Continued*)

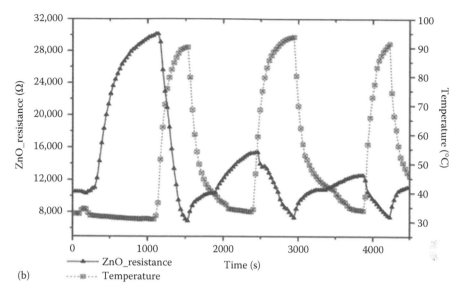

(b)

FIGURE 1.18 (Continued)    (b) Sensitivity: Response of TNT with ZnO sensor: The concentration of TNT vapors was varied gradually, and the response was recorded and integrated with heating. (Data from Reddy, C.V.B. et al., *IEEE Trans. Nanotechnol.*, 11, 701–706, 2012.)

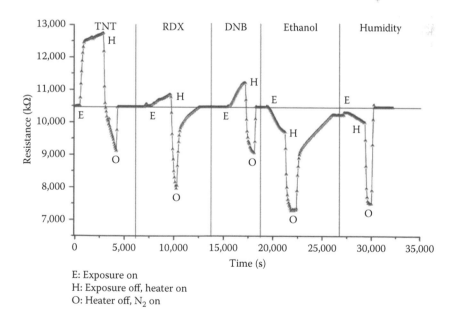

E: Exposure on
H: Exposure off, heater on
O: Heater off, N$_2$ on

FIGURE 1.19    Differential change in resistance is positive for TNT (first panel), whereas it is negative for RDX, DNB, ethanol, and water vapor, respectively. (Data from Reddy, C.V.B. et al., *IEEE Trans. Nanotechnol.*, 11, 701–706, 2012.)

ZnO nanowires/nanorods have proven as excellent amalgamation tools where we seek to unite top-down approach of fabricating devices with the bottom-up approaches, especially in volatile organic compound sensing applications. Piezoelectric properties of ZnO nanowires have been demonstrated earlier [44], showing a promise of using ZnO as a material for generating electrical energy harnessing day-to-day mechanical movements.

## REFERENCES

1. G. Steinlesberger, M. Engelhardt, G. Schindler, W. Steinhogl, A. von Glasow, K. Mosig, and E. Bertagnolli, Electrical assessment of copper damascene interconnects down to sub-50 nm feature sizes, 2002, *Microelectron. Eng.*, 64(1–4), 409–416.
2. W. Steinhogl, G. Schindler, G. Steinlesberger, M. Traving, and M. Engelhardt, Comprehensive study of the resistivity of copper wires with lateral dimensions of 100 nm and smaller, 2005, *J. Appl. Phys.*, 97, 023706.
3. The Semiconductor Industry Association, 2009, *International Technology Roadmap for Semiconductors.* Austin, TX: International SEMATECH.
4. E. S. Moyer, K. Chung, M. Spaulding, T. Deis, R. Boisvert, C. Saha, and J. Bremmer, Ultra low dielectric constant silsesquioxane based resin [ILDs], 1999, *IEEE International Conference on Interconnect Technology*, San Francisco, CA, pp. 196–197.
5. K. J. Chao, P. H. Liu, A. T. Cho, K. Y. Huang, Y. R. Lee, and S. L. Chang, Preparation and characterization of low-k mesoporous silica films, 2004, *Stud. Surf. Sci. Catal.,* Elsevier, 154, Part 1, 94–101.
6. Y. Shacham-Diamond, D. A. Hoffstetter, and W. G. Oldham, Reliability of copper metallization on silicon-dioxide, 1991, *VLSI Multilevel Interconnection Conference*, IEEE, Santa Clara, CA, pp. 109–115.
7. N. Awaya, H. Inokawa, E. Yamamoto, Y. Okazaki, M. Miyake, Y. Arita, and T. Kobayashi, Evaluation of a copper metallization process and the electrical characteristics of copper-interconnected quarter-micron CMOS, 1996, *IEEE Trans. Electron Dev.*, 43(8), 1206–1212.
8. S. Rawal, D. P. Norton, T. J. Anderson, and L. McElwee-White, Properties of W–Ge–N as a diffusion barrier material for Cu, 2005, *Appl. Phys. Lett.*, 87, 111902.
9. P. Kapur, J. P. McVittie, and K. C. Saraswat, Technology and reliability constrained future copper interconnects. I. Resistance modeling, 2002, *IEEE Trans. Electron Dev.*, 49(4), 590–597.
10. F. Da Cruz, K. Driaf, C. Berthier, J.-M. Lameille, and F. Armand, Study of a self-assembled porphyrin monomolecular layer obtained by metal complexation, 1999, *Thin Solid Films*, 349(1/2), 155–161.
11. A. Krishnamoorthy, K. Chanda, S. P. Murarka, G. Ramanath, and J. G. Ryan, Self-assembled near-zero-thickness molecular layers as diffusion barriers for Cu metallization, 2001, *Appl. Phys. Lett.*, 78, 2467–2469.
12. N. Mikami, N. Hata, T. Kikkawa, and H. Machida, Robust self-assembled monolayer as diffusion barrier for copper metallization, 2003, *Appl. Phys. Lett.,* 83(25), 5181–5183.
13. A. M. Caro, G. Maes, G. Borghs, and C.M. Whelan, Screening self-assembled monolayers as Cu diffusion barriers, 2008, *Microelectron. Eng.*, 85(10), 239–242.
14. M. A. Khaderbad, R. Pandharipande, V. Singh, S. Madhu, M. Ravikanth, and V. Ramgopal Rao, Porphyrin self-assembled monolayer as a copper diffusion barrier for advanced CMOS technologies, 2012, *IEEE Trans. Electron Dev.*, 59(7), 1963–1969.
15. E. P. Gusev, V. Narayanan, and M. M. Frank, Advanced high-k dielectric stacks with polySi and metal gates: Recent progress and current challenges, 2006, *IBM J. Res. Dev.*, 50(4), 387–410.

16. H. D. B. Gottlob, T. Echtermeyer, M. Schmidt, T. Mollenhauer, J. K. Efavi, T. Wahlbrink, M.C. Lemme et al., 0.86-nm CET gate stacks with epitaxial $Gd_2O_3$ high-$k$ dielectrics and FUSI NiSi metal electrodes, 2006, *IEEE Electron Dev. Lett.*, 27(10), 814–816.

17. C.-H. Jan, M. Agostinelli, H. Deshpande, M. A. El-Tanani, W. Hafez, U. Jalan, L. Janbay et al., RF CMOS technology scaling in High-k/metal gate era for RF SoC applications, 2010, *IEEE International Electron Devices Meeting*, San Francisco, CA, pp. 27.2.1–27.2.4.

18. D. Gu, R. Sistiabudi, and S. K. Dey, Modification of work function of Ti by selfassembled monolayer molecules on $SiO_2$/p-Si, 2005, *J. Appl. Phys.*, 97, 123710.

19. B. de Boer, A. Hadipour, M. M. Mandoc, T. van Woudenbergh, and P. W. M. Blom, Tuning the metal work function with self assembled monolayers, 2005, *Adv. Mater.*, 17(5), 621–625.

20. K. Asadi, F. Gholamrezaie, E. C. P. Smits, P. W. M. Blom, and B. D. Boer, Manipulation of charge carrier injection into organic field-effect transistors by self-assembled monolayers of alkanethiols, 2007, *J. Mater. Chem.*, 17(19), 1947–1953.

21. G. Heimel, L. Romaner, E. Zojer, and J.-L. Brédas, Toward control of the metal–organic interfacial electronic structure in molecular electronics: A first-principles study on selfassembled monolayers of π-conjugated molecules on noble metals, 2007, *Nano Lett.*, 7(4), 932–940.

22. N. V. Venkataraman, S. Zürcher, A. Rossi, S. Lee, N. Naujoks, and N. D. Spencer, Spatial tuning of the metal work function by means of alkanethiol and fluorinated alkanethiol gradients, 2009, *J. Phys. Chem. C*, 113(14): 5620–5628.

23. M. A. Khaderbad, U. Roy, M. Yedukondalu, M. Rajesh, M. Ravikanth, and V. Ramgopal Rao, Variable interface dipoles of metallated porphyrin self-assembled monolayers for metal-gate work function tuning in advanced CMOS technologies, 2010, *IEEE Trans. Nanotechnol.*, 9(3), 335–337.

24. Y. Zhu, S. Murali, W. Cai, X. Li, J. W. Suk, and J. R. Potts, Graphene and graphene oxide: Synthesis, properties, and applications, 2010, *Adv. Mater.*, 22(35), 3906–3924.

25. J. C. Meyer, A. K. Geim, M. I. Katsnelson, K. S. Novoselov, T. J. Booth, and S. Roth, The structure of suspended graphene sheets, 2007, *Nature*, 446(7131), 60–63.

26. Y. Si and E. T. Samulski, Synthesis of water soluble graphene, 2008, *Nano Lett.*, 8(6), 1679–1682.

27. D. R. Dreyer, S. Park, C. W. Bielawski, and R. S. Ruoff, The chemistry of graphene oxide, 2010, *Chem. Soc. Rev.*, 39(1), 228–240.

28. M. Pumera. Graphene-based nanomaterials for energy storage, 2011, *Energy Environ. Sci.*, 4(3), 668–674.

29. C. Soldano, A. Mahmood, and E. Dujardin, Production, properties and potential of grapheme, 2010, *Carbon*, 48(8), 2127–2150.

30. S. Wang, P. K. Ang, Z. Wang, A. L. L. Tang, J. T. L. Thong, and K. P. Loh, High mobility, printable, and solution-processed graphene electronics, 2009, *Nano Lett.*, 10(1), 92–98.

31. P. H. Wobkenberg, G. Eda, D. S. Leem, J. C. De Mello, D. D. C. Bradley, M. Chhowalla et al., Reduced graphene oxide electrodes for large area organic electronics, 2011, *Adv. Mater.*, 23(13), 1558–1562.

32. X. Wang, X. Li, L. Zhang, Y. Yoon, P. K. Weber, H. Wang et al., N-Doping of graphene through electrothermal reactions with ammonia, 2009, *Science*, 324(5928), 768–771.

33. R. Nouchi, M. Shiraishi, and Y. Suzuki, Transfer characteristics in graphene fieldeffect transistors with Co contacts, 2008, *Appl. Phys. Lett.*, 93(15), 152104-3.

34. Y. Abe, T. Hasegawa, Y. Takahashi, T. Yamada, and Y. Tokura, Control of threshold voltage in pentacene thin-film transistors using carrier doping at the charge transfer interface with organic acceptors, 2005, *Appl. Phys. Lett.*, 87(15), 153506-3.

35. M. F. Calhoun, J. Sanchez, D. Olaya, M. E. Gershenson, and V. Podzorov, Electronic functionalization of the surface of organic semiconductors with self-assembled monolayers, 2008, *Nat. Mater.*, 7(1), 84–89.

36. B. Lee, Y. Chen, F. Duerr, D. Mastrogiovanni, E. Garfunkel, E. Y. Andrei et al., Modification of electronic properties of graphene with self-assembled monolayers, 2010, *Nano Lett.*, 10(7), 2427–2432.

37. X. Cheng, Y. Y. Noh, J. Wang, M. Tello, J. Frisch, R. P. Blum et al., Controlling electron and hole charge injection in ambipolar organic field-effect transistors by self-assembled monolayers, 2009, *Adv. Funct. Mater.*, 19(15), 2407–2415.

38. P. Stoliar, R. Kshirsagar, M. Massi, P. Annibale, C. Albonetti, D. M. deLeeuw et al., Charge injection across self-assembly monolayers in organic field-effect transistors odd-even effects, 2007, *J. Am. Chem. Soc.*, 129(20), 6477–6484.

39. M. A. Khaderbad, T. Verawati, R. Manohar, P. Rohit, M. Sheri, W. Jun, R. Mangalampalli, M. Nripan, G. M. Subodh, and V. R. Rao, Fabrication of unipolar graphene field-effect transistors by modifying source and drain electrode interfaces with zinc porphyrin, 2012, *ACS Appl. Mater. Interf.*, 4(3), 1434–1439.

40. M.A. Khaderbad, C. Youngjin, H. Pritesh, A. Atif, W. Nan, D. Colm, T. Pradyumna, A.J.A. Gehan, V. R. Rao, and A. S. Ashwin, Electrical actuation and readout in a nano-electromechanical resonator based on a laterally suspended zinc oxide nanowire, 2012, *Nanotechnology*, 23(2), 025501.

41. R. Nathawat, M. Patel, P. Ray, N. A. Gilda, M. S. Vinchurkar, and V. R. Rao, ZnO nanorods based ultra sensitive and selective explosive sensor, 2013, In *Proceedings of the IEEE 5th International Nanoelectronics Conference*, IEEE, pp. 40–42.

42. P. Ray and V. R. Rao, ZnO nanowire embedded strain sensing cantilever: A new ultra-sensitive technology platform, 2013, *J. Microelectromech. Syst.*, IEEE, Piscataway, NJ, 22(5): 995–997.

43. N. Kilinc, C. Onur, K. Arif, E. Erhan, O. Sadullah, Y. Yusuf, Z. O. Zafer, and U. Hakan, Fabrication of 1D ZnO nanostructures on MEMS cantilever for VOC sensor application, 2014, *Sensors Actuat. B Chem.*, 202, 357–364.

44. Z. L. Wang and S. Jinhui, Piezoelectric nanogenerators based on zinc oxide nanowire arrays, 2006, *Science*, 312(5771), 242–246.

45. C. V. B. Reddy, M. A. Khaderbad, S. Gandhi, M. Kandpal, S. Patil, K. N. Chetty, K. G. Rajulu, P. C. K. Chary, M. Ravikanth, and V. R. Rao, Piezoresistive SU-8 cantilever with Fe (III) porphyrin coating for CO sensing, 2012, *IEEE Trans. Nanotechnol.*, 11(4), 701–706.

# 2 Synthesis and Assembly of Inorganic and Inorganic–Organic Hybrid Nanomaterials by Microreactor-Assisted Chemical Processes

*Ki-Joong Kim, Chang-Ho Choi,
Seung-Yeol Han, and Chih-Hung Chang*

## CONTENTS

**ABSTRACT** Microreactors have recently gained significant interests as novel tools for producing different types of nanomaterials for a variety of applications in chemical industries, pharmaceuticals, and biotechnology. In this chapter, we describe the current development of microreactor-assisted chemical processes for controlled synthesis and assembly of inorganic and inorganic–organic hybrid nanomaterials, including single-phase nanoparticles of unary, binary, and ternary materials; core–shell nanoparticles; assembly of nanoparticles; and the deposition of nano- and microstructured thin films. Some of the advantages (or issues) comparing to batch chemical reactors are also discussed, along with opportunities for large-scale applications.

## 2.1   INTRODUCTION

Nanoparticles (NPs), or nanocrystals (NCs), defined as chemical substances or materials with 1–100 nm dimensions, have been intensively investigated over the past several decades due to their novel characteristics that are not observed from corresponding bulk materials.[1] As the size of materials becomes small enough to be comparable to the wavelength of the electron, the quantum confinement effect is observed. This effect gives rise to interesting physicochemical and optical properties, which are gaining much attention from researchers.[2–7] The synthetic methods of NPs are very important due to their high relevance to the novel properties of nanomaterials.[8–14] A variety of synthetic routes for NPs have been suggested with the aim of controlling shape and size of the synthesized NPs. Conventional solution-based processes include a precipitation method, sol–gel processing, microemulsion, and solvothermal processing.[15–20] These conventional processes are based on a batch process, where all of the reactants are fed into the reactor and the products are removed after processing is finished. The high-volume batch process has several challenges in obtaining monodispersed NPs. Large gradients associated with the reaction temperature, precursor concentration, and solution pH are of concern in the batch process, which result in heterogeneous NP size and thus broad size distribution. Significant efforts have been made on developing the synthesis methodology for producing controlled NPs in order to overcome these inherent problems of the batch process. Continuous flow reactors, assisted with a microreactor, can be a promising approach to efficiently control the growth stage of NPs and make it more feasible to tailor the morphology of NPs. The advantages of utilizing a continuous flow microreactor-assisted NP synthesis are well established.[21–27] The large surface-area-to-volume ratios in microreactors allow for rapid mixing of reactants and minimize the gradients of precursor concentration and reaction temperature due to the enhanced heat and mass transfer, which facilitate the homogenous nucleation and growth of the NPs. Reducing reaction volume to the microscale can improve production yield and reduce the reactant consumption. Moreover, decoupling the nuclei formation stage from the growth stage makes it more feasible to study the growth mechanism of NPs. The development of microreactor systems has been accompanied by the advance of microreactor fabrication technology. Given the recent development of microfabrication, different types of microreactors have been made more accessible for the synthesis of various functional NPs.[28–33] Furthermore, the continuous flow microreactor opens up the opportunity to conveniently assemble unique functional nanostructures. There are a large number of microreactor-assisted syntheses of nanomaterials, as the microreactor can produce not only reactive fluxes of short-life, intermediate molecules for heterogeneous growth but also NPs and their assemblies for nanostructured surfaces.[26,27,34,35]

In this chapter, we present an overview of the progress in the synthesis of NPs and the assembly of nanostructures using continuous microreactor-assisted chemical processes.

## 2.2 SYNTHESIS OF COLLOIDAL NPs

NPs have attracted a great deal of research interest because their properties can be controlled by varying the particle size and shape. These size- and shape-dependent properties have found applications in many technically important areas such as catalysis, light-emitting diodes,[36,37] lasers,[38,39] and solar cells.[40-44] Each of these applications depends critically on the ability to synthesize *high-quality* NPs as judged by their uniformity in size and shape.[12,44]

High-quality NPs with low defect densities and narrow size distributions can be prepared via the state-of-the-art batch synthesis techniques. These highly successful batch methods suffer from a number of fundamental limitations in control over NP synthesis. In particular, the nature of this batch method leads to local fluctuations in temperature and concentration, and inhomogeneity in mixing, which make precise control of reaction conditions nearly impossible. These problems are magnified as the reaction vessel size is increased, and thus, the technique has a fundamental limitation in scalability.[45] Microreaction technology that allows rapid mixing of reactants and minimizes the gradients of precursor concentration and reaction temperature is desirable for synthesis of high-quality NPs for commercial applications.

In this context, microreactor-assisted chemical processes offer an ideal platform for the synthesis of high-quality NPs because they allow for rapid mixing of reagents and controlled heat and mass transfer. Microreaction technologies offer a much improved control of reaction parameters, and thus allow for higher yields, smaller amounts of by-products, and greater selectivities. By using continuous flow microreactors instead of batch reactors, one could produce high-quality NPs at a high throughput, particularly in parallelized arrays. Two main approaches that have been employed for the synthesis of colloidal NPs including continuous single-phase laminar flow and segmented flow (including slug- and droplet-based processes) are discussed in this section.

### 2.2.1 SINGLE-PHASE FLOW

A wide range of NPs have been synthesized using continuous flow microreactors, including unary (Au,[46-51] Ag,[48,52,55] Cu,[56] Co,[57,58] Pd[59,60]), binary (InP,[61,62] SnTe,[63] CdS,[64,65] CdSe[23,33,55,66-72]), and ternary (CuInSe$_2$)[73] structures in addition to more complicated core–shell structures such as CdSe/ZnS[74-76] and ZnSe/ZnS.[77] Some of the reported NPs synthesized using microreactors are listed in Table 2.1.

Metal NPs have been synthesized successfully using a number of different microreactors. One simple example of a continuous flow reactor consisting of a helical coil in a heated oil bath is shown in Figure 2.1a.[53] The metal precursor is mixed with growth solutions, and the mixture is then directed into temperature-controlled tubing. The capillary tubing, having microscale internal dimensions, improves size and composition control in the NP synthesis, which is due in large part to more precise temperature control. Microreactors used in all cases have been shown to offer significant advantages over similar batch reactors as a result of increased reaction control and the ability to fine-tune the final product.

**TABLE 2.1**

**Overview of the Chemicals for Colloidal NPs Synthesized in a Continuous Flow Microreactors**

| Nanometerials | | Precursors, Stabilizers, and Solvents | Diameter (nm) | Reference |
|---|---|---|---|---|
| Unary | Au | HAuCl$_4$, AA, NaBH$_4$, CTAB | 20–40 | 46 |
| | | HAuCl$_4$, TSC | 35 | 47 |
| | | HAuCl$_4$, NaBH$_4$, CTAB | 40–45 | 48 |
| | | HAuCl$_4$, AA, PVP, KOH | 24–35 | 49 |
| | | HAuCl$_4$, NaBH$_4$, DDT | 2.9–4.9 | 50 |
| | | Au(PPh$_3$)Cl, NaBH$_4$ | 0.8 | 51 |
| | Ag | AgNO$_3$, NaBH$_4$, CTAB | 85–274 | 48 |
| | | AgNO$_3$, NaBH$_4$, TSC, DMF, FDTS, PDMS | 18–46 | 52 |
| | | A$_g$C$_2$F$_5$CO$_2$, TOA | 7.4–8.7 | 53 |
| | | AgNO$_3$, PVP | 29–47 | 54 |
| | | AgNO$_3$, NaBH$_4$, PVP | 12 | 55 |
| | Cu | CuCl$_2$, SB-12, LiBEt$_3$H | 8.9 | 56 |
| | Co | CoCl$_2$, SB-12, LiBEt$_3$H | 3.5–4.7 | 57 |
| | | CoCl$_2$, SB-12, LiBEt$_3$H | 3.8–8.2 | 58 |
| | Pd | PdCl$_2$, SB-12, LiBEt$_3$H | 3.0–5.2 | 59 |
| | | Pd(OAc)$_2$, PBED | 2.7–4.8 | 60 |
| Binary | InP | InCl$_3$, (TMS)$_3$P, OA, OLA, ODE | 6.4 | 61 |
| | | In(MA)$_3$, (TMS)$_3$P, TOP, OTE | 2.0–3.2 | 62 |
| | SnTe | SnCl$_2$, Te, TOP, OA | 3–60 | 63 |
| | CdS | Cd(NO$_3$)$_2$, Na$_2$S, sodium polyphosphate | 3.2–12 | 64 |
| | | Cd(NO$_3$)$_2$, Na$_2$S, sodium polyphosphate | 5.3 | 65 |
| | CdSe | Cd(OH)$_2$, Se, OA, OLA, TOP, squalane | 1.5–2.7 | 66 |
| | | Cd(Ac)$_2$, Se, SA, TOP, TOPO | 2.0–4.5 | 23 |
| | | Cd(Ac)$_2$, Se, OA, TOP, OLA, ODE | 2.5–7.5 | 33 |
| | | Cd(Ac)$_2$, Se, SA, TOP, TOPO | 1.2–8.3 | 67 |
| | | Cd(Ac)$_2$, Se, TOP, TOPO | 0.9–1.5 | 68 |
| | | CdO, Se, OA, TOP, OLA, TOPO, ODE | 2.2–4.0 | 69 |
| | | CdO, Se, OA, TOP, ODE, fomblin | 3.4–3.8 | 70 |
| | | Cd(Acac)$_2$, Se, OA, TOP, OLA, squalane | 2.0–10 | 71 |
| | | Cd(CH$_3$)$_2$, Se, TBP, DDA, TOPO, ODE | 2.4–2.7 | 72 |
| | | CdO, Se, TOP, OA, ODE | 4.0 | 55 |
| Ternary | CuInSe$_2$ | CuCl, InCl$_3$, Se, OA, TOP | 2.6–4.1 | 73 |
| Core– | CdSe/ZnS | CdO, Se, ZDC, OA, TOP, OLA, ODE | 2.5–3.6 | 74 |
| Shell | | Cd(Ac)$_2$, Se, ZDC, SA, TOP, TOPO, ODE | 2.8–4.9 | 75 |
| | | Cd(Ac)$_2$, Se, Zn(C$_2$H$_5$)$_2$, (TMS)$_2$S, TOP, TOPO | – | 76 |
| | ZnSe/ZnS | Zn(Ac)$_2$, Se, S, TOP, OLA, ODE, TBP, PMMA | 3.2–3.9 | 77 |

AA, ascorbic acid; CTAB, trimethylammonium; DDA, dodecylamine; DDT, dodecanethiol; DMF, dimethylformamide; FDTS, trichloro(1$H$,1$H$,2$H$,2$H$-perfluorooctyl)silane; OA, oleic acid; ODE, 1-octadecene; PBED, poly(benzyl ether) dendron; OLA, oleylamine; OTE, octane; PDMS, polydimethylsiloxane; PMMA, poly(methyl methacrylate); PVP, polyvinylpyrrolidone; SA, stearic acid; SB-12, 3-($N$,$N$-dimethyldodecylammonia)propanesulfonate; TBP, tributylphosphine; (TMS)$_3$P, tris(trimethylsilyl) phosphine; (TMS)$_2$S, bis(trimethylsilyl)sulfide; TOA, trioctylamine; TOP, trioctylphosphine; TOPO, trioctylphosphine oxide; TSC, trisodium citrate; ZDC, zinc diethyl dithiocarbamate.

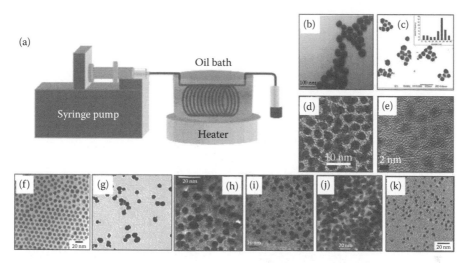

**FIGURE 2.1** Single-phase-based synthesis of unary NPs. (a) Simple continuous flow setup. (Reproduced from Lin, X.Z. et al., *Nano Lett.*, 4, 2227–2232, 2004.) The flow is directed through temperature-controlled tubing, allowing the crystals to grow at a fixed temperature. The extinction of the final product is measured by a flow through spectrometer. Transmission electron microscopic images of (b–e) Au (Reproduced from Weng, C.-H. et al., *J. Micromech. Microeng.*, 18, 035019, 2008, Wagner, J. and Kohler, J.M., *Nano Lett.*, 5, 685–691, 2005, Shalom, D. et al., *Mater. Lett.*, 61, 1146–1150, 2007, Jin, H.D. et al., *Nanotechnology*, 21, 445604, 2010.), (f, g) Ag (Reproduced from Lin, X.Z. et al., *Nano Lett.*, 4, 2227–2232, 2004, Wu, C. and Zeng, T., *Chem. Mater.*, 19, 123–125, 2007.), (h, i) Co (Reproduced from Song, Y. et al., *Chem. Mater.*, 18, 2817–2827, 2006, Song, Y. et al., *Langmuir*, 25, 10209–10217, 2009.), (j) Cu (Reproduced from Song, Y. et al., *J. Phys. Chem. B*, 109, 9330–9338, 2005.), and (k) Pd (Reproduced from Torigoe, K. et al., *J. Nanopart. Res.*, 12, 951–960, 2010.) NPs.

The syntheses of semiconductor NPs (or quantum dots) in microreactor systems have also been widely studied because of the potential advantages over conventional batch reactors. Table 2.1 summarizes the examples of semiconductor NP synthesis by microreactors. Semiconducting NP research to date has primarily focused on II–IV systems such as CdSe due to the relative simplicity of the synthetic routes. Initial studies realized continuous synthesis of CdSe NPs at atmospheric pressure using single-phase laminar flow capillary reactors as shown in Figure 2.2a.[69] These devices have the advantage of not requiring any microfabrication procedure. For example, commercially available silica tubing can easily be immersed in an oil bath for good temperature control, which resulted in CdSe NPs with a narrow size distribution and fairly spherical morphology (Figure 2.2b). Figure 2.2c shows the absorption and emission characteristics and how a variety of residence times influenced the particle size of CdSe NPs: longer reaction times led to larger NPs with red-shifted optical spectra and emission spectra that are narrow indicating excellent monodispersity and high-quality crystal.

Nightingale and de Mello successfully synthesized InP NPs using a chip-based microreactor.[61] The authors used a simple two-in/one-out Y-shaped reactor fabricated from glass suitable for temperatures up to ~300°C. They found that the Y-shaped

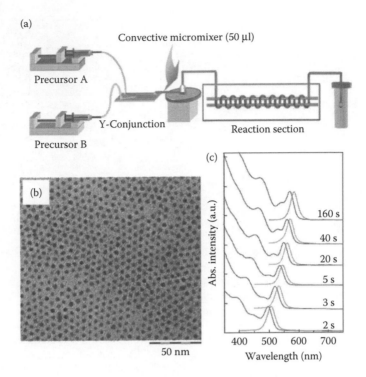

**FIGURE 2.2** (a) A schematic of the capillary microreactor. Two precursor solutions were pumped into a Y-shaped junction and then into a convective mixer, before entering a heated section of capillary where particle nucleation and growth occurred. (b) A typical TEM image of the resultant CdSe NPs obtained at 280°C with a residence time of 160 s. (c) TEM images of (b) 2.2 nm and (c) absorption and emission characteristics of CdSe NP spectra synthesized using different residence times. (Reproduced from Yang, H. et al., *Cryst. Growth Des.*, 9, 1569–1574, 2009.)

microreactor could readily provide precise control over the reaction temperature, the mean reaction time, and the precursor ratios, yielding particles that are of comparable quality to those obtained using macroscale batch chemistry. A more recent study reported a continuous three-stage microreactor system that separates the mixing, aging, and subsequent injection stages of InP NP synthesis. The microreactor system operates at high temperature (320°C) and high pressure (65 bar) enabling the use of solvents operating in the supercritical regime for high diffusivity resulting in the production of high-quality InP NPs.[62]

Very recently, Jin and Chang[73] have reported the first synthesis of ternary I–III–VI$_2$ (CuInSe$_2$) chalcopyrite semiconductor NPs using a continuous hot injection microreactor (Figure 2.3). Using a simple micro T-mixer, they verified that microreactors can be applied to ternary semiconducting materials and achieved CuInSe$_2$ NPs with different sizes, which could be controlled by simply changing the ratios of coordinating solvents. They also reported continuous synthesis of SnTe as both rod-shaped and spherical NPs with uniform size distributions.[63]

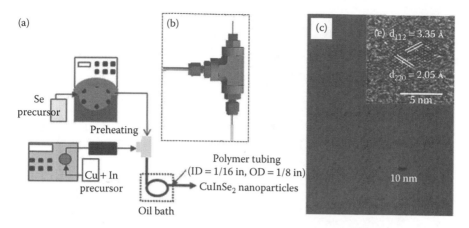

**FIGURE 2.3** Schematic illustration of (a) the continuous flow microreactor system and (b) the continuous hot injection microreactor. (c) TEM images of $CuInSe_2$ NPs with an average diameter of 3.5 nm. Inset shows the high-resolution image of a 3.7 nm-diameter $CuInSe_2$ NPs. (Reproduced from Jin, H.D. and Chang, C.-H., *J. Nanopart. Res.*, 14, 1180, 2012.)

## 2.2.2 SEGMENTED FLOW

Microreactors are a fascinating technology for the highly controlled synthesis of semiconductor NPs. The main limitation of the single-phase reaction mixtures is the wider residence time distributions through axial dispersion, leading to broader particle size distributions than desired. In addition, synthesized NP deposition on the walls of the channel during the synthesis could lead to clogging of the device.[78] Use of segmented flow offers important advantages over single-phase reaction flow, including enhanced reaction control and more reliable operation due to the reduction of deposition and fouling. Segmented flow is generated by introducing an additional immiscible fluid (gas or liquid) into the channel, causing the reaction phase to spontaneously divide into a succession of discrete *slugs* or *droplets*. Application of segmented flows for continuous synthesis of narrowly distributed CdSe NPs has been demonstrated for liquid–gas[79] and liquid–liquid[70] segmented flows.

Yen et al.[79] have used a gas–liquid segmented flow microreactor fabricated in silicon–Pyrex incorporating distinct temperature zones for the synthesis of high-quality CdSe NPs (Figure 2.4). Each precursor was introduced separately into a heated microreactor, mixed and then segmented with argon gas, generating well mixed nanoliter-sized reactors in series. They could control the size of CdSe NPs by changing the precursor flow rates. Compared with a single-phase flow, slug flow consistently yielded particles with sharper absorption spectra (Figure 2.4d) and much narrower emission spectra (Figure 2.4e).

Particle deposition due to physical contact of the synthesized NPs with the walls is a common issue for single-phase microreactors. Droplet flow has one very important advantage over slug flow: as the reaction mixture is isolated from the channel walls, the risk of fouling is minimized. This was demonstrated by Chan et al.,[70] with high-temperature synthesis of CdSe NPs in a glass-based microfluidic droplet reactor with a jet injector (Figure 2.5). They were able to controllably synthesize high-quality CdSe NPs at temperatures up to 300°C, and were able to avoid particle

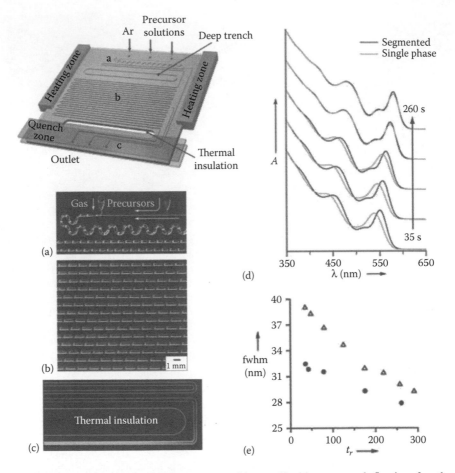

**FIGURE 2.4** Silicon–Pyrex microreactor with gas–liquid segmented flowing for the synthesis of CdSe NPs. (a) Heated inlet region (mixing), (b) main channel region (aging), and (c) cooled outlet region (quenching). (d) Absorbance and emission spectra of the CdSe NPs synthesized using either continuous or slug flow. (e) Photoluminescence full width at half maximum (fwhm) of CdSe NPs synthesized in gas–liquid segmented (●) and single-phase (△) flow. (Reproduced from Yen, B.K.H. et al., *Angew. Chem. Int. Edit.*, 44, 5447–5451, 2005.)

deposition over the 5 h active lifetime of their reactors, whereas analogous conditions in a single-phase continuous flow reactor showed clear layers of particle deposition after just 20 min. The avoidance of particle deposition in droplet flow reactors, which ensures stable flow conditions and consistency of product, is highly beneficial for the practical application of flow reactors.[80] A key issue using microwave in the synthesis of NCs is deposition of the synthesized NCs on the reactor tube wall resulting in sparking. A gas–liquid segmented flow is also used to avoid this problem and has been demonstrated to provide better control of the sizes and shapes of ternary CuInSe$_2$ NCs in comparison with more conventional synthetic methods.[81] The use of this system allows for finely tuned parameters to achieve a high level of control over the reaction by separating the nucleation and growth stages.

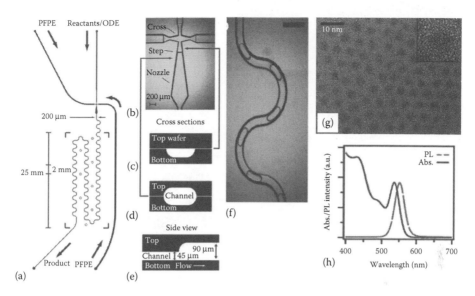

**FIGURE 2.5** Droplet-based liquid–liquid microreactor with droplet jet injector for the synthesis of CdSe NPs. (a) Channel schematic showing dimensions and boundaries of Kapton heater (square brackets). (b) Optical micrograph of droplet injection crosses. The narrowest point is 160 μm wide. (c) Lateral D-shaped cross section of channel etched on the bottom wafer only. (d) Cross section of ellipsoidal channel etched on both top and bottom wafers. (e) Axial cross section showing the 45 μm steps up in channel height. (f) Droplet images in main channel. (g) TEM image of CdSe NPs synthesized at 290°C. Inset shows the high-resolution TEM image of a 3.4 nm-diameter NPs. (h) Online photoluminescence spectrum and off-line absorption spectrum of CdSe NPs. (Reproduced from Chan, E.M., et al., *J. Am. Chem. Soc.*, 127, 13854–13861, 2005.)

Another good example of the use of droplet-based microreactor is tunable growth of Au NPs reported by Duraiswamy and coworkers in 2009 as shown in Figure 2.6.[46] They controlled the size and shape of the NPs at different temperatures using a droplet flow within a microchannel and showed that it is possible to modify the surface plasmons and hence tune the optical absorption spectra. The dimensions of the Au rods could be controlled by varying the chemical composition of the reaction mixture, with aspect ratios as high as 4 being obtained. The segmentation flow studies above provide the driving force for improvements in reaction yield and size distribution by enhanced mixing and reduced residence time distributions. The droplet-based microreactor particularly can be used to manipulate droplets over a wide range of flow rates with the ability to form droplets at a low viscosity ratio of the two immiscible liquids at high capillary numbers and temperatures.

## 2.2.3 Multistep Synthesis of Fluorescent Core–Shell NPs

Significant research has been done to study the size-dependent optical properties of NPs over the past two decades. A significant fraction of these organically passivated

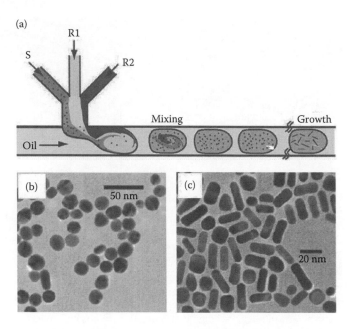

**FIGURE 2.6** (a) Droplet-based synthesis of Au NPs synthesized in microreactor. TEM images of (b) Au nanospheres and (c) Au nanorods. (Reproduced from Duraiswamy, S. and Khan, S.A., *Small*, 5, 2828–2834, 2009.)

core NPs exhibits surface-related trap states acting as fast recombination of photo-generated charge carriers, thereby reducing the fluorescence quantum yield. The fluorescence efficiency and stability of various types of NPs could be improved by the appropriate choice of a shell of a second semiconductor material; it is also possible to tune the emission wavelength in a larger spectral window than with each material alone.[82] The most studied core–shell structure to date is CdSe/ZnS NPs, since first described by Hines and Guyot-Sionnest, who overcoated 3 nm CdSe NPs with one to two monolayers of ZnS using a batch process.[83]

Microreaction technologies had proven to be beneficial for producing core–shell NPs with well-controlled shell thickness. Several multistep continuous synthetic methods of core–shell CdSe/ZnS (or ZnSe/ZnS) for the synthesis of highly luminescent ZnS-capped CdSe (or ZnSe) with a narrow size distribution (Table 2.1) have been reported. Luan et al.[74] demonstrated a facile method for the synthesis of CdSe/ZnS core/shell structures showing clean green luminescence spectra via a process in a microreactor with a short residence time and low reaction temperature. CdSe NPs and Zn sources were mixed by a convective mixer before entering a heated polytetrafluoroethylene capillary for the coating process. They achieved homogeneous coatings of the ZnS shell with fairly wide operation parameters, such as residence times and temperatures. All of these CdSe NPs with different ZnS shell thicknesses could

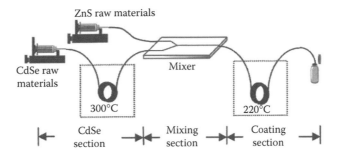

**FIGURE 2.7** An example schematic illustration of CdSe/ZnS core–shell NP synthetic process using a continuous-flow thermoplastic-based microfluidic reaction system. (Reproduced from Wang, H. et al., *Chem. Commun.*, 48–49, 2004.)

be generated by simply varying the flow rate and capillary length. Specifically, Wang et al.[75] used a chip-based microreactor and two separate oil baths to maintain defined temperatures for the nucleation and growth of CdSe core and ZnS shell, respectively. Subsequently, the same group demonstrated that highly luminescent ZnS-capped CdSe with a quantum yield of over 50% and narrow size distribution can be produced using the commercial single-molecular precursor of ZnS in a microreactor (Figure 2.7).[76]

## 2.2.4 Synthesis of Metal Oxide NPs

Nanostructured metal oxides are one of the most important materials in many applications such as energy generation and storage, and environmental, electronic, chemical, and biological systems. Abou-Hassan et al.[84] reported the synthesis of $Fe_2O_3$ NPs using a hydrodynamic focusing flow that enables the formation of a stable central jet in which the nuclei formation of ferrihydrite occurs (Figure 2.8). The separation of nucleation and growth controls the particle growth more effectively. Ferrihydrite-$Fe_2O_3$ nucleated in microreactor R1, and these seeds were delivered to the aging microreactor for the growth of goethite NPs. An aging time over 15 min is necessary to obtain goethite $Fe_2O_3$ platelike nanostructures. The anisotropic structure is a result of aggregation of nanodots.

Cottam et al.[85] synthesized one-dimensional anatase $TiO_2$ nanostructures using microfluid chips. The nanostructures possess broom-like aggregates of anatase $TiO_2$ nanorods. Compared to $TiO_2$ nanorods prepared by a batch reaction, the reaction rate is significantly higher in microfluidic chips. The rapid mixing conditions in the microchannels give rise to significantly higher heat transfer and uniform reactant concentration, increasing the reaction rate. Synthesis of p-type ZnO NPs is known to be a challenging task because of the self-compensation and low solubility of

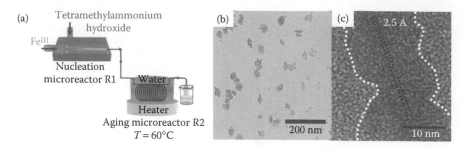

**FIGURE 2.8** (a) Scheme of the microreactor, (b) transmission electron microscopic (TEM) image of platelike $Fe_2O_3$ nanostructures, and (c) high-resolution TEM image of a typical platelike $Fe_2O_3$ nanostructure. (Reproduced from Abou-Hassan, A. et al., *Angew. Chem. Int. Edit.*, 48, 2342–2345, 2009.)

dopants.[86,87] Kim et al.[88] successfully prepared flowerlike assemblies of p-type ZnO NPs using a continuous flow microreactor. The p-type ZnO NPs were synthesized by incorporation of Na into the ZnO lattice by adding NaOH as the precipitating agent. X-ray photoelectron spectroscopic and high-resolution transmission electron microscopic analysis confirmed the intercalation of Na ions into the ZnO structure. Photocatalytic activities of p-type Na-doped ZnO are significantly improved over undoped ZnO, which is attributed to the increased surface defect sites that constitute the oxygen vacancy and the substitution of Na for Zn site.

Frenz et al.[89] synthesized $Fe_2O_3$ NPs using droplet-based microreactors (Figure 2.9a). Individual droplets contain equal quantities of reactants and hence offer better control of NP growth conditions. The nucleation and growth of NPs take place after individual droplets containing different reactants combine together. The segmented flow facilitates the production of monodispersed $Fe_2O_3$ NPs (Figure 2.9b).

Khan et al.[28] used the segmented flow for the synthesis of colloidal $SiO_2$ NPs (Figure 2.10). Four different reactant solutions pump into the four inlets and mix well

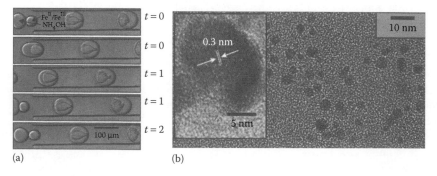

**FIGURE 2.9** (a) Formation of $Fe_2O_3$ precipitates after coalescence of reactants in droplet-based microreactor and (b) TEM image of $Fe_2O_3$ NPs. (Reproduced from Frenz, L. et al., *Angew. Chem. Int. Edit.*, 47, 6817–6820, 2008.).

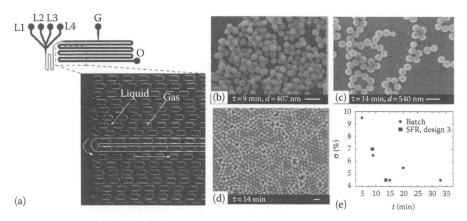

**FIGURE 2.10** (a) Segmented flow in microreactor: uniform, noncoalescing gas–liquid flow. Sequence of Scanning electron microscope (SEM) images corresponding to various resident times: (b) $\tau = 9$ min, (c) $\tau = 14$ min, (d) low-magnification SEM of sample, (e) plot of standard deviation ($\sigma$) of mean diameter as a function of residence time. SFR, segmented flow reactor. Scale bar: 1 μm. (Reproduced from Khan, S.A. et al., *Langmuir*, 20, 8604–8611, 2004.)

in the microreactor. Gas, injected into the microreactor, generates the gas–liquid segmented flow. Colloidal $SiO_2$ NPs with a narrow particle size distribution comparable to the state-of-the-art batch results were obtained as shown in Figure 2.10b–d.

## 2.3   SELF-ASSEMBLY OF NANOSTRUCTURES

Fabrication of an NC superlattice via self-assembly is important for the development of NC-based electronic and optoelectronic devices such as displays, solar cells, and light-emitting diodes. In a self-assembly process, the building blocks organize themselves into a superstructure as driven by the energetics of the system. The major driving force for self-assembly is the interaction between the building blocks. Microfluidics provides a number of unique advantages, such as ultrasmall material consumption and precise control over molecular diffusion and crystal nucleation. Abou-Hassan's group[90] has successfully assembled the multifunctional fluorescent, plasmonic, and magnetic nanostructures Au and $\gamma$-$Fe_3O_4$ NPs onto $SiO_2$ NPs using multistep microfluidic device. A colloidal dispersion of $SiO_2$ NPs and a solution of citrate-coated Au NPs were injected into the first microreactor ($\mu$R1), and then a stable suspension of the obtained $SiO_2$-Au nanostructures was continuously injected in one inlet of the second microreactor ($\mu$R2). At the same time, stable citrate-coated $\gamma$-$Fe_3O_4$ NPs were injected in the second inlet of $\mu$R2 (Figure 2.11). By varying the flow rate of the Au (or $\gamma$-$Fe_3O_4$) and $SiO_2$ NPs, they could control the Au (or $\gamma$-$Fe_3O_4$) coverage on the $SiO_2$ surfaces. By contrast, $\gamma$-$Fe_3O_4$ on $SiO_2$-Au nanostructures was found to depend on the residence time where short residence times showed a very poor number of Au NPs covering the surface of $SiO_2$, and the $SiO_2$ surface was completely covered with the $\gamma$-$Fe_3O_4$ in case of $\gamma$-$Fe_3O_4$-$SiO_2$ and $\gamma$-$Fe_3O_4$-Au-$SiO_2$ assemblies.

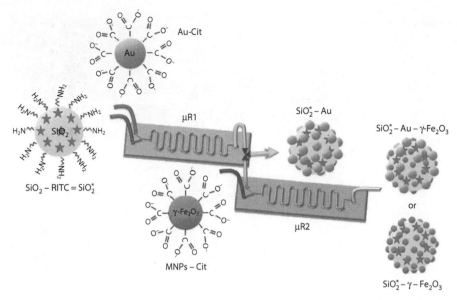

**FIGURE 2.11** A two-step microfluidic synthetic procedure for the assembly of multifunctional NPs/fluorescent SiO$_2$ sphere assemblies. RITC, rhodamine isothiocyanate; Cit, citrate; MNPs, Maghemite γ-Fe$_2$O$_3$ nanoparticles. (Reproduced from Hassan, N. et al., *Angew. Chem. Int. Edit.*, 52, 1994–1997, 2013.)

The Talapin group[91] successfully grew large superlattices from metallic, semiconducting, and magnetic NCs in the microfluidic plugs using both *evaporation-driven* and *destabilization-driven* approaches (Figure 2.12). In this module, they assembled long-range ordered three-dimensional colloidal NC superlattices within nanoliter microfluidic droplets via the destabilization-driven approach, which uses slow destabilization of a colloidal solution by contacting the NC solution with a precipitant that slowly diffuses into the NCs. By properly tuning of the precipitant/solvent ratio, they demonstrated impressive plug-to-plug reproducibility of the crystallization conditions and found the conditions leading to large NC superlattices with well-developed facets of the selected NCs.

Self-assembled noble metal-decorated core–shell particles (Pt/Fe$_3$O$_4$/SiO$_2$) using a continuous flow microreactor were explored by the Jensen group.[92] Highly dispersed Pt NPs with a narrow size distribution were coated on Fe$_3$O$_4$/SiO$_2$ particles when ethylene glycol was used as a reducing agent. Precise control of the inner-to-outer flow rate ratio in the microfludic device led to controlled emulsion sizes with a very uniform distribution and the particles are converted into spherical assemblies with micron sizes. The catalytic activity for the oxidation of 4-isopropyl benzaldehyde showed an excellent activity and selectivity compared with commercial noble metal catalysts.

Choi et al.[93] synthesized stable monodispersed colloidal ZnO NPs and their assemblies in an aqueous medium using a continuous flow microreactor (Figure 2.13). Assembly of ZnO NPs were obtained via a competition between the Dean vortices

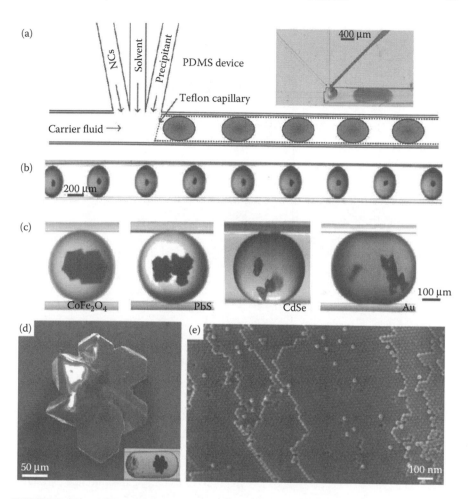

**FIGURE 2.12** (a) Microfluidic device used for NC self-assembly. The nanoliter plugs were formed by mixing the NC colloidal solution with the precipitant. The inset shows an optical photograph of a typical polydimethylsiloxane device during an experiment. (b) Optical image of an array of plugs containing PbS NC superlattices. (c) Optical images of plugs containing superlattices formed from various NCs: 11 nm $CoFe_2O_4$, 10 nm PbS, 3.5 nm CdSe, and 7 nm Au (left to right). (d) SEM image of a faceted superlattice of 11 nm $CoFe_2O_4$ NCs grown by incubation of plugs containing NCs in the ethanol/toluene solution. The inset shows an optical image of the plug containing exactly the same superlattice. (e) High-resolution SEM image of a superlattice self-assembled from 20 nm $CoFe_2O_4$ NCs. PDMS, polydimethyl siloxane. (Reproduced from Bodnarchuk, M.I. et al., *J. Am. Chem. Soc.*, 133, 8956–8960, 2011.)

and the electrostatic forces. Because the magnitude of Dean vortices is determined by the solution flow rates, assemblies of ZnO NPs with different structures could be generated at different flow rates. Most of the NPs are dispersed stably without showing aggregation at a slow flow rate, whereas high flow rates create the assemblies (Figure 2.13d–f).

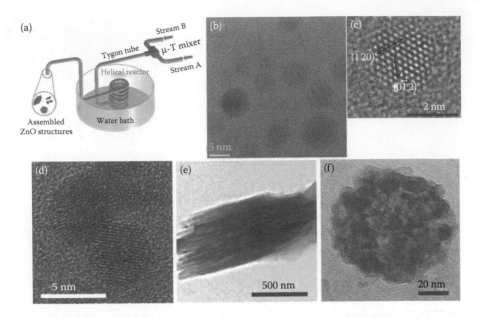

**FIGURE 2.13** (a) Scheme of microreactor-assisted nanoparticles deposition process, (b) monodispersed ZnO NPs, (c) HRTEM image of ZnO NPs, (d) dispersed ZnO NPs, (e) tactoid assembly, and (f) spherical assembly. (Reproduced from Choi, C.-H. et al., *Cryst. Eng. Commun.*, 15, 3326–3333, 2013.)

## 2.4 METAL–ORGANIC FRAMEWORKS BY CONTINUOUS FLOW MICROREACTORS

Metal–organic framework (MOF) materials composed of metal ions, or clusters, as a center and organic molecule ligands as linkers have recently garnered much attention due to their excellent properties for gas purification, separation, and gas storage.[94–96] Several hundred different types of MOF materials have been synthesized using a concept known as reticular design.[97] MOF materials are typically synthesized utilizing solvothermal batch methods.[98] The major drawback is that the reaction under batch solvothermal conditions typically requires long reaction times, sometimes up to several days. In order to reduce the reaction time from days to minutes, a few synthesis methods have been developed recently.

Continuous flow droplet-based microfluidic devices offer a high level of mixing operations, and hence are promising for the development of a wide range of porous MOF assemblies. Ameloot et al.[99] synthesized uniform hollow Cu-BTC microspheres via interfacial reaction conditions using a continuous droplet microreactor (Figure 2.14a). They demonstrated how the intrinsically hybrid nature of MOFs enables self-completing growth, which is directly related to the micropore size of the MOF crystallites forming the capsule wall. Microdroplets can also produce novel heterostructures, such as core–shell particles, by confining the MOF precursor solutions within individual droplets that are suspended in an immiscible fluid. Faustini et al.[100] have used this technique for preparation of $Fe_3O_4$ NPs encapsulated by zeolitic

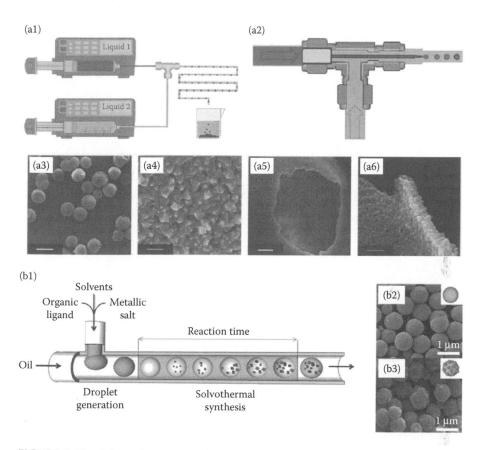

**FIGURE 2.14** Schematic representation of the continuous synthesis systems used for the production of MOFs. (a1) Droplet technique for the formation of hollow MOF capsules. (a2) Detailed emulsification step in T-junction and SEM images with different scales: (a3) 500 nm, (a4) 25 nm, (a5) 2 nm, and (a6) 2 nm. (b1) Schematic representation of the continuous droplet microfluidic synthesis of $Fe_3O_4$@ZIF-8 catalyst. SEM images of (b2) magnetic $Fe_3O_4$ particles and (b3) core–shell $Fe_3O_4$@ZIF-8 catalyst. (*Continued*)

imidazolate framework (ZIF-8) using polystyrene sulfonate (Figure 2.14b). The $Fe_3O_4$@ZIF-8 particles were proven to be an efficient catalyst for a model Knoevenagel condensation reaction of benzaldehyde and ethyl cyanoacetate to yield ethyl (E)-α-cyanocinnamate. Kim et al.[101] have demonstrated the utility of a continuous flow microreactor consisting of a micro T-mixer for instantaneous mixing of reactants at high pressure (Figure 2.14c) and its ability to control the physical properties of Cu-BTC MOF materials. The particle size and the crystal phase of these materials could be tuned by changing the relative ratios of solvents. Carné-Sánchez et al.[102] used a specialized nozzle to spray precursor solutions to create atomized droplets which instantly react to form hollow nanocrystalline MOF in a continuous process (Figure 2.14d). They suggest that this spray-drying process enables the construction of multicomponent nanoMOF superstructures and the encapsulation of guest species within these superstructures.

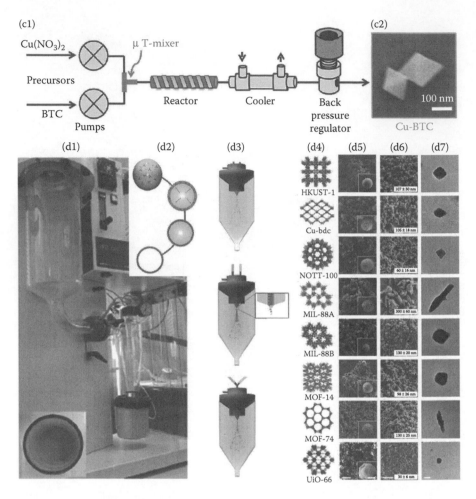

**FIGURE 2.14 (Continued)** Schematic representation of the continuous synthesis systems used for the production of MOFs. (c1) Microreactor-assisted system and (c2) SEM image of nanoscale Cu-BTC. (Reproduced from Ameloot, R. et al., *Nat. Chem.*, 3, 382–387, 2011; Faustini, M. et al., *J. Am. Chem. Soc.*, 135, 14619–14626, 2013; Kim, K.-J., *Chem. Commun.*, 49, 11518–11520, 2013.) (d1) Photograph of the spray dryer. (d2) Proposed spherical super-structure formation process. (d3) Schematic diagrams of the spray-drying processes. (d4) Series of MOF superstructures. SEM (d5, d6) and TEM (d7) images of the synthesized MOF super-structures and discrete nanoMOF crystals. Scale bars: 10 μm (d5), 500 nm (d6), and 50 nm (d7). (Reproduced from Carné-Sánchez, A. et al., *Nat. Chem.*, 5, 203–211, 2013.)

## 2.5   DEPOSITION OF NANOSTRUCTURED THIN FILMS BY CONTINUOUS FLOW MICROREACTORS

In this section, the deposition of nanostructured films using a microreactor-assisted chemical process, referred to herein as microreactor-assisted solution deposition (MASD), is introduced and compared to the static batch process.

It is necessary to optimize the chemical bath deposition (CBD) process by controlling the homogeneous particle formation and the heterogeneous molecular surface reaction with high growth rate and conversion efficiencies in addition to eco-friendly processing. MASD (also referred to as continuous flow microreactor deposition) was developed to meet those demands by many research groups led by Chang, Ryu, Baxter, and Paul.[27,35,103–110,116–118] The various names of microreactor-assisted processes are represented as MASD in this section. To control the film growth mechanism, either homogeneous or heterogeneous reactions have to be well controlled by providing precursors continuously and reducing thermal/chemical diffusion times in a micro-channel. Moreover, a novel deposition technique is demanded for creating high-quality and uniform films through a microreactor system. Consequently, the MASD is able to minimize the drawbacks of conventional chemical solution deposition by controlling the operating variables of linear velocity, mean residence time, thermal/chemical distributions, concentration gradients, and reagent mixing conditions. Diverse novel deposition techniques, such as fixed-tilt, rotating, flow cell, and rod coating, have been combined with microreactor systems for investigating, developing, and improving the mechanism of film deposition and scale-up, shown in Figure 2.15. Excellent results have been achieved by using these novel deposition techniques so far. However, more advanced techniques could be adopted for improving film deposition.

## 2.5.1 Effect of MASD over CBD Process on Surface Morphology and Film Quality

Dense and smooth nanocrystalline thin films using NPs as building blocks are able to be produced via the MASD process because the reactants are continuously refreshed and the effect of homogeneously nucleated particles agglomerating on the surface of the growing film is minimized. By contrast, a conventional static batch reactor process would produce films from large aggregates that exhibit coarse morphologies, which could then form a mixture of heterogeneously grown particles along with homogeneously grown, largely aggregated particles.

Prior works for CdS thin-film deposition from CBD and MASD processes done by Chang's group have clearly demonstrated the advantages of MASD.[26,35,103,105] A comparison between CdS thin-film depositions by a conventional static batch reactor (CBD) and an MASD process has been reported and the selected results are shown in Figure 2.16. The CdS film deposited by the MASD process is smooth and dense, and demonstrates continuous morphology, although the static batch reactor process resulted in discrete CdS islands on the order of hundreds of nanometers (Figure 2.16a and b).[106] Transmission electron microscopic images suggest that the batch reactor process generates particles through the homogeneous reaction and the MASD process results in dense films via molecule-by-molecule heterogeneous surface reaction. The X-ray diffraction (XRD) patterns show that the MASD process creates more uniform, highly oriented CdS films and that the batch process creates films that are of a more polycrystalline nature as shown in Figure 2.16c. Outstanding film morphology of dense CuSe thin film deposited by MASD process over CBD is also reported by the Ryu's group (shown in Figure 2.16d and e).[107] Based on the

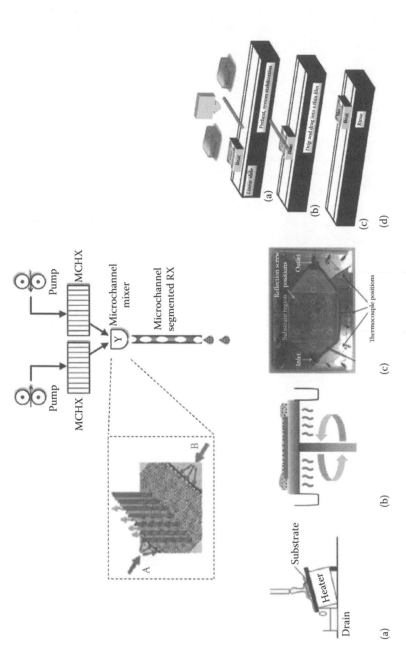

**FIGURE 2.15** Schematic diagram of film deposition techniques using microreactor-assisted chemical process: (a) fixed-tilt sample holder, (b) rotating sample holder, (c) flow cell, and (d) rod-coating process. MCHX, micro-channel heat exchanger. (Reproduced from Chang, Y. J. et al., *Electrochem. Solid-State Lett.*, 9, G174–G177, 2006; Paul, B.K. et al., *Cryst. Growth Des.*, 12, 5320–5328, 2012; Chang, Y.J. et al., *Electrochem. Solid-State Lett.*, 12, H244–H247, 2009; Su, Y.W. et al., *Thin Solid Films*, 532, 16–21, 2013.)

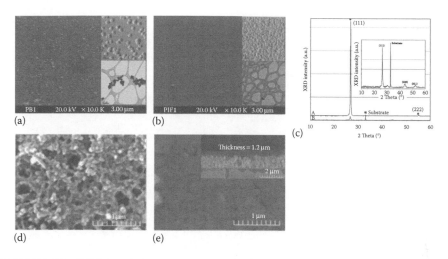

**FIGURE 2.16** SEM, Atomic force microscopic (AFM), and TEM images of CdS films from CBD (a) and MASD (b) processes. (c) XRD patterns of CdS thin films deposited by a continuous-flow (inset of a batch reactor). (Reproduced from Mugdur, P.H. et al., *J. Electrochem. Soc.*, 154, D482–D488, 2007.) SEM images of CuSe thin films by CBD (d) and MASD (e) processes. (Reproduced from Kim, C.R. et al., *Mol. Cryst. Liquid Cryst.*, 532, 455–463, 2010.)

excellent achievement of CuSe thin-film deposition via MASD process, $CuInS_2$ and $CuInSe_2$ thin films were also successfully deposited through MASD process.[108–109] These results indicate that MASD process could open a new avenue for next-generation solution-based film deposition process.

## 2.5.2 Effect of MASD over CBD Process on Growth Rate

The MASD process also provides a higher growth rate compared to a typical static batch reactor process, and the higher growth rates of MASD processes follow a linear trend. The linear growth rate of MASD conditions is due in part to the constant growth conditions provided by continuously replenishing the precursor solution. By contrast, the stagnation of growth in the static batch reactor deposition condition at longer deposition times can be attributed to a decrease in the concentration of precursors. The films from MASD have a higher density because of the higher nucleation density that corresponds to higher reactant concentrations. High growth rates of various material thin films via the MASD process have been demonstrated. Su et al.[105] reported high growth rates of dense CdS thin films on fluorine-doped tin oxide (FTO)-coated glass. The thickness of 251.7 nm CdS thin film on top of an FTO layer was deposited by an MASD process (Figure 2.17a). The average growth rate of CdS thin film from the MASD process was calculated around 25.2 nm min$^{-1}$. This growth rate is significantly higher than the values from typical static batch reactor processes, which have reported highly textured columnar CdS films with average deposition rates of 2.5 and 1.67 nm min$^{-1}$ from $CdSO_4$/EDA and $CdCl_2$/$NH_3$, respectively.[110] The MASD process was extended by Ryu's group into $CuInS_2$[108] and $CuInSe_2$[109] thin

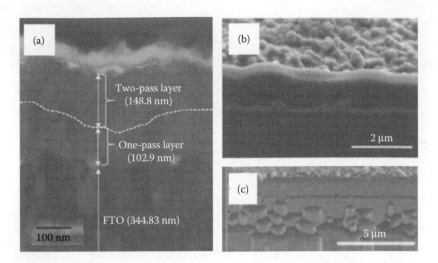

**FIGURE 2.17** (a) CdS (Reproduced from Su, Y.W. et al., *Thin Solid Films*, 532, 16–21, 2013.), (b) CuInS$_2$ (Reproduced from Park, M.S. et al., *Curr. Appl. Phys.*, 10, S379–S382, 2010.), and (c) CuInSe$_2$ (Reproduced from Kim, C.R. et al., *Curr. Appl. Phys.*, 10, S383–S386, 2010.) thin films from MASD process.

films, and they were able to enhance the growth rates: 1 μm thickness of CuInS$_2$ films in 5 min and 2.5 μm thickness of CuInSe$_2$ films in 7 min via MASD processes (Figure 2.17b and c).

### 2.5.3 EFFECT OF MASD OVER CBD ON CONTROL OF NANOSTRUCTURES

The ability of MASD to tailor ZnO nanostructures by simply adjusting physical process parameters such as the flow rate of solutions and the rotating speed of substrate was demonstrated by Choi and Chang.[111] In contrast to CBD process, the MASD enables to control the reaction kinetics as a function of the flow rate of solution, and thus controlled reactive species can be continuously delivered onto a substrate for nanostructure formation. The type of reactive species can vary upon the flow rate of solution, including zinc molecular precursors, colloidal ZnO NPs, and assembly of ZnO NPs. These selectively formed species result in the formation of ZnO nanostructures with various morphologies. The nanostructure, including an amorphous layer (Figure 2.18a), flowerlike structured film (Figure 2.18b), vertical ZnO nanowire (NW) arrays (Figure 2.18c), and crystalline ZnO film (Figure 2.18d), is obtained. The authors found that zinc molecular precursors serve as building blocks for amorphous ZnO film growth under high rotating speed of substrate. The assembly of ZnO NPs served as the seed layer for the subsequent growth of flowerlike ZnO nanostructure by consumptions of zinc molecular precursors. The amorphous ZnO film was also used as the seed layer for the growth of vertical ZnO NW arrays. Because the colloidal ZnO NPs and molecular precursors were continuously delivered onto the amorphous seed layer as building blocks for the growth of vertical ZnO NW arrays, the growth was controlled by the deposition period, not limited by the depletion of

**FIGURE 2.18** SEM images of ZnO nanostructures fabricated by the MASD process: (a) Amorphous film, (b) flowerlike structured film, (c) vertical ZnO NW arrays, and (d) polycrystalline film. (Reproduced from Choi, C.-H. and Chang, C.-H., *Cryst. Growth Des.*, 14, 4759–4767, 2014.)

precursors, offering wide range of NW dimensions. Continuous polycrystalline ZnO layer was also formed in a similar manner by delivering the assembly of ZnO NPs onto the amorphous seed layer.

## 2.5.4 Effect of MASD over CBD Process on Scale-Up

It is expected that the MASD process will be easier to scale up than conventional batch solution-based processes while also offering the possibility to minimize the environmental impact of nanomanufacturing practices through solvent-free mixing and integrated separation. Point-of-use synthesis has the potential to eliminate the need to store and transport hazardous NPs. For these reasons, MASD has been pursued in many solution-processing applications.[23,28,53,70,72,112,113] Various materials and deposition techniques through MASD process for film formation have been demonstrated and achieved with significant evolution. Few groups led by Chang, Paul, and Baxter have demonstrated large-scale film deposition using a microreactor system similar to that shown in Figure 2.15.

McPeak et al.[114–116] have demonstrated the deposition of ZnO and CdZnS thin films with a plug flow reactor under the name continuous flow microreactor process. Although this approach was able to form ZnO and CdZnS films by taking advantage of the microreactor system, it still has to be optimized for film deposition over large areas with good uniformity. Paul et al.[104] have developed a deflected plate flow cell technique for coating large areas with uniform CdS thin films incorporated with the MASD process, shown in Figure 2.15c. The deflected plate flow cell was designed for eliminating dead volume, avoiding depletion of reagents by balancing reactant concentration and residence time within the deposition region, and minimizing the variation of flow distribution. Recently, Ramprasad et al.[38] have demonstrated a route to process scale-up for depositing CdS films with high uniformity operating near room temperature under atmospheric pressures. The schematic of a pilot rod coater deposition unit was shown in Figure 2.15d. The rod coating technique adapted to the MASD process is simple and cost-effective to implement. This approach is favorably comparable to commercial roll coating systems, which are well known as a typical thin-film coating process. This excellent result opens a potential avenue of the

MASD process for large area deposition of various thin films with excellent uniformity and thicknesses ranging from 70 to 230 nm.

The significant improvements seen with MASD processes have been achieved by addressing the problems of typical chemical solution processes with microreactor systems. It is expected that these microreactor processes could hold the potential to deposit large area films with excellent uniformity at low cost, low processing temperature, and simple operational demand. However, this technique still has room to improve film uniformity for large area film deposition. Therefore, greater precision and further efforts are demanded to manage and address the remaining problems.

## 2.6  CONCLUSION AND OUTLOOK

In this chapter, we described the recent development of the microreactor-assisted chemical process in synthesis, assembly, and film formation of NPS. The high surface-area-to-volume ratios within microreactor systems permit shorter diffusional distances allowing for rapid mixing and precise control of reaction conditions, which led to improved homogeneity of particle size distributions and reproducible control of particle size. They also allow the potential for simplifying the scale-up of nanomaterials synthesis with reduced formation of associated wastes. High throughput can also be directly scaled by continuous flow operation of multiple microreactors in parallel. In addition, MASD processing combines the merits of microreaction technology with solution-phase NP synthesis and film deposition. This process takes advantage of microreaction technology and the large library of near room-temperature, liquid-phase nanomaterial synthesis recipes to produce and assemble nanofilms at the point of deposition. As mentioned above, microreactor-assisted chemical processes have distinct advantages over traditional batch processes that motivate their adoption by the broader scientific community. Therefore, continuous flow microreactors have the potential to become standard technology and a great deal of effort has provided for significant growth in developing microreaction technologies for nanoscale chemical synthesis.

## ACKNOWLEDGMENT

The authors express their gratitude to Peter B. Kreider at the department of chemical engineering, Oregon State University, Corvallis, Oregon, for his valuable comments and suggestions.

## REFERENCES

1. K. E. Drexler, *Nanosystems: Molecular Machinery, Manufacturing, and Computation*, Wiley, New York, 1992.
2. J. Chen, B. J. Wiley, and Y. Xia, One-dimensional nanostructures of metals: Large-scale synthesis and some potential applications, *Langmuir*, 2007, 23, 4120–4129.
3. Y. N. Xia, P. D. Yang, Y. G. Sun, Y. Y. Wu, B. Mayers, B. Gates, Y. D. Yin, F. Kim, and Y. Q. Yan, One-dimensional nanostructures: Synthesis, characterization, and applications, *Advanced Materials*, 2003, 15, 353–389.
4. N. R. Jana, L. Gearheart, S. O. Obare, and C. J. Murphy, Anisotropic chemical reactivity of gold spheroids and nanorods, *Langmuir*, 2002, 18, 922–927.

5. C. H. Leung and K. S. Song, Model of excitonic mechanism for defect formation in alkali halides, *Physical Review B*, 1978, 18, 922–929.
6. R. J. Chimentao, I. Kirm, F. Medina, X. Rodriguez, Y. Cesteros, P. Salagre, and J. E. Sueiras, Different morphologies of silver nanoparticles as catalysts for the selective oxidation of styrene in the gas phase, *Chemical Communications*, 2004, 846–847.
7. L. A. Dick, A. D. McFarland, C. L. Haynes, and R. P. Van Duyne, Metal film over nanosphere (MFON) electrodes for surface-enhanced raman spectroscopy (SERS): Improvements in surface nanostructure stability and suppression of irreversible loss, *Journal of Physical Chemistry B*, 2002, 106, 853–860.
8. K. Esumi, K. Matsuhisa, and K. Torigoe, Preparation of rodlike gold particles by uv irradiation using cationic micelles as a template, *Langmuir*, 1995, 11, 3285–3287.
9. F. Kim, J. H. Song, and P. D. Yang, Photochemical synthesis of gold nanorods, *Journal of the American Chemical Society*, 2002, 124, 14316–14317.
10. C. B. Murray, D. J. Norris, and M. G. Bawendi, Synthesis and characterization of nearly monodisperse cde (e = sulfur, selenium, tellurium) semiconductor nanocrystallites, *Journal of the American Chemical Society*, 1993, 115, 8706–8715.
11. T. Hyeon, S. S. Lee, J. Park, Y. Chung, and H. Bin Na, Synthesis of highly crystalline and monodisperse maghemite nanocrystallites without a size-selection process, *Journal of the American Chemical Society*, 2001, 123, 12798–12801.
12. J. Park, K. J. An, Y. S. Hwang, J. G. Park, H. J. Noh, J. Y. Kim, J. H. Park, N. M. Hwang, and T. Hyeon, Ultra-large-scale syntheses of monodisperse nanocrystals, *Nature Materials*, 2004, 3, 891–895.
13. F. Dumestre, B. Chaudret, C. Amiens, P. Renaud, and P. Fejes, Superlattices of iron nanocubes synthesized from $Fe[N(SiMe_3)_2]_2$, *Science*, 2004, 303, 821–823.
14. J. Joo, H. B. Na, T. Yu, J. H. Yu, Y. W. Kim, F. X. Wu, J. Z. Zhang, and T. Hyeon, Generalized and facile synthesis of semiconducting metal sulfide nanocrystals, *Journal of the American Chemical Society*, 2003, 125, 11100–11105.
15. J. P. Jolivet, C. Froidefond, A. Pottier, C. Chaneac, S. Cassaignon, E. Tronc, and P. Euzen, Size tailoring of oxide nanoparticles by precipitation in aqueous medium. A semi-quantitative modelling, *Journal of Materials Chemistry*, 2004, 14, 3281–3288.
16. J. R. Zhang and L. Gao, Synthesis and characterization of nanocrystalline tin oxide by sol-gel method, *Journal of Solid State Chemistry*, 2004, 177, 1425–1430.
17. I. R. Mangani, C. W. Park, and J. Kim, Synthesis and electrochemical properties of layered $Li[Li_{(1/3-x/3)}Cr_xMn_{(2/3-2x/3)}]O_2$ prepared by sol-gel method, *Electrochimica Acta*, 2006, 52, 1451–1456.
18. I. Capek, Preparation of metal nanoparticles in water-in-oil (w/o) microemulsions, *Advances in Colloid and Interface Science*, 2004, 110, 49–74.
19. M. J. Rosemary and T. Pradeep, Solvothermal synthesis of silver nanoparticles from thiolates, *Journal of Colloid and Interface Science*, 2003, 268, 81–84.
20. Y. Sun, N. G. Ndifor-Angwafor, D. J. Riley, and M. N. R. Ashfold, Synthesis and photoluminescence of ultra-thin zno nanowire/nanotube arrays formed by hydrothermal growth, *Chemical Physics Letters*, 2006, 431, 352–357.
21. A. J. DeMello, Control and detection of chemical reactions in microfluidic systems, *Nature*, 2006, 442, 394–402.
22. S. Krishnadasan, R. J. C. Brown, A. J. deMello, and J. C. deMello, Intelligent routes to the controlled synthesis of nanoparticles, *Lab on a Chip*, 2007, 7, 1434–1441.
23. H. Nakamura, Y. Yamaguchi, M. Miyazaki, H. Maeda, M. Uehara, and P. Mulvaney, Preparation of cdse nanocrystals in a micro-flow-reactor, *Chemical Communications*, 2002, 2844–2845.
24. R. K. Shah, H. C. Shum, A. C. Rowat, D. Lee, J. J. Agresti, A. S. Utada, L.-Y. Chu et al., Designer emulsions using microfluidics, *Materials Today*, 2008, 11, 18–27.

25. L.-Y. Chu, A. S. Utada, R. K. Shah, J.-W. Kim, and D. A. Weitz, Controllable monodisperse multiple emulsions, *Angewandte Chemie International Edition*, 2007, 46, 8970–8974.
26. Y. J. Chang, Y. W. Su, D. H. Lee, S. O. Ryu, and C. H. Chang, Investigate the reacting flux of chemical bath deposition by a continuous flow microreactor, *Electrochemical and Solid State Letters*, 2009, 12, H244–H247.
27. S. Y. Han, Y. J. Chang, D. H. Lee, S. O. Ryu, T. J. Lee, and C. H. Chang, Chemical nanoparticle deposition of transparent ZnO thin films, *Electrochemical and Solid State Letters*, 2007, 10, K1–K5.
28. S. A. Khan, A. Gunther, M. A. Schmidt, and K. F. Jensen, Microfluidic synthesis of colloidal silica, *Langmuir*, 2004, 20, 8604–8611.
29. A. R. Abate, D. Lee, T. Do, C. Holtze, and D. A. Weitz, Glass coating for PDMS microfluidic channels by sol-gel methods, *Lab on a Chip*, 2008, 8, 516–518.
30. K. Jahnisch, V. Hessel, H. Lowe, and M. Baerns, Chemistry in microstructured reactors, *Angewandte Chemie International Edition*, 2004, 43, 406–446.
31. R. Knitter and M. A. Liauw, Ceramic microreactors for heterogeneously catalysed gas-phase reactions, *Lab on a Chip*, 2004, 4, 378–383.
32. C. Iliescu, B. Chen, and J. Miao, On the wet etching of pyrex glass, *Sensors and Actuators A: Physical*, 2008, 143, 154–161.
33. S. Marre, J. Park, J. Rempel, J. Guan, M. G. Bawendi, and K. F. Jensen, Supercritical continuous-microflow synthesis of narrow size distribution quantum dots, *Advanced Materials*, 2008, 20, 4830–4834.
34. S. Liu, C.-H. Chang, B. K. Paul, and V. T. Remcho, Convergent synthesis of polyamide dendrimer using a continuous flow microreactor, *Chemical Engineering Journal*, 2008, 135, S333–S337.
35. S. Ramprasad, Y.-W. Su, C.-H. Chang, B. K. Paul, and D. R. Palo, Cadmium sulfide thin film deposition: A parametric study using microreactor-assisted chemical solution deposition, *Solar Energy Materials and Solar Cells*, 2012, 96, 77–85.
36. N. Tessler, V. Medvedev, M. Kazes, S. Kan, and U. Banin, Efficient near-infrared polymer nanocrystal light-emitting diodes, *Science*, 2002, 295, 1506–1508.
37. A. L. Rogach, N. Gaponik, J. M. Lupton, C. Bertoni, D. E. Gallardo, S. Dunn, N. L. Pira et al., Light-emitting diodes with semiconductor nanocrystals, *Angewandte Chemie International Edition*, 2008, 47, 6538–6549.
38. M. Kazes, D. Y. Lewis, Y. Ebenstein, T. Mokari, and U. Banin, Lasing from semiconductor quantum rods in a cylindrical microcavity, *Advanced Materials*, 2002, 14, 317–321.
39. V. I. Klimov, S. A. Ivanov, J. Nanda, M. Achermann, I. Bezel, J. A. McGuire, and A. Piryatinski, Single-exciton optical gain in semiconductor nanocrystals, *Nature*, 2007, 447, 441–446.
40. S. E. Habas, H. A. S. Platt, M. F. A. M. van Hest, and D. S. Ginley, Low-cost inorganic solar cells: From ink to printed device, *Chemical Reviews*, 2010, 110, 6571–6594.
41. I. Gur, N. A. Fromer, M. L. Geier, and A. P. Alivisatos, Air-stable all-inorganic nanocrystal solar cells processed from solution, *Science*, 2005, 310, 462–465.
42. A. Shavel, D. Cadavid, M. Ibáñez, A. Carrete, and A. Cabot, Continuous production of Cu2ZnSnS4 nanocrystals in a flow reactor, *Journal of the American Chemical Society*, 2011, 134, 1438–1441.
43. R. S. Selinsky, Q. Ding, M. S. Faber, J. C. Wright, and S. Jin, Quantum dot nanoscale heterostructures for solar energy conversion, *Chemical Society Reviews*, 2013, 42, 2963–2985.
44. Y. Yin and A. P. Alivisatos, Colloidal nanocrystal synthesis and the organic–inorganic interface, *Nature*, 2005, 437, 664–670.
45. A. Abou-Hassan, O. Sandre, and V. Cabuil, Microwave-assisted synthesis of colloidal inorganic nanocrystals, *Angewandte Chemie International Edition*, 2011, 50, 11312–11359.

46. S. Duraiswamy and S. A. Khan, Droplet-based microfluidic synthesis of anisotropic metal nanocrystals, *Small*, 2009, 5, 2828–2834.
47. C.-H. Weng, C.-C. Huang, C.-S. Yeh, H.-Y. Lei, and G.-B. Lee, Synthesis of hexagonal gold nanoparticles using a microfluidic reaction system, *Journal of Micromechanics and Microengineering*, 2008, 18, 035019.
48. J. Boleininger, A. Kurz, V. Reuss, and C. Sonnichsen, Microfluidic continuous flow synthesis of rod-shaped gold and silver nanocrystals, *Physical Chemistry Chemical Physics*, 2006, 8, 3824–3827.
49. J. Wagner and J. M. Kohler, Continuous synthesis of gold nanoparticles in a microreactor, *Nano Letters*, 2005, 5, 685–691.
50. D. Shalom, R. C. R. Wootton, R. F. Winkle, B. F. Cottam, R. Vilar, A. J. deMello, and C. P. Wilde, Synthesis of thiol functionalized gold nanoparticles using a continuous flow microfluidic reactor, *Materials Letters*, 2007, 61, 1146–1150.
51. H. D. Jin, A. Garrison, T. Tseng, B. K. Paul, and C.-H. Chang, High-rate synthesis of phosphine-stabilized undecagold nanoclusters using a multilayered micromixer, *Nanotechnology*, 2010, 21, 445604.
52. M. Carboni, L. Capretto, D. Carugo, E. Stulz, and X. Zhang, Microfluidics-based continuous flow formation of triangular silver nanoprisms with tuneable surface plasmon resonance, *Journal of Materials Chemistry C*, 2013, 1, 7540–7546.
53. X. Z. Lin, A. D. Terepka, and H. Yang, Synthesis of silver nanoparticles in a continuous flow tubular microreactor, *Nano Letters*, 2004, 4, 2227–2232.
54. C. Wu and T. Zeng, Size-tunable synthesis of metallic nanoparticles in a continuous and steady-flow reactor, *Chemistry of Materials*, 2007, 19, 123–125.
55. A. M. Nightingale, S. H. Krishnadasan, D. Berhanu, X. Niu, C. Drury, R. McIntyre, E. Valsami-Jones, and J. C. deMello, A stable droplet reactor for high temperature nanocrystal synthesis, *Lab on a Chip*, 2011, 11, 1221–1227.
56. Y. Song, E. E. Doomes, J. Prindle, R. Tittsworth, J. Hormes, and C. S. S. R. Kumar, Investigations into sulfobetaine-stabilized cu nanoparticle formation: Toward development of a microfluidic synthesis, *Journal of Physical Chemistry B*, 2005, 109, 9330–9338.
57. Y. Song, H. Modrow, L. L. Henry, C. K. Saw, E. E. Doomes, V. Palshin, J. Hormes, and C. S. S. R. Kumar, Microfluidic synthesis of cobalt nanoparticles, *Chemistry of Materials*, 2006, 18, 2817–2827.
58. Y. Song, L. L. Henry, and W. Yang, Stable amorphous cobalt nanoparticles formed by an in situ rapidly cooling microfluidic process, *Langmuir*, 2009, 25, 10209–10217.
59. Y. Song, C. S. S. R. Kumar, and J. Hormes, Synthesis of palladium nanoparticles using a continuous flow polymeric micro reactor, *Journal of Nanoscience and Nanotechnology*, 2004, 4, 788–793.
60. K. Torigoe, Y. Watanabe, T. Endo, K. Sakai, H. Sakai, and M. Abe, Microflow reactor synthesis of palladium nanoparticles stabilized with poly(benzyl ether) dendron ligands, *Journal of Nanoparticle Research*, 2010, 12, 951–960.
61. A. M. Nightingale and J. C. deMello, Controlled synthesis of III–V quantum dots in microfluidic reactors, *ChemPhysChem*, 2009, 10, 2612–2614.
62. J. Baek, P. M. Allen, M. G. Bawendi, and K. F. Jensen, Investigation of indium phosphide nanocrystal synthesis using a high-temperature and high-pressure continuous flow microreactor, *Angewandte Chemie International Edition*, 2011, 50, 627–630.
63. H. D. Jin and C.-H. Chang, Continuous synthesis of SnTe nanorods, *Journal of Materials Chemistry*, 2011, 21, 12218–12220.
64. J. B. Edel, R. Fortt, J. C. deMello, and A. J. deMello, Microfluidic routes to the controlled production of nanoparticles, *Chemical Communications*, 2002, 1136–1137.
65. L.-H. Hung, K. M. Choi, W.-Y. Tseng, Y.-C. Tan, K. J. Shea, and A. P. Lee, Alternating droplet generation and controlled dynamic droplet fusion in microfluidic device for cds nanoparticle synthesis, *Lab on a Chip*, 2006, 6, 174–178.

66. B. K. H. Yen, N. E. Stott, K. F. Jensen, and M. G. Bawendi, A continuous-flow microcapillary reactor for the preparation of a size series of cdse nanocrystals, *Advanced Materials*, 2003, 15, 1858–1862.

67. H. Nakamura, A. Tashiro, Y. Yamaguchi, M. Miyazaki, T. Watari, H. Shimizu, and H. Maeda, Application of a microfluidic reaction system for CdSe nanocrystal preparation: Their growth kinetics and photoluminescence analysis, *Lab on a Chip*, 2004, 4, 237–240.

68. S. Krishnadasan, J. Tovilla, R. Vilar, A. J. deMello, and J. C. deMello, On-line analysis of CdSe nanoparticle formation in a continuous flow chip-based microreactor, *Journal of Materials Chemistry*, 2004, 14, 2655–2660.

69. H. Yang, W. Luan, S.-T. Tu, and Z. M. Wang, High-temperature synthesis of CdSe nanocrystals in a serpentine microchannel: Wide size tunability achieved under a short residence time, *Crystal Growth & Design*, 2009, 9, 1569–1574.

70. E. M. Chan, A. P. Alivisatos, and R. A. Mathies, High-temperature microfluidic synthesis of CdSe nanocrystals in nanoliter droplets, *Journal of the American Chemical Society*, 2005, 127, 13854–13861.

71. B. K. H. Yen, A. Gunther, M. A. Schmidt, K. F. Jensen, and M. G. Bawendi, A microfabricated gas-liquid segmented flow reactor for high-temperature synthesis: The case of CdSe quantum dots, *Angewandte Chemie International Edition*, 2005, 117, 5583–5587.

72. E. M. Chan, R. A. Mathies, and A. P. Alivisatos, Size-controlled growth of cdse nanocrystals in microfluidic reactors, *Nano Letters*, 2003, 3, 199–201.

73. H. D. Jin and C.-H. Chang, Synthesis of CuInSe2 nanocrystals using a continuous hot-injection microreactor, *Journal of Nanoparticle Research*, 2012, 14, 1180.

74. W. Luan, H. Yang, N. Fan, and S.-T. Tu, Synthesis of efficiently green luminescent CdSe/ZnS nanocrystals via microfluidic reaction, *Nanoscale Research Letters*, 2008, 3, 134–139.

75. H. Wang, H. Nakamura, M. Uehara, Y. Yamaguchi, M. Miyazaki, and H. Maeda, Highly luminescent CdSe/ZnS nanocrystals synthesized using a single-molecular ZnS source in a microfluidic reactor, *Advanced Functional Materials*, 2005, 15, 603–608.

76. H. Wang, X. Li, M. Uehara, Y. Yamaguchi, H. Nakamura, M. Miyazaki, H. Shimizu, and H. Maeda, Continuous synthesis of CdSe–ZnS composite nanoparticles in a microfluidic reactor, *Chemical Communications*, 2004, 48–49.

77. B. H. Kwon, K. G. Lee, T. J. Park, H. Kim, T. J. Lee, S. J. Lee, and D. Y. Jeon, Continuous in situ synthesis of ZnSe/ZnS Core/shell quantum dots in a microfluidic reaction system and its application for light-emitting diodes, *Small*, 2012, 8, 3257–3262.

78. S. Marrea and K. F. Jensen, Synthesis of micro and nanostructures in microfluidic systems, *Chemical Society Reviews*, 2010, 39, 1183–1202.

79. B. K. H. Yen, A. Günther, M. A. Schmidt, K. F. Jensen, and M. G. Bawendi, A microfabricated gas–liquid segmented flow reactor for high-temperature synthesis: The case of cdse quantum dots, *Angewandte Chemie International Edition*, 2005, 44, 5447–5451.

80. A. M. Nightingale and J. C. demello, Segmented flow reactors for nanocrystal synthesis, *Advanced Materials*, 2013, 25, 1813–1821.

81. K.-J. Kim, R. P. Oleksak, E. B. Hostetler, D. A. Peterson, P. Chandran, D. M. Schut, B. K. Paul, G. S. Herman, and C.-H. Chang, Continuous microwave-assisted gas-liquid segmented flow reactor for controlled nucleation and growth of nanocrystals, *Crystal Growth & Design*, 2014, 14, 5349–5355.

82. P. Reiss, M. Protiere, and L. Li, Core/shell semiconductor nanocrystals, *Small*, 2009, 2, 154–168.

83. M. A. Hines and P. Guyot-Sionnest, Synthesis and characterization of strongly luminescing ZnS-capped CdSe nanocrystals, *Journal of Physical Chemistry*, 1996, 100, 468–471.

84. A. Abou-Hassan, O. Sandre, S. Neveu, and V. Cabuil, Synthesis of goethite by separation of the nucleation and growth processes of ferrihydrite nanoparticles using microfluidics, *Angewandte Chemie International Edition*, 2009, 48, 2342–2345.

85. B. F. Cottam, S. Krishnadasan, A. J. deMello, J. C. deMello, and M. S. P. Shaffer, Accelerated synthesis of titanium oxide nanostructures using microfluidic chips, *Lab on a Chip*, 2007, 7, 167–169.
86. B. Xiang, P. Wang, X. Zhang, S. A. Dayeh, D. P. R. Aplin, C. Soci, D. Yu, and D. Wang, Rational synthesis of p-type zinc oxide nanowire arrays using simple chemical vapor deposition, *Nano Letters*, 2007, 7, 323–328.
87. P.-J. Li, Z.-M. Liao, X.-Z. Zhang, X.-J. Zhang, H.-C. Zhu, J.-Y. Gao, K. Laurent, Y. Leprince-Wang, N. Wang, and D.-P. Yu, Electrical and photoresponse properties of an intramolecular p-n homojunction in single phosphorus-doped ZnO nanowires, *Nano Letters*, 2009, 9, 2513–2518.
88. K.-J. Kim, P. B. Kreider, C. Choi, C.-H. Chang, and H.-G. Ahn, Visible-light-sensitive na-doped p-type flower-like ZnO photocatalysts synthesized via a continuous flow microreactor, *RSC Advances*, 2013, 3, 12702–12710.
89. L. Frenz, A. El Harrak, M. Pauly, S. Begin-Colin, A. D. Griffiths, and J.-C. Baret, Droplet-based microreactors for the synthesis of magnetic iron oxide nanoparticles, *Angewandte Chemie International Edition*, 2008, 47, 6817–6820.
90. N. Hassan, V. Cabuil, and A. Abou-Hassan, Continuous multistep microfluidic assisted assembly of fluorescent, plasmonic, and magnetic nanostructures, *Angewandte Chemie International Edition*, 2013, 52, 1994–1997.
91. M. I. Bodnarchuk, L. Li, A. Fok, S. Nachtergaele, R. F. Ismagilov, and D. V. Talapin, Three-dimensional nanocrystal superlattices grown in nanoliter microfluidic plugs, *Journal of the American Chemical Society*, 2011, 133, 8956–8960.
92. S.-K. Lee, X. Liu, V. S. Cabeza, and K. F. Jensen, Synthesis, assembly and reaction of a nanocatalyst in microfluidic systems: A general platform, *Lab on a Chip*, 2012, 12, 4080–4084.
93. C.-H. Choi, Y.-W. Su, and C.-H. Chang, Effects of fluid flow on the growth and assembly of ZnO nanocrystals in a continuous flow microreactor, *Cryst. Eng. Commun.*, 2013, 15, 3326–3333.
94. S. J. Dalgarno, P. K. Thallapally, L. J. Barbour, and J. L. Atwood, Engineering void space in organic van der waals crystals: Calixarenes lead the way, *Chemical Society Reviews*, 2007, 36, 236–245.
95. J. R. Li, R. J. Kuppler, and H. C. Zhou, Selective gas adsorption and separation in metal-organic frameworks, *Chemical Society Reviews*, 2009, 38, 1477–1504.
96. O. Shekhah, J. Liu, R. A. Fischer, and C. Woll, MOF thin films: Existing and future applications, *Chemical Society Reviews*, 2011, 40, 1081–1106.
97. O. M. Yaghi and M. O'Keeffe, Design of solids from molecular building blocks: Golden opportunities for solid state chemistry, *Journal of Solid State Chemistry*, 2000, 152, 1–2.
98. G. Ferey, Hybrid porous solids: Past, present, future, *Chemical Society Reviews*, 2008, 37, 191–214.
99. R. Ameloot, F. Vermoortele, W. Vanhove, M. B. J. Roeffaers, B. F. Sels, and D. E. De Vos, Interfacial synthesis of hollow metal-organic framework capsules demonstrating selective permeability, *Nature Chemistry*, 2011, 3, 382–387.
100. M. Faustini, W.-S. Ahn, J. Kim, G.-Y. Jeong, J. Y. Kim, H. R. Moon, and D.-P. Kim, Microfluidic approach toward continuous and ultrafast synthesis of metal-organic framework crystals and hetero structures in confined microdroplets, *Journal of the American Chemical Society*, 2013, 135, 14619–14626.
101. K.-J. Kim, Y. J. Li, P. B. Kreider, C.-H. Chang, N. Wannenmacher, P. K. Thallapally, and H.-G. Ahn, High-rate synthesis of Cu–BTC metal-organic frameworks, *Chemical Communications*, 2013, 49, 11518–11520.
102. A. Carné-Sánchez, I. Imaz, M. Cano-Sarabia, and D. Maspoch, A spray-drying strategy for synthesis of nanoscale metal-organic frameworks and their assembly into hollow superstructures, *Nature Chemistry*, 2013, 5, 203–211.

103. Y. J. Chang, P. H. Mugdur, S. Y. Han, A. A. Morrone, S. O. Ryu, T. J. Lee, and C.-H. Chang, Nanocrystalline CdS MISFETs fabricated by a novel continuous flow microreactor, *Electrochemical and Solid-State Letters*, 2006, 9, G174–G177.

104. B. K. Paul, C. L. Hires, Y.-W. Su, C.-H. Chang, S. Ramprasad, and D. Palo, A uniform residence time flow cell for the microreactor-assisted solution deposition of CdS on an FTO-glass substrate, *Crystal Growth & Design*, 2012, 12, 5320–5328.

105. Y. W. Su, S. Ramprasad, S.-Y. Han, W. Wang, S.-O. Ryu, D. R. Palo, B. K. Paul, and C.-H. Chang, Dense CdS thin films on fluorine-doped tin oxide coated glass by high-rate microreactor-assisted solution deposition, *Thin Solid Films*, 2013, 532, 16–21.

106. P. H. Mugdur, Y.-J. Chang, S.-Y. Han, Y.-W. Su, A. A. Morrone, S. O. Ryu, T.-J. Lee, and C.-H. Chang, A comparison of chemical bath deposition of CdS from a batch reactor and a continuous-flow microreactor, *Journal of the Electrochemical Society*, 2007, 154, D482–D488.

107. C. R. Kim, S.-Y. Han, C.-H. Chang, T. J. Lee, and S. O. Ryu, A study on copper selenide thin films for photovoltaics by a continuous flow microreactor, *Molecular Crystals and Liquid Crystals*, 2010, 532, 455–463.

108. M. S. Park, S.-Y. Han, E. J. Bae, T. J. Lee, C.-H. Chang, and S. O. Ryu, Synthesis and characterization of polycrystalline CuInS$_2$ thin films for solar cell devices at low temperature processing conditions, *Current Applied Physics*, 2010, 10, S379–S382.

109. C. R. Kim, S.-Y. Han, C.-H. Chang, T. J. Lee, and S. O. Ryu, Synthesis and characterization of CuInSe$_2$ thin films for photovoltaic cells by a solution-based deposition method, *Current Applied Physics*, 2010, 10, S383–S386.

110. M. Kokotov, Y. Feldman, A. Avishai, M. DeGuire, and G. Hodes, Chemical bath deposition of CdS highly-textured, columnar films, *Thin Solid Films*, 2011, 519, 6388–6393.

111. C.-H. Choi and C.-H. Chang, Aqueous synthesis of tailored ZnO nanocrystals, nanocrystal assemblies, and nanostructured films by physical means enabled by a continuous flow microreactor, *Crystal Growth & Design*, 2014, 14, 4759–4767.

112. S. H. Lee, H. J. Lee, D. Oh, S. W. Lee, H. Goto, R. Buchmaster, T. Yasukawa et al., Control of the ZnO nanowires nucleation site using microfluidic channels, *Journal of Physical Chemistry B*, 2006, 110, 3856–3859.

113. S. A. Khan and K. F. Jensen, Microfluidic synthesis of titania shells on colloidal silica, *Advanced Materials*, 2007, 19, 2556–2560.

114. K. McPeak and J. Baxter, ZnO nanowires grown by chemical bath deposition in a continuous flow microreactor, *Crystal Growth & Design*, 2009, 9, 4538–4545.

115. K. McPeak and J. Baxter, Microreactor for high-yield chemical bath deposition of semiconductor nanowires: ZnO nanowire case study, *Industrial & Engineering Chemistry Research*, 2009, 48, 5954–5961.

116. K. McPeak, B. Opasanont, T. Shibata, D.-K. Ko, M. A. Becker, S. Chattopadhyay, H. P. Bui et al., Microreactor chemical bath deposition of laterally graded Cd$_{1-x}$Zn$_x$S thin films: A route to high-throughput optimization for photovoltaic buffer layers, *Chemistry of Materials*, 2013, 25, 297–306.

# 3 Studying Biologically Templated Materials with Atomic Force Microscopy

*Andrew J. Lee and Christoph Walti*

## CONTENTS

**ABSTRACT** Biologically templated materials differ considerably from their more traditional counterparts in both their properties as well as in the way they are fabricated. They are generally constructed by exploiting self-assembly properties intrinsic to many biological molecules, are organized hierarchically, and feature highly non-uniform properties at the lengthscales of nanometers. One of the many challenges facing the development of such materials is the understanding of their assembly processes, which in turn requires sophisticated tools for studying their properties at relevant length and timescales. Within the context of biologically templated materials, arguably the atomic force microscope (AFM) is the tool of choice, due to its versatility, spatial and temporal resolutions, and ability to manipulate directly biomolecular complexes.

This chapter aims to give an introduction to the underlying mechanics of the AFM and how the direct interaction of a physical probe can be utilized to extract meaningful topographical and mechanical quantities from biological components and bionanomaterials. This background theory is related directly to the ascertainable spatial resolutions, providing insight to systems such as the spatial addressing of DNA nano-architectures with the *E. coli* protein Recombinase A (RecA). More recent developments, enabling exquisite force sensitivity and improved spatiotemporal resolutions, are discussed and examined in relation to nano-mechanical characterization and real-time observations of biological interactions. Finally, the characteristics of the AFM are related to the direct construction of nano-materials, highlighting the AFM as a versatile nano-manipulator.

## 3.1  INTRODUCTION

Materials have always played an important role in progressing our civilization. In many cases, the development of new classes of materials, together with capability to exploit these to create novel and more sophisticated devices, has triggered the birth of a new age—historically, for example, the Stone Age or the Iron Age. More recently, the enormous advances of the semiconductor industry over the past few decades have relied heavily on rapid advancements in materials science, which, in turn, relied on the progress of nanotechnology. The general idea of nanotechnology and its exploitation for materials purposes has originally been proposed over half a century ago, often credited to Richard Feynman and his groundbreaking talk "There's plenty of room at the bottom" in 1959.[1] However, progress was impeded by the lack of tools to manipulate and visualize small structures, a challenge Feynman already referred to in his talk.

It took more than two decades until a family of tools, the scanning probe microscopes (SPMs), which eventually enabled the visualization and later even the direct manipulation of matter at the nanoscale, was introduced with the development of the scanning tunneling microscope (STM) in 1981 by Gerd Binnig and Heinrich Rohrer.[2] The range of tools has rapidly expanded ever since and now includes Kelvin probe force microscopy (KPFM),[3] magnetic force microscopy (MFM),[4] chemical force microscopy (CFM),[5] scanning near-field optical microscopy (SNOM),[6] scanning ion-conductance microscopy (SCIM),[7] scanning thermal microscopy (SThM),[8] and atomic force microscopy (AFM),[9] to name but a few. Today, SPMs have become a fixture in modern nanotechnology.

Despite the extensive range of tools, by far the most versatile workhorse for materials science in the SPM arsenal is the AFM. Developed in 1986 in an attempt to extend the use of SPMs to nonconductive samples, the AFM directly addresses a physical probe to a sample's surface, responding to minute intermolecular forces between the tip and the sample.[9] The AFM's operation can be likened to that of reading Braille and is found to be applicable to a very large range of different types of matter, and it can operate under vacuum, in ambient conditions, or under liquids.

In Feynman's vision of nanotechnology,[1] construction at the nanoscale was anticipated through top-down manipulation of atoms. Indeed, it is this view of connecting directly down to the nanoscale that has led the way for micro- and nanoelectronics manufacturing for several decades. However, more recently, bottom-up approaches have been developed to assemble sophisticated complexes from individual building blocks.[10,11] Although the smallest building blocks for any assembly are the individual constituent atoms, the focus has largely been upon harnessing macromolecules with appropriate assembly functionalities as building blocks. In particular, biological molecules have received considerable attention for the bottom-up construction of sophisticated nanoscale complexes owing to the inherent fidelities and specificities of their interactions and the hierarchical nature through which they self-assemble. By exploiting and adapting biological molecules— such as proteins, nucleic acids, and lipids—it is thus possible to construct novel, complex, and even functional systems at nanoscale dimensions in a massively parallel manner. Materials created using these principles are often referred to as bionanomaterials.

However, to capitalize fully on this potential, it is necessary to interrogate directly the mechanical properties of the biological molecules and the resulting assemblies, to observe their interactions, and even to manipulate them directly. AFM, in particular in connection with recently developed high-speed and high-resolution imaging modes for imaging biological systems, is arguably one of the most suitable and versatile tools to study such systems with relevant spatial as well as temporal resolution. It is within this context that this text discusses the ability of AFM to study a range of relevant aspects of biologically templated materials with great versatility.

In this chapter, the basic physics that underpins the operation of the AFM will be introduced, and how this relates to some of the specific quantities we wish to extract from measurements of bionanomaterials will be discussed. The AFM with respect to high-spatial resolution imaging and developments which have taken place to increase the temporal resolutions, sufficient to follow the specific interactions of biological molecules, will be highlighted. Furthermore, the ability to apply and sense nanoscale forces will be discussed to highlight the nanomechanical properties that can be extracted. Finally, how this exquisite force control can be applied to the direct manipulation of molecules for the orchestration of directed assemblies at the nanoscale will be described.

## 3.2  ATOMIC FORCE MICROSCOPE

The original AFM,[9] as seen in Figure 3.1, was based on the simple concept of moving a sharp tip at the end of a cantilever across the sample surface while recording the deflection of the cantilever. The cantilever was maintained in constant contact with the sample surface during imaging and this method is nowadays referred to as contact mode (Figure 3.2b).[9,12]

The AFM can be broken down into a few basic constituents, a cantilever and probe, a system for detecting motion of the cantilever, and several piezoelectric elements arranged to coordinate movement of either the tip or the sample in $X$, $Y$, and $Z$. These elements can occur in various arrangements such as a sample scanning setup, seen in Figure 3.2a.

In the original work, the deflections of the cantilever as a result of tracking the topology of the surface, were measured by the tunneling current occurring between the cantilever and an STM tip positioned at a fixed location above.[9] The conceptual

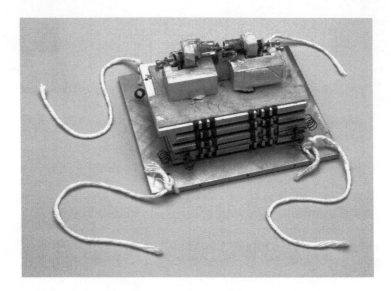

**FIGURE 3.1**  The first atomic force microscope designed by C. F. Quate, G. Binning, and C. Gerber. The core components of the microscope can be seen perched on top. All other components, including the stacked metal plates, are providing the crucial vibration isolation of the instrument. The whole instrument was suspended from each corner using the attached cords for further isolation from vibration. (Image reprinted from Copyright of Science Museum/ Science & Society Picture Library, London.)

simplicity and the stunning imaging resolution that was achieved by this concept— lateral and vertical resolutions of 30 Å and 1 Å, respectively—led to a large variety of investigations being published in rapid succession, imaging a multitude of objects in air and liquid, and under vacuum. The latter environments often resulted in significant enhancements to the spatial resolution due to removal of the capillary forces present in ambient environments owing to the accumulated water layers between the AFM tip and the surface.[13] This will be discussed in more detail below. At this stage, AFM imaging was restricted mostly to hard surfaces, whereas the examination of biological samples was limited because of the high lateral forces imposed by the tip during imaging, causing deformation and disruption of the substrate.

A first solution to this problem was introduced by Martin et al. in 1987 in the form of a noncontact imaging mode.[14] Here, the cantilever was maintained within 1–10 nm from the surface and oscillated at its resonance frequency. Perturbations in the oscillation amplitude were used to form an image (Figure 3.2d). In this imaging mode, the vertical and lateral forces imposed on the surface are negligible, making the imaging of delicate biological samples a possibility. However, this came at the price of significantly reduced spatial resolution.

Modern AFMs can be operated in a variety of different imaging modes, utilizing different ways of how the probe interacts with the sample, taking into account the specific requirements of the surface to be imaged. In particular, the introduction of dynamic imaging modes led to significant advances in the imaging of soft and in particular biological samples. Here, instead of maintaining a constant height as in noncontact mode

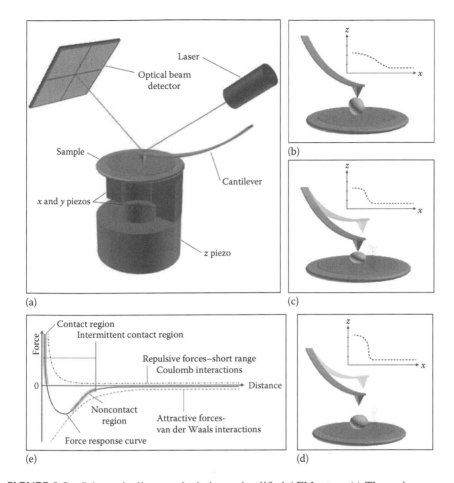

**FIGURE 3.2** Schematic diagram depicting a simplified AFM setup. (a) The main components of an AFM are depicted in a sample scanner arrangement, with the $x$, $y$, and $z$ piezos actuating the movement of the sample relative to the fixed position of the cantilever. Alternatively, a tip scanning setup can be used with the piezos actuating the movement of the cantilever relative to the sample. A laser beam is reflected from the back of the cantilever to an optical beam detector (OBD). Minute cantilever deflections are amplified as linear movements of the laser across the OBD. The conventional modes of cantilever actuation are depicted with associated image traces, (b) contact mode, (c) tapping mode, and (d) noncontact mode. (e) A force–distance curve depicting the force experienced by the cantilever upon approach to the surface. The imaging modes are indicated in their relevant region of the interaction curve.

and measuring the changes in amplitude as a result of the topography of the sample, the vertical position of the cantilever is adjusted dynamically such that a particular tip–sample interaction property, for example, the amplitude of the resonant oscillations of the cantilever, is maintained. This is achieved using a feedback loop, and the output of the feedback loop is then used to form a topographical representation of the surface.

The most commonly used dynamic mode is an intermittent contact mode, commonly referred to as *tapping mode* (Figure 3.2c).[15] Again, the cantilever is excited at its resonance frequency and is subsequently brought into proximity with the surface. In contrast to noncontact mode, the cantilever tip lightly taps the surface at the extent of each oscillation, which leads to small distortion in the oscillation. Instead of maintaining a constant height and measuring the changes in amplitude as a result of topography, the amplitude is maintained and a feedback loop is used to adjust the vertical position of the cantilever relative to the sample surface via the actuation of a $z$-piezo, to keep the distortion in oscillation constant. This method imposes higher vertical forces to the sample surface compared to noncontact mode imaging, but lateral forces remain low. Importantly, it provides greater spatial resolution over its true noncontact counterpart.[13] Shortly after the first demonstration in ambient environments, Hansma et al. demonstrated tapping mode AFM in fluids in 1994,[16,17] which is arguably one of the major breakthroughs that opened up AFM techniques to the world of biological molecules. Additional enhancements to the spatial and especially the temporal resolutions of dynamic AFM modes have occurred over the last decade, increasing the applicability of the AFM in the study of biological molecules. As this chapter is concerned with the AFM imaging of biologically templated materials, we will focus mainly on dynamic imaging modes, although some of the theoretical aspects will also be applicable for general AFM operation.

Irrespective of the particular mode of imaging used, the accurate monitoring of the deflection of the cantilever is of key importance. As mentioned above, in the original work by Binning et al., the deflection of the cantilever was monitored by measuring the tunneling current occurring between the cantilever and an STM tip positioned above.[9] An alternative detection method utilized piezoresistive layers applied directly to the top surface of the cantilever.[18] However, the most widely adopted method of cantilever detection was established in 1988 by Meyer et al.[19] Here, a laser beam is focused on the back of the cantilever. The reflected beam is then directed onto the center of a four-quadrant area of photodiodes. For the nondeflected cantilever, half the laser beam now hits the upper photodiodes and the other half the lower photodiodes (Figure 3.2a). Upon deflection of the cantilever, this distribution is shifted and hence can be used to quantify the deflection. This optical system enables convenient and high-precision measurements of deflection as the small changes in deflection of the cantilever are amplified into large linear movements across the surface of the optical beam detector (OBD), allowing for small height features on the surface to be registered, according to the following equation:

$$\Delta Z \approx 2\frac{D}{L}\Delta z \tag{3.1}$$

where:
 $\Delta Z$ is the change in position of the deflected laser beam on the OBD
 $\Delta z$ is the change in $z$-position of the free end of the cantilever
 $D$ is the cantilever–OBD distance
 $L$ is the cantilever length

### 3.2.1 Background Theory

Despite the wide adoption of AFM, a solid understanding of the tip–sample interaction and cantilever dynamics has taken a long time to be established. Moreover, no single description has thus far accounted for all imaging conditions. It is this limitation in theoretical understanding that has hampered rapid advancement in AFM techniques, especially with respect to minimizing interaction forces while maintaining spatial resolution. However, recent innovations have enabled the development of advanced imaging regimes, such as peak force tapping (PFT), which address these challenges (see Section 3.4.2). The following discussion examines the physical interactions occurring between the tip and the sample that underpin the formation of AFM images.

### 3.2.1.1  Tip–Sample Interactions

When considering the formation of an AFM image, we must consider the combination of attractive and repulsive forces that cumulatively act upon the tip to form

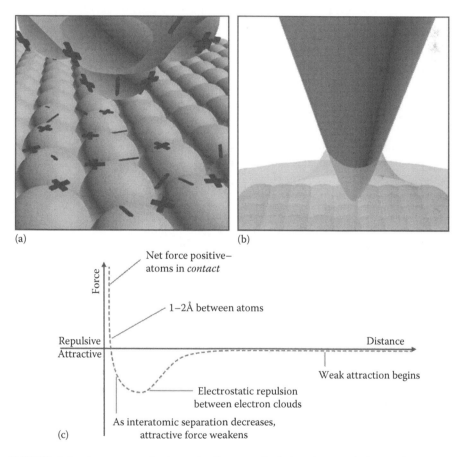

**FIGURE 3.3**  A montage of schematic diagrams depicting tip–sample interactions. (a) A surface and a tip are represented as ensembles of half-spheres with permanent and induced dipoles with the relative charges indicated. VdW forces act between the dipoles of both the surface and the tip giving rise to attractive interactions. (b) A tip is depicted with an associated capillary neck as a consequence of an ambient water layer. (c) Interaction forces experienced between the tip and the sample as a function of distance.

the observable quantity (Figure 3.3). These interaction forces are, however, difficult to disentangle and understanding each contribution in isolation is not trivial. The interaction between the tip of an AFM cantilever and a substrate is mediated by an array of electromagnetic forces, examined in brief below. Other more comprehensive reviews exist in the literature, for example, by Garcia.[13]

First and foremost, we must consider van der Waals (VdW) forces, which are long-range attractive interactions between atoms and/or molecules arising from electric dipole interactions. In the case of tip–sample interaction, we must consider the sum of all dipoles within both the sample and the tip (Figure 3.3a). Surfaces are rarely flat on the nanometer scale, and hence we approximate the surface as an array of half-spheres. When considering the tip as a half-sphere a distance $d$ away from the surface, the VdW force is given by[20]

$$F_{\text{VdW}} = -\frac{H}{6d^2}\frac{RR_s}{R+R_s} \qquad (3.2)$$

where:
  $R_s$ is the effective radius of the surface roughness
  $R$ is the radius of the tip
  $H$ is the Hamaker constant

As will be discussed later, this view of a nanoscale surface defined by interconnected half-spheres, in part, defines the limit in spatial resolution achievable when imaging with another half-spherical object.

In addition, a range of repulsive forces, including forces arising from Pauli exclusion and ionic repulsion, must be considered. However, these interactions can be described sufficiently well by phenomenological contact mechanics models, which take into account the load (applied force) and the elastic properties of the materials involved (both the tip and the surface). Hertz first described the contact and subsequent deformation of two spherical objects under an applied force in 1881.[21] Subsequent revised models by Derjaguin, Muller, and Toporov (DMT)[22] and Johnson, Kendall, and Roberts (JKR) elaborate on the relationships between deformation and load by additionally considering an adhesion force. Upon first encounter, the JKR and DMT models appear contradictory; however, they describe opposing ends of the elastic modulus spectra.[23] Thus, the DMT model is relevant for low adhesion forces and stiff contact moduli when applied to small tip radii.[23] By contrast, the JKR model successfully describes large tip radii with a contact that is of low stiffness and has high adhesive forces. It is important to note that the former considers adhesions forces acting outside of the contact area, whereas the latter does not.[13,23,24] In addition, these models also account for some nonconservative forces such as surface adhesion hysteresis and viscoelasticity.[13]

When operating the AFM under ambient conditions, a water meniscus forms spontaneously between the tip and the sample surface, owing to capillary condensation (Figure 3.3b). The two principle radii of curvature of the meniscus, $r_1$ and $r_2$, are linked to the vapor pressure $P$ of the liquid as follows[25]:

$$\frac{1}{r_1} + \frac{1}{r_2} = \frac{1}{r_k} = \frac{R_g T}{\gamma_L V_m} \log \frac{P}{P_{sat}} \tag{3.3}$$

where:

$r_k$ is the Kelvin radius
$R_g$ is the gas constant
$T$ is the temperature
$P_{sat}$ is the vapor pressure at saturation
$\gamma_L$ is the surface tension of the liquid
$V_m$ is the molar volume of the liquid

From this equation, it can be seen that a meniscus will form when the tip reaches a distance from the surface equal to that of the Kelvin radius.

In general, analytical solutions of the tip–sample-associated capillary force $F_{cap}$ do not exist, but Israelachvili provides a solution in the approximation where the tip is modeled as a half-sphere and the surface is considered flat[26]:

$$F_{cap} = -\frac{4\pi R \gamma_L \cos\theta}{1 + \left(\dfrac{d}{d_0}\right)} \tag{3.4}$$

where:

$d$ is the distance between the tip and the surface
$R$ is the tip radius
$d_0$ is the height of the meniscus
$\theta$ is the contact angle at the surface

We note that the force reaches a maximum when the tip is in contact with the surface ($d = 0$). Although the above equation is only valid for a specific approximation, the general cases are not dissimilar. It has been shown both numerically and experimentally that capillary force values can reach 100 nN, clearly becoming the dominating force in any ambient imaging regime compared to other typical tip–sample interaction forces which range from high piconewton to low nanonewton.

The prevalence of a dominating capillary force under ambient conditions can cause problems for stability and resolution, in particular when imaging soft specimens. This force can be eliminated altogether when the entire tip–sample interaction occurs within a liquid medium. However, other forces such as solvation and electrostatic double-layer forces come into play owing to the presence of the solid–liquid interface at the tip surface.

When submerged in polar solvents, solid surfaces become charged through the absorption of ions or charged molecules, or by the ionization of the surface atoms themselves. This excess charge at the surface is compensated by an accumulation of an equal number of opposite charges in the solution close to the surface. The thin layer of liquid containing the extra charges is generally referred to as the electrostatic double layer (EDL), which can extend into the solution from 1 nm to hundreds of nanometers, depending on the ionic strength of the solution. The EDL is highly dynamic, in constant battle between the entropy of ions becoming diffuse throughout the solution and the electrostatic attraction of the surface.

An EDL force is encountered when a second surface is introduced into the liquid and brought into close enough proximity to the first surface such that their respective EDLs overlap and hence interact. The EDL force is generally repulsive if both surfaces are either positively or negatively charged upon immersion in the polar solvent.

When imaging under liquid, the EDL and the VdW forces are usually dominating. These forces form the basis of the Derjaguin–Landau–Verwey–Overbeek theory, and an approximation of the resulting force in the case where the tip is modeled as a half-sphere and the surface is considered flat was given by Butt et al. for the case of $d \gg \lambda_D$, the Debye length, that is, the thickness of the EDL:[27]

$$F_{DLVO} = -\frac{4\pi R}{\in \in_0} \sigma_t \sigma_s \lambda_D \exp\left(\frac{-d}{\lambda_D}\right) - \frac{HR}{6d^2} \tag{3.5}$$

where:

$\in$ and $\in_0$ are the dielectric constants of the medium and the vacuum, respectively

$\sigma_t$ and $\sigma_s$ are the surface charges of the tip and the sample, respectively

We note that the thickness of the EDL varies strongly with the ionic strength of the medium, and hence the strength of the EDL force can be tuned accordingly. In doing so, access to high spatial resolutions is achievable.[28,29] Furthermore, an appropriate balance between a stable imaging environment, a suitable sample binding environment, and biological reaction conditions is required.[30]

In addition, one must account for solvation and hydration forces that occur due to the confined nature of liquid molecules between the tip and the sample surface.[31,32] It must also be considered that there is a dependency upon the mechanical and chemical properties of the substrate itself and its imaging environment.

### 3.2.2 CANTILEVER DYNAMICS

So far, we have only considered the complex interactions between the surface and the tip. We must, however, be mindful of the role of the cantilever onto which the tip is mounted, in particular for dynamic mode imaging which is the main interest of this chapter. The cantilever can generally be modeled as a perturbed harmonic oscillator. This has been discussed extensively in the literature and an excellent overview is given by Garcia.[13]

The way in which a cantilever behaves when excited is defined by three characteristics: its resonant frequencies $f_n$, its quality factor $Q$, and its spring constant $k$. When considering a standard beam cantilever with the dimensions $w$, $h$, and $l$, and only considering the most commonly used mode of oscillation, that is, the transverse mode where the cantilever oscillates vertically, the above characteristics are given by[13]

$$f_n = \frac{1}{4\pi} \lambda_n^2 \frac{h}{l^2} \sqrt{\frac{E}{3\rho}} \tag{3.6}$$

where:

$E$ is the Young's modulus

$\rho$ is the mass density of the cantilever material

The coefficients $\lambda_n$ are defined by $\cos \lambda_n \cosh \lambda_n = -1$, which leads to $\lambda_0 = 1.875$ and $\lambda_1 = 4.694$ for the lowest two resonances of the cantilever

The spring constant $k$ is given by

$$k = \frac{Ewh^3}{4l^3} \tag{3.7}$$

The $Q$-factor describes any damping that affects a resonating cantilever and therefore is largely defined by the surrounding medium (air, liquid, or vacuum). This dimensionless parameter can range from a few thousand in vacuum to a few hundred in air, and even lower in liquid, and effectively describes the *sharpness* of the principle resonance peak:

$$Q = \frac{f_0}{\Delta f} \tag{3.8}$$

where $\Delta f$ is the full width half maximum of the resonance peak.

We note that these equations are valid in vacuum, and that both resonance frequencies and $Q$-factor can change considerably when the cantilever is oscillated in a medium.

The desire to visualize biologically templated materials, as well as eventually the actual assembly processes at the nanoscale, requires imaging in aqueous buffers which mimic their native environments.[33,34] This further complicates cantilever dynamics owing to the changes in resonance frequencies as a result of the fluid environment (rather than vacuum) and viscous damping forces. Furthermore, when a cantilever is oscillated in a fluid environment, there are two additional factors to consider: first, the interaction of the liquid with the cantilever, and second, the interaction at the sample–liquid–tip interface, where long-range forces such as EDL forces can become significant. However, it is important to remember that by imaging under liquid the capillary force, which can be dominating in ambient environments, is removed.[13,17]

For simplicity, we only consider the case of imaging in a water environment, as EDLs formed by ions in aqueous buffers further complicate tip–sample interactions, and thus cantilever dynamics, and hence must be tuned appropriately depending on the system that is investigated. As mentioned above, in water there is a reduction in $Q$ and $f_n$ of the cantilever. For example, when considering a beam cantilever, the primary resonance frequency is typically reduced by a factor of 3–5 when operated in water compared with air. This decrease is caused by fluid boundary layers forming around the cantilever, which must be displaced during each oscillation cycle. Moreover, the viscous properties of the medium impart hydrodynamic damping to the cantilever, which is responsible for the reduction of $Q$.[35,36]

As the oscillating cantilever is brought toward the surface, the fluid boundary layers form a greater proportion of the cantilever–sample separation distance, thus increasing the fluid shear, which in turn causes further damping. In most cases, it is acceptable to neglect such shifts as these occur between the tip and the sample only, whereas the prevailing hydrodynamic effects act upon the entire cantilever body.[37] Additionally, higher harmonics can complicate the cantilever oscillations in liquid owing to the significantly reduced $Q$, whereas these are negligible in air.[38]

## 3.3   SPATIAL RESOLUTION

Defining the spatial resolution of an AFM is different to that of conventional radiation-based microscopes. The AFM forms images of surfaces in three dimensions and hence requires us to distinguish between lateral ($x$ and $y$) and vertical ($z$) resolutions. The main contributors that determine vertical spatial resolution are the mechanical and electrical noise in the feedback system, and thermal fluctuations of the cantilever. By contrast, factors such as the tip radius, the aspect ratio, the distance over which the surface forces decay, and the compliance of the sample govern the lateral spatial resolution. Note that unlike with most radiation-based microscopy techniques, the mechanical properties of the sample play a defining role in the spatial resolution obtainable.

### 3.3.1   VERTICAL RESOLUTION LIMITS

The ability to distinguish small step heights is directly limited by the inherent noise of the imaging signal, either amplitude or frequency.[39] In the case of amplitude modulation, the most commonly used method, the ratio of the noise in the amplitude signal, $\delta A$, to the gradient of the amplitude versus tip–surface-distance defines the vertical noise inherent to the image:

$$\delta h_n = \frac{\delta A}{\left| \dfrac{dA}{dd} \right|} \tag{3.9}$$

For soft samples such as biological molecules, the gradient is typically 0.2–0.5, and hence $\delta h_n$ is typically between 2 $\delta A$ and 5 $\delta A$.[39]

There are two major factors that contribute to noise in the cantilever signal, and these can be considered independent of each other: First, the cantilever experiences thermal instabilities, and second, there is an inherent noise in the detector/feedback loop, dominated primarily by noise in the optical detection of the deflected laser beam. The latter has been reduced significantly over recent years through enhancements to the OBD, improved laser sources, and small cantilever systems. As a result, the thermal instabilities of the cantilever are now the dominating noise source of the cantilever in modern AFM systems.[31,40,41] The thermal vertical noise of the deflection signal can be approximated by:[42]

$$\delta h_{\mathrm{th}} = \sqrt{\frac{4k_{\mathrm{B}}TQB}{\pi f_0 k}} \tag{3.10}$$

where:

$k_B$ is the Boltzmann constant

$B$ is the noise bandwidth

and all other quantities as defined above. Provided that this is the dominating contribution and generally small in modern cantilevers, systems can be built which are sensitive even to the short-range forces proximal to the surface, and therefore offer increased spatial resolution and force sensitivity, as the two are highly intertwined.[31]

### 3.3.2 LATERAL RESOLUTION LIMITS

The lateral resolution is governed by the typical feature sizes of both the tip and the sample, and the resulting AFM image is in fact a convolution of both shapes (Figure 3.4). As a result, unless the tip radius $R$ is significantly smaller than the typical feature sizes of the sample, the lateral sizes of the imaged structures are generally overestimated by AFM. Furthermore, where typical surface features are smaller than the apex of the tip, we can argue that the resulting AFM image is dominated by the tip apex rather than the sample surface. The lateral resolution, $\delta l_r$, can now be defined as the smallest distance at which adjacent features with a height difference of $\Delta h$ on the sample can be resolved, and is given by[43]

$$\delta l_r = \sqrt{2R}\left(\sqrt{\delta h_r} + \sqrt{\delta h_r + \Delta h}\right) \tag{3.11}$$

where $\delta h_r$ is the vertical resolution.

In the context of this chapter, it is justified to set $\delta h_r \approx \delta h_n$. It is evident from the above equation that the lateral resolution depends not only on the tip but also on the

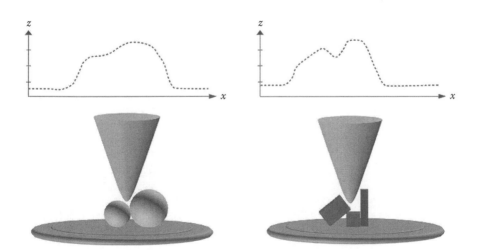

**FIGURE 3.4** Schematic diagram illustrating the resolution limits when imaging objects with a probe of finite size. Sample features smaller than the diameter of the probe cannot be resolved resulting in convolution artifacts, which represent a superimposition of the shape of the tip upon the sample.

vertical resolution of the system, as well as the height of the features. For illustration, if we consider an AFM system with a typical vertical noise level of 0.1 nm and employing a tip with a radius of 5 nm, the lateral resolution $\delta l_r$ when resolving peaks of identical height on the surface is 2 nm, whereas $\delta l_r \approx 4.3$ nm when imaging features with a height difference of 1 nm. Any features occurring closer than $\delta l_r$ will be subjected to tip convolution artifacts (or dilation); this effect is represented in Figure 3.4.

In the above approximation, we considered the tip apex as being the only volume that interacts with the surface. However, this is not always true, in particular when imaging surfaces where the height of the features is larger than the tip radius. In such cases, the sidewalls of the tip will also interact with the surface. To address this problem, the use of single-walled carbon nanotubes as AFM tips has been investigated owing to their excellent aspect ratio and tip radius.[44] However, difficulties in preparing such tips impedes their wide adoption.

The above discussion of lateral resolution assumes noncompliance of both the tip and the sample, that is, neither sample nor tip deformation is taken into account. Although this is reasonable for solid samples, the elastic deformation of soft samples under an applied tip load will significantly alter both the vertical and lateral resolutions achievable.

When considered in relation to the Hertz model,[21] where the applied force, $F$, the tip radius, $R$, and the effective Young's modulus of the sample, $E_{eff}$, are taken into account, the lateral resolution is found to be

$$\delta l_r = 2 \left( \frac{3RF}{4E_{eff}} \right)^{\frac{1}{3}} \tag{3.12}$$

where the effective Young's modulus $E_{eff}$ of the surface–tip interface is given by

$$E_{eff} = \frac{1 - v_s^2}{E_s} + \frac{1 - v_t^2}{E_t} \tag{3.13}$$

where $E_{t,s}$ and $v_{t,s}$ are the Young's modulus and the Poisson ratio of the tip and the sample, respectively. This implies that forces below 1 nN and tip radii of around 1 nm are required to obtain subnanometer resolutions. More importantly, it indicates that such high resolutions are substantially easier to achieve on materials with high Young's moduli.[13,43] We note that several groups have reported AFM images with molecular resolution of biomolecules by increasing $E_{eff}$ by forming rigid structures (e.g., protein crystals).[28,45] By contrast, alternative instrumental developments to circumnavigate the above challenges have been developed to control and reduce actively (PFT, see Section 3.4.2.) and passively (torsional tapping[46–48]) interaction forces giving rise to higher resolutions.

### 3.3.3 APPLICATION TO BIONANOMATERIALS—DNA-TEMPLATED MATERIALS

The use of DNA as a structural element was first envisaged by Nadrian Seeman in 1980 as a method for creating nanoscale arrays with programmable periodicities.[49] DNA is a semiflexible biopolymer assembled from four nucleic acids or bases, which natively exists as a duplex owing to the high fidelity of the Watson and Crick base-pairing interactions between the nucleic acids. Seeman realized that the sequence of bases could be arranged such that topologically more complex structures than linear double-stranded DNA could be formed. Such structures, examples of which are in fact found in nature—hairpin loops and four-way Holliday junctions—have great potential in nonbiological applications such as material science. Through intelligent design of the sequence of bases (primary structure), multiple DNA strands can be designed to bind to one and other in a specific fashion forming a unique and potentially very complex structure (secondary structure). This method results in the creation of branched junction elements which can be added together to create very large homogeneous arrays.[50–52]

Great interest in this field has led to a large number of publications demonstrating 2D and 3D structures of increasing complexity. In addition, the technique was greatly expanded with the introduction of DNA origami by Paul Rothemund in 2006.[53] This new methodology simplified the design process by relieving some of the limitations imposed by the small number of nucleotide interactions available. Here, a large single-stranded DNA template, typically a viral genome, is folded into the desired shape through duplex formation with hundreds of short DNA strands, known as staple strands. This form of structural DNA nanotechnology has greater stability and typically more favorable assembly kinetics. It can easily be extended to enable the assembly of heterogeneous arrangements of multiple DNA origami tiles through additional base-pair interactions in order to form larger structures. Such structures have great potential as scaffolds for DNA-templated materials.

To expand the concept of DNA-templated materials, it is desirable to introduce capabilities for hierarchical spatial functionalization of DNA nanoarchitectures or arrays to produce heterogeneous structural assemblies with nanoscale resolution. This may be approached through chemical modification of DNA bases or sequence-specific interaction with DNA binding or recombination proteins.

A promising approach is to utilize the *Escherichia coli* protein Recombinase A (RecA) for spatially addressing DNA architectures.[54] The RecA protein is involved in DNA repair pathways through homologous recombination. It polymerizes upon single-stranded DNA in the presence of ATP and magnesium to form a nucleoprotein filament. Subsequently, this nucleoprotein complex is able to self-assemble onto a double-stranded DNA at regions of shared homology between the encapsulated single-stranded DNA and the double-stranded DNA (Figure 3.5a). Upon successful assembly, the two homologous strands exchange resulting in the substitution of the new single-stranded DNA molecule with the homologous region of the double-stranded DNA molecule, and then the RecA complex disassociates through the hydrolysis of ATP.

This mechanism can be exploited for DNA-templated material applications, as it offers direct spatial addressing of nanoscale DNA structures. Owing to the spatial positioning being a function of the homology search, no specific binding sites are required to be preprogrammed into the DNA template. This makes this approach highly flexible and fully programmable, allowing the template strand to be addressed at any point and at any time by altering the sequence of the single-stranded DNA

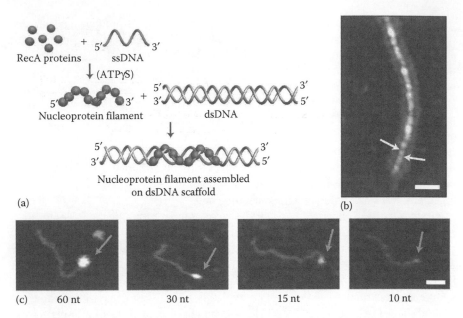

(a)

(b)

(c)        60 nt                    30 nt                    15 nt                    10 nt

**FIGURE 3.5** Application of AFM to RecA-based bionanomaterials. (a) Schematic diagram depicting the formation and homology-based patterning system of RecA nucleoprotein filaments. Monomeric RecA protein binds single-stranded DNA in the presence of ATPγS to form a nucleoprotein filament, which is able to locate a region of sequence homology within a double-stranded DNA substrate forming a stable triple-stranded complex. (b) An AFM image showing a RecA nucleoprotein filament formed upon 890 bp double-stranded DNA. The 10 nm helical pitch is indicated with arrows. Scale bar: 50 nm. (c) A series of white AFM images showing RecA nucleoprotein filaments of decreasing size, formed upon single-stranded DNA from 60 (60 nt) to 10 nucleotides (10 nt) long, patterned at the termini of a double-stranded DNA template (grey arrow). Scale bar: 50 nm. All AFM images are captured using the tapping mode in air. Images adapted from R. Sharma, A. G. Davies, and C. Wälti. Directed assembly of 3-nm-long RecA nucleoprotein filaments on double-stranded DNA with nanometer resolution. *ACS Nano*, 8(4):3322–3330, 2014.

from which the nucleoprotein filament is formed. Furthermore, by substituting ATP with a nonhydrolyzable analog—ATPγS—it is possible to lock the complex in place once assembled, resulting in the stable patterning of the DNA substructure.[55,56]

Previous work by Keren et al. has demonstrated the use of RecA filaments as shadow masks for the production of gaps of controlled size in metallized DNA nanowires.[57] Alternatively, RecA itself can be functionalized or metallized for the production of nanowires,[58] which could ultimately be arranged specifically within a larger DNA structure. The applicability of this method for true nanoscale pattering of DNA templates and the programmable assembly of RecA nucleoprotein filaments as small as 3 nm with single DNA base-pair resolution have been recently demonstrated (Figure 3.5c).[54]

Visualizing the resulting small structures is of great importance to understand and exploit the process; however, given the typical feature sizes and the structural properties of the materials involved, it is a significant challenge. AFM offers the spatial resolution required for assessing RecA structures of all sizes, from revealing the helical pitch of the RecA nucleoprotein filament superstructure (Figure 3.5b) to confirming the fidelity of patterned structures down to 3 nm in size (Figure 3.5c).

Although conventional tapping-mode AFM is able to generate images of biologically templated materials, it is desirable to improve the resolution even further. As has been discussed above, the imaging resolution and force sensitivity of AFM are intrinsically linked, meaning that the tight regulation and control of tip–sample interaction forces can give rise to increased spatial resolutions as a smaller portion of the tip interacts with the sample. In addition, smaller imaging forces are particularly advantages when dealing with soft, fragile samples, such as those of a biological nature.

## 3.4 SENSING AND CONTROLLING FORCES

AFM employs a physical probe that interacts directly with the sample to gain the required information to form an image. This characteristic feature of AFM is, however, not limited to image formation. As discussed above, the typical forces experienced during imaging are on the order of piconewton to nanonewton, and therefore, AFM lends itself to direct sensing of local interaction forces with high spatial and force resolutions. AFM probes can be utilized as nanoscale force sensors to investigate an external force acting upon it, and mechanical characteristics of a sample can be derived through the application of force. Properties such as deformation, energy dissipation, adhesion, and moduli can be mapped with the same spatial resolution as available to topographical imaging.

Furthermore, by functionalizing the AFM tip to introduce specific enhanced affinity for certain parts of the sample, AFM can be applied to the study of inter- and intramolecular forces, including affinity mapping,[59] calculation of binding energies,[60] and examination of the internal mechanics of proteins.[61]

### 3.4.1 FORCE SPECTROSCOPY

The mechanical stability of biological systems is often critically linked with their biological activity, and thus being able to map nanoscale biological mechanics is of great importance in understanding how form and function are linked, and in this endeavor AFM is undoubtedly unsurpassed.

Considering the AFM cantilever as a Hookian spring, the interaction force between the AFM tip and a sample can be given as

$$F = -k\Delta z \qquad (3.14)$$

where $k$ and $\Delta z$ are the cantilever spring constant and deflection, respectively.

Mechanical properties can be extracted by driving the cantilever along the $z$-axis from a position far away into the sample surface (and back) while monitoring the cantilever deflection signal. This is typically referred to as a force–distance curve, however, not the tip–sample distance but only the position of the $z$-piezo (Z) is known, which represents the rest position of the cantilever. The tip–sample separation distance $d$ is affected by the deflection of the cantilever and the elastic properties of the sample. Therefore, $d$ is given by

$$d = Z - \left(\Delta z + \delta_s\right)$$  (3.15)

where $\delta_s$ is the sample deformation.

Hence, the recorded curve contains elastic contributions from the cantilever and the sample, as well as contributions from the tip–sample interactions,[62] including both long-range attractive and short-range repulsive (mechanical contact) forces.

Because of the elastic nature of the cantilever—and in many cases, the sample itself—discontinuities between the approach and the retract parts of the curve occur when considering the attractive interactions. This hysteresis mainly results from the snap to contact dominated by the formation of a capillary neck in ambient conditions, but also includes VdW and hydrophobic contributions. In general, more detailed information on the attractive part of the tip–sample interaction can be obtained upon retraction.

When considering the repulsive contributions, the spring constant of the cantilever plays a major part. Mechanical deformation of the sample occurs as the cantilever is lowered and the tip starts to interact with the sample surface until the cantilever begins to deflect due to the resistive force of the surface. Therefore, it is important to match the compliance of the cantilever and the sample to avoid damage to the sample and optimize force sensitivity, and this is of particular concern where biological samples are investigated. In addition, the tip geometry must also be taken into account, for example, when investigating biological membranes or imaging living cells, a sharp, high aspect ratio tip will lead to high loading forces across a small surface area, and hence can result in the rupturing of the membranes.

In 1996, Mitsui et al. exploited this principle to exert an external force on a protein in an attempt to gain insight into the protein's mechanical stability,[63] and shortly afterward this method was utilized to study the folding and unfolding pathways of the model protein system titin.[64] However, these early experiments, which focussed directly on unfolding single protein molecules, yielded complicated data signals arising from a complex combination of actual protein unfolding and nonspecific tip–sample interactions.[61] To overcome this challenge, recombinant polyprotein constructs that contain multiple protein units linked by short peptide domains are used to deconvolute the various contributions to the signal. A detailed discussion of this approach is given by Hoffmann and Dougan.[61]

In general, force spectroscopy by AFM is carried out using two different approaches (Figure 3.6). In force extension measurements, the AFM tip is lowered into the solution containing the surface-immobilized proteins such that an arbitrary number of proteins can attach to the tip and is then retracted from the surface at a constant velocity (typically 50–1000 nm/s). With increasing separation, the force experienced by the tip increases and at a particular force an individual domain within the protein will unfold, causing a sudden elongation of the effective length of the polypeptide chain, and an associated reduction of the cantilever deflection (Figure 3.6a

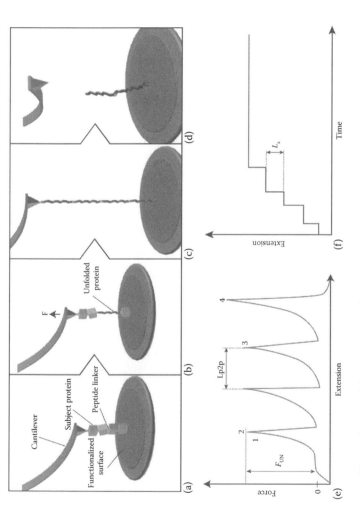

**FIGURE 3.6** A comparison of force-clamp and force-extension force spectroscopy. A typical force extension experiment is shown schematically in panels (a–d), and the corresponding force extension profile is depicted in (e). A series of linked protein units (grey cubes) is sandwiched between a cantilever and a substrate (a). A force is applied by withdrawing the cantilever from the surface causing it to deflect (point 1 in [e]). Eventually a protein unit unfolds (panel [b] and point 2 in [e]) restoring the cantilever to its resting position. When all the protein units have unfolded (panel [c] and point 3 in [e]), the polyprotein detaches from either the substrate or the cantilever ([d] and point 4 in [e]). The unfolding force ($F_{UN}$) and the protein length extension ($L_{P2p}$) are indicated in the trace in (e). Alternatively, force-clamp experiments produce stepped profiles as a function of time (f), and the size of the steps ($L_s$) directly correlates to the number of amino acids released at each unfolding event.

and b). The resulting force–extension curves typically resemble repeated sawtooth patterns, with peaks that correspond to the rupture forces of particular internal protein domains (Figure 3.6e). Once all the domains have unfolded, a detachment peak is observed (Figure 3.6c and d).

Alternatively, the force clamp method applies a constant tip–protein force while recording the $z$-position of the cantilever as a function of time. As a protein domain unfolds, the polypeptide chain extends, and the cantilever position changes in a stepwise fashion (Figure 3.6f). The resulting curve is typically staircase-like, where step heights relate directly to the number of amino acids released per unfolding event. Force clamp experiments can be conducted across a range of forces and along several trajectories by engineering the tip and surface such that specific residues at different regions of the protein are attached to the tip and the surface, respectively.[65]

### 3.4.2 Mapping Mechanical Properties

Force curves are not limited to single points, and can in fact be collected across an entire surface analogous to, for example, imaging by tapping mode where height information is collected for each point across a surface. This process of mapping the mechanical properties across an entire surface is known as force–volume mapping. Traditional AFM instruments can collect force curves at a rate of up to 10 Hz, and typically each curve is treated independent of the previous one, which can lead to overshooting and increased sample load.

This approach of measuring force at each point across the sample surface has recently been adapted and developed further to enable ultra-high-resolution imaging. In this approach, generally referred to as PFT (Figure 3.7), the applied force is continuously analyzed and the $z$-position of the cantilever is adjusted via a fast feedback loop such that the maximum force applied to the sample remains constant. We note that PFT effectively decouples the resonance of the cantilever from its response by oscillating at a set low frequency, typically fixed at 2 kHz.[66,67] Furthermore, the $z$-piezo is actuated with a sinusoidal waveform, contrary to traditional force–distance curves, which are derived using linear ramps. This results in the tip velocity approaching zero at the point of contact, enabling a more controlled interaction.[67,68]

Owing to the increased speed of modern controller electronics, the system is able to modulate the peak force applied to the sample surface per oscillation cycle to a self-adjusting set point at which sample deformation begins. In practice, the probe is able to interact with the sample with forces of 10s of piconewton for merely 100 μs, with the system deriving multiple force–distance curves per pixel at a typical rate of 200 Hz. In addition, the signal-to-noise ratio is drastically increased by algorithmically removing any parasitic components from the deflection signal at each surface contact.[67]

This approach of direct force control enables imaging at very low interaction forces and hence leads to increased spatial resolution compared to traditional modes, in particular for soft polymer and biological samples. In addition, PFT avoids excessive sample deformation,[69,70] minimizes tip wear,[68] and maintains stable imaging as the set points are not susceptible to drift, which is a particular benefit when imaging in fluid. Another important benefit of this mode is the fact that the mechanical

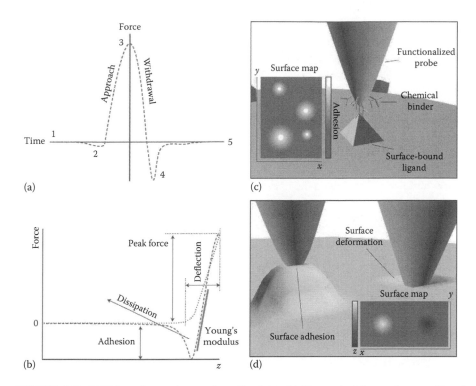

**FIGURE 3.7**    PFT imaging regime and applications. (a) The typical force–time profile of PFT. As the tip–sample distance is decreased, the cantilever goes from being at rest (point 1) to sensing long-range attractive forces (point 2). Upon further reduction of the separation, the cantilever experiences mechanical loading until a set threshold is reached (point 3). An adhesion force is experienced by the cantilever upon retraction (point 4), and when overcome by the restoring force of the cantilever, the system returns to its rest (point 5). An associated force–distance profile is depicted (b) showing both the approach (dotted line) and retract (dashed line) curves. Some characteristic quantities that can be derived from the interaction are indicated. The sensitivity of this method has applications in recognition mapping (c) and nanomechanical mapping (d).

properties are extracted at each pixel, which enables comprehensive statistical analysis of the sample allowing true quantitative mapping (Figure 3.7c).[66,69]

Additional variations of force mapping of biological samples include the direct investigations of ligand–receptor interactions.[71] An early example of this method is discussed by Gaub et al., measuring the interaction forces of individual biotin–avidin complexes[72,73] and DNA hybridization.[72] When coupled with force–volume or PFT, this method can be used to examine the spatial distribution of specific receptors, typically across a biologically active surface (Figure 3.7d).[59,60]

### 3.4.3    High-Resolution Imaging of Biologically Templated Materials

As discussed, the application of small forces gives access to superior resolutions upon fragile samples, in particular those of biological nature. It is now possible to image bionanomaterials with sufficient resolution to probe their internal structures and identify perturbations as a consequence of hierarchical interactions such as the structural manipulation of DNA by proteins. This high spatial resolution offers insight into the exact role each biological component plays in a particular assembly, crucial information when attempting to manipulate their interaction to form bionanomaterials.

If we consider the previous discussion of RecA nucleoprotein filaments, a clear advancement in spatial resolution can be seen in Figures 3.8a and 3.5b. Through the controlled application of only 100 pN of force to the sample, we are able to image the subunit structure of nucleoprotein filaments absorbed onto mica. Interestingly, we can see a clear discontinuity in the helical pitch of the nucleoprotein filament, indicating that it has polymerized upon the encapsulated DNA from multiple locations.

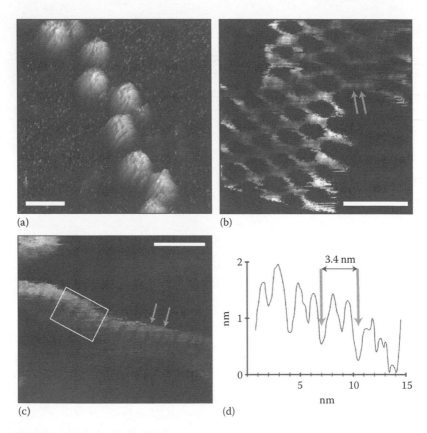

**FIGURE 3.8** High-resolution AFM images. (a) The subunits of a 3.5 kbp RecA nucleoprotein filament are clearly resolved, indicating the helical structure of the complex. A dislocation is observed, suggesting that polymerization of the RecA protein occurred from multiple sites. Scale bar: 20 nm. (b) Internal structure of a DNA origami, indicating the weave of crossovers between adjacent helices integral to the structure. The helical pitch of DNA can clearly be resolved (arrows). Scale bar: 15 nm. (c) The B-form DNA helical pitch clearly resolved upon Lambda bacteriophage DNA (arrows). Both the minor and major grooves are resolved as indicated in the averaged trace (d) across the region highlighted by the white box. Larger scale writhe is observed, indicating supercoiling of the biopolymer. Scale bar: 15 nm. All images are captured in aqueous buffer using PFT.

As discussed, DNA is one of the most promising molecules for use in bionano-materials application due to the accuracy and fidelity of the Watson and Crick base pairing, and its structural characteristics as a semiflexible polymer.

Similarly, observing the helical pitch of linear DNA molecules (Figure 3.8c) is now becoming relatively routine, and is even possible in more complex DNA geometries found in structural DNA nanotechnology and DNA origami (Figure 3.8b). In Figure 3.8b and c and the trace—taken across the white box in Figure 3.8c—in Figure 3.8d, the major grooves (arrows) and minor grooves are clearly resolved, showing the characteristic 3.4 nm helical pitch of B-form DNA. Additional periodicity with longer unit size can be observed owing to supercoiling of the DNA molecule; this can be seen in the overall height difference along the long axis of the DNA molecule.

Direct imaging with this level of detail can offer valuable insight into the internal forces induced in biologically templated structures, or forces exerted by additional protein components in hierarchical assemblies. For example, the helical pitch of DNA when encapsulated by RecA-based nucleoprotein filaments is approximately 5.2 nm, which represents an increase of 50% over native double-stranded DNA. Where this occurs at a region of homology on a DNA template, the resulting stress has to be relaxed outside the patterned region of the DNA molecule. Although this is relatively straightforward in arrangements comprising only linear DNA molecules in solution, in more complex geometries or where two nucleoprotein filaments are in close proximity upon the same DNA molecule, the rotational freedom will be reduced, potentially resulting in unwanted stresses within the bionanomaterial. An understanding of how RecA nucleoprotein filaments impact on the underlying DNA scaffold is critical for advancing the field.

## 3.5 TEMPORAL RESOLUTION

It can be argued that one of the great limitations of AFM for the study of biological systems until very recently has been its lack in ability to image at sufficient speed to monitor biological processes in real time. However, in the past few years, a number of instruments have become available which make this possible. Before these recent advances are discussed, the theoretical aspects governing temporal resolution of dynamic AFM under fluid are briefly considered.

The highest possible imaging rate is fundamentally defined by the feedback bandwidth of the control loop that maintains the tapping force during a scan, with further limitations implied by the sample itself.[34,74–79]

The image acquisition rate, $r_{aq}$, is given by

$$r_{aq} = \frac{v_t}{2An} \tag{3.16}$$

where:
$v_t$ is the velocity of the tip ($x$-axis)
$A$ is the scan area
$n$ is the density of scan lines ($y$-axis)

If we consider a sample whose typical features are distributed along the $x$-axis with a periodicity $\eta$, the bandwidth $v_B$ of the $z$-axis feedback loop must be sufficient to trace the sample surface, that is,

$$v_B \geq \frac{v_t}{\eta} = \frac{2r_{aq}An}{\eta} \tag{3.17}$$

and hence

$$r_{aq} \leq \frac{v_B \eta}{2An} \tag{3.18}$$

Provided that the feedback loop acquires a phase delay of $\pi/4$ at the bandwidth frequency, the time domain delay resulting from the feedback loop is at least $\Delta\tau = 1/8v_B$. Most AFM systems operate in closed-loop feedback configuration, and it is reasonable to assume that the total time delay doubles as a result of the closed loop,[34] that is, $\Delta\tau_{cl} \approx 2\Delta\tau$, and hence

$$v_B \leq \frac{1}{16\Delta\tau_{total}} \tag{3.19}$$

where $\Delta\tau_{total}$ is the total time delay in the feedback which is equal to or bigger than $\Delta\tau_{cl}$ Hence, the maximum acquisition rate is given by

$$r_{aq}^{max} = \frac{\eta}{32\Delta\tau_{total}An} \tag{3.20}$$

For illustration, in order to achieve an acquisition rate of five frames per second at a scan line density of 10 nm$^{-1}$ for a sample of 50 nm × 50 nm with a typical feature periodicity of 1 nm, the total time delay in the feedback loop must not exceed 0.25 μs, and a minimum feedback bandwidth of 250 kHz must be achieved.

### 3.5.1 INSTRUMENTATION PERSPECTIVE

Early efforts to increase the imaging speed of conventional AFMs have focused on contact mode techniques. It was identified that the mechanical response of the microscope is limited by the time delay in the piezo tube that actuates displacements in $x$, $y$, and $z$. In order to increase the bandwidth limitations of these piezo stacks, feedback actuation was attempted by the deposition of piezoelectric films directly on the cantilevers.[80] Subsequent developments saw the attempted use of these integrated piezoelectric films in dynamic modes.[81,82]

Through these methods, it was possible to increase the imaging speed by an order of magnitude, although it was limited by complicated signals that arose from integrating the piezoelectric films into the full length of the cantilever itself.[12] Furthermore, limitations in fabrication sizes of such cantilevers resulted in high spring constants and poor resonance frequencies,[83,84] which were further compounded in liquid.[85] Efforts to develop large arrays of individually addressable cantilevers were simultaneously attempted.

Further progress in fast-scan AFM was achieved by Miles et al.[86,87] by developing novel approaches to constant mode. In their method, a tuning fork, resonating at 100 kHz, was employed as the $x$ scanner, enabling frame rates of up to 14 frames per second in ambient conditions when imaging semicrystalline polymer samples.

Ando's and Hansma's groups simultaneously reported attempts to increase the feedback bandwidth of dynamic modes. A flurry of inventions followed, including small cantilevers with dimensions of less than 10 μm and corresponding optical detection systems to utilize these small cantilevers,[17,88–90] active damping $z$ scanners,[91,92] dynamic proportional–integral–derivative controllers,[93] and fast data acquisition systems.[94–96]

### 3.5.2  IMPORTANCE OF SURFACE PREPARATION

In particular, for high-speed AFM, it is essential to remember that in addition to instrument capabilities, the sample preparation also plays a defining role in what can be achieved with AFM techniques. AFM does not require sample staining, fixation, or labeling, which is often characteristic of radiation-based microscopy techniques. However, as this method employs a physical probe, the deposition of molecules on a flat and rigid support surface—typically Muscovite mica—is required.

The exact method of sample preparation is dictated by the intended experimental procedures and can be broadly split between those that allow for maximum spatial resolution and those that allow for dynamic events to be observed *in situ*. In simplified terms, we can say that the difference between these two objectives is a function of the strength with which the molecules are bound to the support surface. Tightly bound molecules allow for greater spatial resolutions when imaging but lack the flexibility to allow subsequent interactions to occur *in situ*. Here, only preparations that allow biological processes to be observed with high-speed AFM will be discussed.

In order to observe biological process in real time, we require that the biological molecules largely retain their degrees of freedom when confined within 2D on the support surface. In short, they must be sufficiently free, that is, unconstrained by the interaction with the support surface, to undergo interactions, yet sufficiently fixed to resist the instantaneous interaction forces imposed by the scanning AFM probe. This seems to be a contradiction impossible to overcome.

Let us first consider the net charges on the surface of both the biological molecules and the support substrate. Where mica is utilized, it holds a net negative surface charge owing to the dissociation of $K^+$ ions from its aluminum phyllosilicate lattice upon submersion in aqueous fluids. Biological polymers such as nucleic acids hold a net negative charge, whereas proteins tend to have more complex surface charge distributions, which may be split into localized specific domains. It is worth noting that the charge on the biological molecules varies significantly with pH as their isoelectric point is generally not too far from neutral pH, and hence changes in the pH of the imaging buffer can result in altering the net charge of the molecule from negative to positive, or vice versa. However, the useful pH range is typically constrained as the biological molecules are typically only fully functional in narrow pH ranges.

The interactions that occur between nucleic acids, such as DNA, and proteins are the most commonly investigated by AFM owing to their relevance in biology and bionanotechnology. For such AFM experiments, the DNA molecules are usually bound to the mica surface, and subsequently the proteins are imaged while interacting with the surface associated DNA. In order to bind the negatively charged DNA molecules to mica, previous studies have chemically modified the surface, such as with APTES ((3-aminopropyl)triethoxysilane), resulting in a net positive surface charge.[97] However, this typically results in an associated increase in surface roughness. Alternative approaches include the cycling of buffers[98–100] or electric fields[101,102] to switch between loose and bound states to image molecular interactions in stages; however, these lack the ability of true *in situ* observation. Additionally, novel support structures have been fabricated using DNA origami to isolate the biological interaction through indirect surface attachment.[103] Although all these methods have proven successful to varying degrees, they all put significant constraints on the experimental setup or do not provide the required functionality.

We have recently demonstrated the ability to tune the surface translational freedom of DNA molecules such that their surface association is sufficient to resist the lateral forces imposed by the tip during high-speed AFM imaging, while simultaneously providing enough mobility for nucleo-protein interactions to occur within physiologically relevant buffering conditions.[104]

It is well established that divalent cations can be used to substitute the dissociated $K^+$ ions in the mica lattice. Transition metal cations, such as $Ni^{2+}$, $Co^{2+}$, and $Zn^{2+}$, bind irreversibly to mica, whereas biological relevant cations such as $Mg^{2+}$ bind transiently, continually exchanging with $K^+$ and $H^+$ ions in the imaging buffer. This difference in binding strength can be exploited to tune the overall surface interaction of the DNA molecules with the mica. $Ni^{2+}$ provides strong localized attachment points, whereas $Mg^{2+}$, as it binds only transiently, provides weak and diffuse attachment.

Mica surfaces can be preincubated with $Ni^{2+}$ as these ions bind irreversibly to the mica, but the $Mg^{2+}$ has to be present within the imaging buffer. By carefully controlling the ratio of $Ni^{2+}$ preincubation concentrations to the concentration of $Mg^{2+}$ in the imaging buffer, it is possible to tune the surface translational mobility of DNA as a function of $Ni^{2+}$ concentration, where the $Mg^{2+}$ concentration is kept at a physiologically relevant concentration.[104] An example of how the surface association of the DNA can be tuned from very strongly bound to very weakly bound is shown in Figure 3.9.

### 3.5.3 Observation of Biological Events in Real Time

Time-resolved AFM investigations have been successfully carried out over the past two decades, offering critical insight into many areas of biological study. To date, many model systems have been observed in action, such as RNA

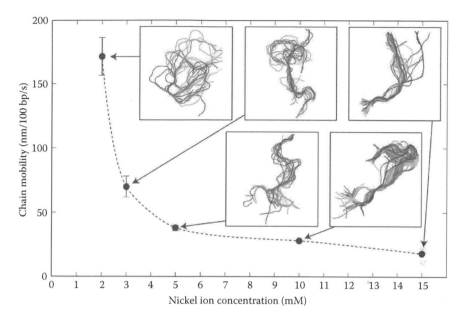

**FIGURE 3.9**    Surface translational mobility of 3.5 kbp DNA as a function of $Ni^{2+}$ concentration used for preincubation. The mobility of the DNA molecules is observed to decrease with increasing $Ni^{2+}$ concentration. An example DNA evolution profile is given for each condition in support of the observed trend in translational mobility. The dashed line through the data points is intended as a guide to the eye.

polymerase,[105,106] bacteriophage Lambda Cro protein,[107] DNA photolyase,[108] nucleosomes,[109] and restriction enzymes,[110] to name but a few.

Despite the success of these studies at revealing important information on the dynamics of biological molecules, the limited temporal resolutions of traditional AFM have impeded true real-time observations. Biological reactions take place on the order of seconds, and it is only recently that AFM technology was developed far enough to enable studies at multiple frames per second.

With the development of high-speed AFM over the last decade, the ability to follow biological processes with subsecond resolution while maintaining single-molecule spatial resolution is now becoming possible. One of the most astounding examples, work by the Ando group, demonstrates the observation of the hand-over-hand walking motion of the motor protein myosin V, confirming the long postulated mode of action (Figure 3.10).[111] Additional work by the same group has demonstrated the direct observation of the response of bacteriorhodopsin to light[112] and the response of rotorless F1-ATPase to ATP.[113]

Furthermore, the Sugiyama group has demonstrated an array of dynamic studies of DNA structural transitions situated within DNA origami reference frames, including folding of G-quadruplex and identifying its intermediate states,[114–116] and the transition of B to Z form of DNA.[117] Moreover, they have utilized the same system to observe directly the action of Holliday junction-resolving enzymes,[118] T7 RNA polymerase,[119] and the site-specific recombination protein Cre.[120]

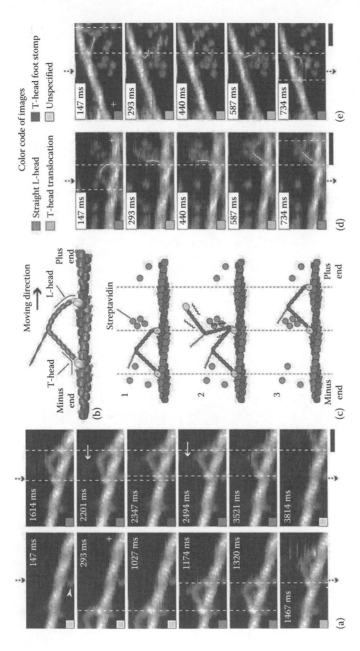

**FIGURE 3.10** High-speed AFM observation of Myosin V walking along actin. (a) A sequence of HS-AFM images showing the progressive movement of Myosin V along an actin filament in the presence of 1 mM ATP. The arrowhead indicates a molecule of streptavidin, which forms part of the sample preparation. The arrows indicate the coiled-coil tail of the Myosin V. Scale bar: 30 nm. (b) and (c) Schematic diagrams depicting the structure of Myosin V and walking mechanism, respectively. (d) and (e) A sequence of HS-AFM images depicting the hand-over-hand motion of Myosin V in 1 and 2 mM ATP, respectively. The center of mass is indicated in (a), (d), and (e) with dashed lines and a plus denotes the swinging lever of the motor protein is indicated by a white line. Scale bars: 50 nm (d) and 30 nm (e). The center of mass is indicated in (a), (d), and (e) with dashed lines and a plus denotes the positive end of the actin filament. The swinging lever of the motor protein is indicated by a white line. (Reprinted by permission from Macmillan Publishers Ltd. *Nature* Kodera, N. et al., 468(7320), 72–76, Copyright 2010.)

## 3.6   DIRECT MANIPULATION AND CONSTRUCTION

With accurate control of a physical probe, it becomes possible to deposit, remove, and move molecules directly on a surface. Macroscale methods of construction utilize similar principles such as additive manufacturing or the bulk machining of materials to form functional structures. In this section, scanning probe lithography techniques are briefly reviewed, which are—at least originally—derived from their microscopical counterparts.

### 3.6.1   MECHANICAL MANIPULATION—DESTRUCTIVE

Although the sharp tip on the end of the flexible cantilever is primarily utilized for imaging and force spectroscopy, it can also be employed to manipulate the substrate. By applying a large loading force to the substrate, one can intentionally scratch the surface as would be achieved at the macroscale using a scalpel blade; this is unimaginatively known as nanoshaving (Figure 3.11a). The applied force, geometry of the tip, and material hardness (relative to the tip hardness) all govern the overall profile of the scratch and this method is typically conducted using diamond-like carbon tips. This process can be used, for example, to scrape away polymer resists to expose the underlying substrate for subsequent depositions of metals or other materials. In addition, it can be further extended to work with self-assembled monolayers (SAMs), such as alkane thiolates that spontaneously absorb onto gold surfaces through a terminal thiol group. By applying lateral force, the AFM tip can be used to scrape away regions of the monolayer leaving the gold substrate selectively exposed. This method can therefore be utilized to pattern one SAM within another by immersing the substrate with the selectively exposed gold areas to a second SAM solution.[121,122] Applications of this technique include the creation of patterns with alternative protein-binding and protein-repelling regions. This methodology is known as nanografting, and is capable of achieving sub-50 nm resolutions (Figure 3.11b).

### 3.6.2   NANO WRITING—CONSTRUCTIVE

AFM-mediated nanolithography is not necessarily a destructive process, but can be used to lay down selectively new material as well. In its simplest form, dip pen nanolithography (DPN) can be described as a nanoscale fountain pen, where a probe is dipped in ink and then used to write arbitrary patterns upon a surface (Figure 3.11c). Typically, the resolution is between 20 nm and 10s of μm, depending upon the viscosity of the ink utilized and the chemistry of the surface to be patterned.

Although DPN is derived from AFM, it does not directly depend on it. The process is governed by the relative humidity of the environment, the viscosity of the *ink* and consequently the meniscus formed between the surface and the tip. The writeable feature size has little dependence upon the applied force and therefore, DPN can be operated independent of optical cantilever detection systems or complicated feedback loops.

Broadly speaking, there are two types of DPN based upon the type of molecule deposited as ink, those based on molecular inks, and those based on liquid inks. The former were the first described, utilizing small molecules that are delivered to a substrate

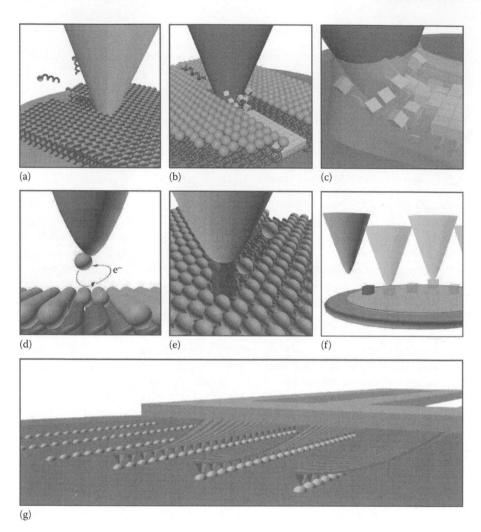

**FIGURE 3.11** Nano manipulation. A summary of tip-based nano manipulation modes, including (a) nanoshaving, (b) nanografting, (c) DPN, (d) electrochemical modification of a substrate with C-AFM, (e) photolithography conducted by NSOM, and (f) the direct manipulation of individual molecules and atoms by STM. (g) Large multiplexed cantilever arrays can enhance the throughput of nano manipulation techniques, as depicted for DPN.

via a water meniscus. The feature size is controlled by varying the meniscus size as a function of relative humidity and by varying the tip dwell time. It is critical to match the deposition rate to the diffusion rate of the molecule utilized, which is different for each component. Additionally, it is important to note that this methodology means that the ink is restricted to being written to a specific substrate where the binding chemistry is matched, for example, alkanethiols onto gold or silanes onto silicon substrates.

By contrast, liquid inks are not restricted to dedicated surface chemistries and can be written onto virtually any substrate. These are composed of materials that can be dissolved in a liquid or that are liquids themselves, including but not restricted to proteins, peptides, DNA, lipids, hydrogels, sol–gels, and conductive inks. The viscosity of the ink is the dominating factor in governing resolution, with higher viscosities delivering greater control over feature size and deposition rate.

DPN can be multiplexed by increasing the number of cantilevers that can simultaneous address the surface, potentially applying different inks, which allows, for example, the production of multiplexed protein or DNA arrays for biosensing applications (Figure 3.11g). Salaita et al. demonstrated the simultaneous multiplexed writing by DPN using a 2D array of 55,000 cantilevers for such applications.[123]

A combination of DPN with nanoshaving, termed nanopen reader and writer, has been used to introduce 4 $nm^2$ patches of octadecanethiol directly into a decanethiol monolayer.[124] Here, an octadecanethiol solution is used as an ink, and as soon as the preformed decanethiol SAM is removed through nanoshaving, the new SAM molecule, octadecanethiol, is written into the resulting cavities.

In alternative approaches, scanning probe-based techniques have been used for data storage applications with potential advantages such as ultrahigh areal density, potentially up to terabytes per square inch, and fast read and write speeds, comparable with flash memory.[125–129] Multiple regimes of operations have been demonstrated over the past two decades, including thermomechanical,[126,130] localized phase changing of a polymer media,[131–133] magnetic,[134] thermo-enhance-magnetic,[135,136] electrical,[137] ferroelectric,[138] and optical.[139] These different approaches have all been demonstrated in large multiplexed arrays employing large numbers of individually addressable tips to read and write data in a highly parallel nature.

### 3.6.3 Modifying Local Surface Chemistry

Alternative approaches for modifying the sample surface locally using AFM-derived techniques include methods where the tip is used to apply an electrical current (e.g., conductive AFM [C-AFM]) (Figure 3.11d) or emit light on to the sample (e.g., near-field scanning optical microscopy [NSOM]) (Figure 3.11e). C-AFM allows the simultaneous acquisition of topographical information and application of an electrical current. In this context, C-AFM has be used to oxidize locally the functional groups of SAMs, for example, transforming the hydrophobic $CH_3$ terminal groups into hydrophilic hydroxyl groups.[140] Such constructive processing, followed by the use of selective chemistries to modify the SAM further, has been used to demonstrate the directed fabrication of gold clusters,[124] nanoparticles,[141] and silver islands.[140]

Photolithographic processes are generally restricted in resolution by the diffraction limit, that is, the smallest features that can be defined are on the order of the wavelength of the light that is being used. However, this is only true in the far field, that is, where the distance of the light source to the sample is significantly larger than the wavelength of the light. This limitation can be circumvented by making use of near-field effects, that is, by exploiting the radiation very close to the source of the light. This is realized by NSOMs, where an AFM-like tip is used as a light source. This can be conducted in two distinct ways. Radiation, for example, a laser beam, can be focussed onto the tip apex, which then acts as a light source. The light emitted from the tip apex, which is confined between the tip and the sample surface reminiscent of tip-enhanced Raman spectroscopy, can then be used for photolithography in the near field.[142,143] Alternatively, apertures can be fabricated into the tips from which the light is emitted. Despite significant challenges associated with the manufacturing of regular apertured tips, tip heating, and interference artifacts, aperture-mode NSOM has been demonstrated by Leggett et al. who employed a UV laser to the patterning of photo-cleavable SAMs producing features of less than 50 nm.[144–146] Despite the advantages that this approach conveys, it remains limited—as all SPM techniques are—by the serial nature of the technique. In order to address this challenge, large arrays of tips can be used in parallel to increase the simultaneous write area of the technique.[147,148]

### 3.6.4 BUILDING WITH ATOMS

Scanning probe techniques can also be used to directly manipulate and move atoms around on a sample surface (Figure 3.11f). Eigler's group at IBM were the first to demonstrate this capability, and they demonstrated the positioning of individual iron atoms using an STM to confine the electronic states within a copper surface, known as a quantum corral.[149] IBM have since gone on to demonstrate the world's smallest stop motion movie called *A Boy and His Atom*. Here, carbon monoxide molecules were individually placed upon a copper surface using an STM as part of IBMs continued development of atomic memory and data storage applications. Although these impressive demonstrations are undertaken in solid-state systems rather than biologically templated systems, they highlight the increasing potential for SPM techniques to position directly isolated molecules, which could eventually extend to the directed assembly of discrete biological systems.

## 3.7 CONCLUSIONS AND PERSPECTIVES

Since its invention, the AFM has proven to be an extremely powerful tool for the characterization of materials at the nanoscale. The exceptional versatility of this form of microscopy has made it into a key tool for the interrogation of biological molecules. Although the operating principles of most, if not all, aspects of AFM are conceptually straightforward, the theoretical foundations for understanding the various different interactions involved have proven to be much more demanding to establish. In this chapter, the basic underlying theory of dynamic AFM modes has been discussed, although only to a limited extent, and readers trying to acquire a more detailed picture are referred to the literature cited in the text.

The interrogation of a substrate by a sharp tip in mechanical contact with the surface means that the spatial resolution is directly governed by the physical properties of both the tip and the sample, as well as being a function of the force applied by the tip to the surface. By carefully optimizing tip geometry and in particular the mechanical properties of the cantilevers, the forces acting on the sample can be reduced and thus significant improvements in resolution were demonstrated over the past decades employing a range of different approaches. However, this was particularly challenging when imaging soft samples such as biological molecules and biologically templated materials. The most notable recent advancement was achieved through significant developments in feedback electronics, enabling very fast and sensitive control over the applied interaction forces.

Furthermore, as AFM instruments collect data in a serial fashion, the temporal resolution is governed principally by the response time of the mechanical and electrical components of the instrument's feedback loop. A number of key advancements to the temporal resolutions have been highlighted in this chapter, indicating substantial developments in AFM instrumentation. In particular, recent developments of high-speed AFM is beginning to allow the interrogation of biological mechanisms over relevant timescales, a revolution that will undoubtedly change our understanding of some of the most fundamental processes in biology and further enable their application in bionanotechnology.

In contrast to radiation-based microscopies, AFM requires the biological molecules to be deposited on an ultra-flat and rigid support surface. As a result, the sample preparation plays a major role in all AFM investigations. This is of particular importance when investigating dynamic biological processes using HS-AFM techniques, as a fine balance between strong enough attachment to the surface to enable imaging with a mechanical tip, which applies a certain force to the molecules upon imaging, and the requirement to minimize the interaction of the molecules with the surface to allow the molecules to move as freely as possible, has to be struck.

Although AFM techniques have advanced considerably over the past few decades, they are far from reaching saturation, and many more developments will follow. For the advancement of biologically templated materials, where materials with novel properties are being developed by, for example, exploiting functionality of biological molecules, it is of fundamental importance that these biological mechanisms are understood in detail. Considering the potential of biologically templated materials, and of course of the vast range of other science or engineering challenges that benefit from such advances, further and more powerful tools to interrogate these underpinning mechanisms will be developed over the coming decades and beyond, and it can be anticipated that AFM techniques will play a very prominent role for the foreseeable future.

## REFERENCES

1. R. P. Feynman. There's plenty of room at the bottom. In *Annual Meeting of the American Physical Society*, Caltech, PA, 1959.
2. G. Binnig and H. Rohrer. Scanning tunneling microscopy. *IBM Journal of Research and Development*, 30(4):279–293, 1986.
3. M. Nonnenmacher, M. P. O'Boyle, and H. K. Wickramasinghe. Kelvin probe force microscopy. *Applied Physics Letters*, 58:2921–2923, 1991.
4. U. Hartmann. Magnetic force microscopy. *Annual Review of Materials Science*, 29:53–87, 1999.
5. A. Noy, D. V. Vezenov, and C. M. Lieber. Chemical force microscopy. *Annual Review of Materials Science*, 27:381–421, 1997.
6. D. W. Pohl, W. Denk, and M. Lanz. Optical stethoscopy: Image recording with resolution λ/20. *Applied Physics Letters*, 44(7):651, 1984.
7. P. K. Hansma, B. Drake, O. Marti, S. A. C. Gould, and C. B. Prater. The scanning ion-conductance microscope. *Science*, 243(4891):641–643, 1989.

8. C. C. Williams and H. K. Wickramasinghe. Scanning thermal profiler. *Applied Physics Letters*, 49(23):1587, 1986.

9. G. Binnig, C. F. Quate, and C. H. Gerber. Atomic force microscope. *Physical Review Letters*, 56(9):930–933, 1986.

10. S. Zhang. Building from the bottom up. *Materials Today*, 6(5):20–27, 2003.

11. S. Zhang. Fabrication of novel biomaterials through molecular self-assembly. *Nature Biotechnology*, 21(10):1171–1178, 2003.

12. N. C. Santos and M. A. R. B. Castanho. An overview of the biophysical applications of atomic force microscopy. *Biophysical Chemistry*, 107(2):133–149, 2004.

13. R. García. *Amplitude Modulation Atomic Force Microscopy*. John Wiley & Sons, Weinheim, Germany, 2011.

14. Y. Martin, C. C. Williams, and H. K. Wickramasinghe. Atomic force microscope-force mapping and profiling on a sub 100-Å scale. *Journal of Applied Physics*, Weinheim, Germany, 61(10):4723, 1987.

15. Q. Zhong, D. Inniss, K. Kjoller, and V.B. Elings. Fractured polymer/silica fiber surface studied by tapping mode atomic force microscopy. *Surface Science Letters*, 290:L688–L692, 1993.

16. H. G. Hansma, K. A. Browne, M. Bezanilla, and T. C. Bruiceg. Bending and straightening of DNA induced by the same ligand: Characterization with the atomic force microscope. *Biochemistry*, 33(28):8436–8441, 1994.

17. T. E. Schaffer, J. P. Cleveland, F. Ohnesorge, D. A. Walters, and P. K. Hansma. Studies of vibrating atomic force microscope cantilevers in liquid. *Journal of Applied Physics*, 80(7):3622, 1996.

18. M. Tortonese, R. C. Barrett, and C. F. Quate. Atomic resolution with an atomic force microscope using piezoresistive detection. *Applied Physics Letters*, 62(8):834, 1993.

19. G. Meyer and N. M. Amer. Novel optical approach to atomic force microscopy. *Applied Physics Letters*, 53(12):1045, 1988.

20. C. Argento and R. H. French. Parametric tip model and force-distance relation for Hamaker constant determination from atomic force microscopy. *Journal of Applied Physics*, 80(11):6081, 1996.

21. H. Hertz. Über die Berührung fester elastischer Körper. *Journal für die reine und angewandte Mathematik*, 92:156–171, 1881.

22. B. V. Derjaguin, V. M. Muller, and Y. U. P. Toporov. Effect of contact deformations on the adhesion of particles. *Progress in Surface Science*, 45(1–4):131–143, 1994.

23. K. L. Johnson and J. A. Greenwood. An adhesion map for the contact of elastic spheres. *Journal of Colloid and Interface Science*, 333(192):326–333, 1997.

24. K. L. Johnson, K. Kendall, and A. D. Roberts. Surface energy and the contact of elastic solids. *Proceedings of the Royal Society of London. Series A, Mathematical and Physical*, 324(1558):301–313, 1971.

25. R. Fisher and J. Israelachvili. Experimental studies on the applicability of the Kelvin equation to highly curved concave menisci. *Journal of Colloid and Interface Science*, 80(2):528–541, 1981.

26. J. Israelachvili. *Intermolecular and Surface Forces*. Elsevier, London, 3rd edition, 2011.

27. H. J. Butt, B. Cappella, and M. Kappl. Force measurements with the atomic force microscope: Technique, interpretation and applications. *Surface Science Reports*, 59(1–6):1–152, 2005.

28. D. J. Müller, D. Fotiadis, S. Scheuring, S. A. Müller, and A. Engel. Electrostatically balanced subnanometer imaging of biological specimens by atomic force microscope. *Biophysical Journal*, 76(2):1101–1111, 1999.

29. J. Sotres and A. M. Baró. AFM imaging and analysis of electrostatic double layer forces on single DNA molecules. *Biophysical Journal*, 98(9):1995–2004, 2010.

30. V. J. Morris, A. R. Kirby, and A. P. Gunning. *Atomic Force Microscopy for Biologists.* Imperial College Press, London, 2nd edition, 2010.
31. T. Fukuma and S. P. Jarvis. Development of liquid-environment frequency modulation atomic force microscope with low noise deflection sensor for cantilevers of various dimensions. *Review of Scientific Instruments*, 77(4):043701, 2006.
32. J. I. Kilpatrick, S. H. Loh, and S. P. Jarvis. Directly probing the effects of ions on hydration forces at interfaces. *Journal of the American Chemical Society*, 135(7):2628–2634, 2013.
33. C. A. J. Putman, K. O. Van der Werf, B. G. De Grooth, N. F. Van Hulst, and J. Greve. Tapping mode atomic force microscopy in liquid. *Applied Physics Letters*, 64(18):2454, 1994.
34. T. Ando. High-speed atomic force microscopy coming of age. *Nanotechnology*, 23(6):062001, 2012.
35. S. Basak, A. Raman, and S. V. Garimella. Hydrodynamic loading of microcantilevers vibrating in viscous fluids. *Journal of Applied Physics*, 99(11):114906, 2006.
36. S. Basak and A. Raman. Dynamics of tapping mode atomic force microscopy in liquids: Theory and experiments. *Applied Physics Letters*, 91(6):064107, 2007.
37. C. P. Green and J. E. Sader. Frequency response of cantilever beams immersed in viscous fluids near a solid surface with applications to the atomic force microscope. *Journal of Applied Physics*, 98(11):114913, 2005.
38. A. Raman, J. Melcher, and R. Tung. Cantilever dynamics in atomic force microscopy. *Nano Today*, 3(1):20–27, 2008.
39. F. J. Giessibl. Advances in atomic force microscopy. *Reviews of Modern Physics*, 75(3):949–983, 2003.
40. T. Fukuma, K. Kobayashi, K. Matsushige, and H. Yamada. True molecular resolution in liquid by frequency-modulation atomic force microscopy. *Applied Physics Letters*, 86(19):193108, 2005.
41. T. Fukuma. Wideband low-noise optical beam deflection sensor with photothermal excitation for liquid-environment atomic force microscopy. *Review of Scientific Instruments*, 80(2):023707, 2009.
42. T. Ando, T. Uchihashi, and T. Fukuma. High-speed atomic force microscopy for nano-visualization of dynamic biomolecular processes. *Progress in Surface Science*, 83(7–9):337–437, 2008.
43. L. Martínez, M. Tello, M. Díaz, E. Román, R. Garcia, and Y. Huttel. Aspect-ratio and lateral-resolution enhancement in force microscopy by attaching nanoclusters generated by an ion cluster source at the end of a silicon tip. *Review of Scientific Instruments*, 82(2):023710, 2011.
44. R. M. Stevens. New carbon nanotube AFM probe technology. *Materials Today*, 12(10):42–45, 2009.
45. C. Möller, M. Allen, V. Elings, A. Engel, and D. J. Müller. Tapping-mode atomic force microscopy produces faithful high-resolution images of protein surfaces. *Biophysical Journal*, 77(2):1150–1158, 1999.
46. N. Mullin and J. Hobbs. Torsional resonance atomic force microscopy in water. *Applied Physics Letters*, 92(5):053103, 2008.
47. N. Mullin, C. Vasilev, J. D. Tucker, C. N. Hunter, C. H. M. Weber, and J. K. Hobbs. Torsional tapping atomic force microscopy using T-shaped cantilevers. *Applied Physics Letters*, 94(17):173109, 2009.
48. N. Mullin and J. Hobbs. Direct imaging of polyethylene films at single-chain resolution with torsional tapping atomic force microscopy. *Physical Review Letters*, 107(19):1–5, 2011.
49. N. C. Seeman. Nucleic acid junctions and lattices. *Journal of Theoretical Biology*, 99(2):237–247, 1982.

50. N. C. Seeman. Structural DNA nanotechnology: Growing along with nano letters. *Nano Letters*, 10(6):1971–1978, 2010.

51. N. C. Seeman. Nanomaterials based on DNA. *Annual Review of Biochemistry*, 79:65–87, 2010.

52. C. Lin, Y. Liu, and H. Yan. Designer DNA nanoarchitectures. *Biochemistry*, 48(8):1663–1674, 2009.

53. P. W. K. Rothemund. Folding DNA to create nanoscale shapes and patterns. *Nature*, 440(7082):297–302, 2006.

54. R. Sharma, A. G. Davies, and C. Wälti. Directed assembly of 3-nm-long RecA nucleo-protein filaments on double-stranded DNA with nanometer resolution. *ACS Nano*, 8(4):3322–3330, 2014.

55. R. Sharma, A. G. Davies, and C. Wälti. RecA protein mediated nano-scale patterning of DNA scaffolds. *Journal of Nanoscience and Nanotechnology*, 11(12):10629–10632, 2011.

56. R. Sharma, A. G. Davies, and C. Wälti. Nanoscale programmable sequence-specific patterning of DNA scaffolds using RecA protein. *Nanotechnology*, 23(36):365301, 2012.

57. K. Keren, M. Krueger, R. Gilad, G. Ben-Yoseph, U. Sivan, and E. Braun. Sequence-specific molecular lithography on single DNA molecules. *Science*, 297(5578):72–75, 2002.

58. T. Nishinaka, A. Takano, Y. Doi, M. Hashimoto, A. Nakamura, Y. Matsushita, J. Kumaki, and E. Yashima. Conductive metal nanowires templated by the nucleoprotein filaments, complex of DNA and RecA protein. *Journal of the American Chemical Society*, 127(22):8120–8125, 2005.

59. P. Kaur, F. Qiang, A. Fuhrmann, R. Ros, L. O. Kutner, L. A. Schneeweis, R. Navoa et al. Antibody-unfolding and metastable-state binding in force spectroscopy and recognition imaging. *Biophysical Journal*, 100(1):243–250, 2011.

60. A. P. Gunning, S. Chambers, C. Pin, A. L. Man, V. J. Morris, and C. Nicoletti. Mapping specific adhesive interactions on living human intestinal epithelial cells with atomic force microscopy. *FASEB Journal*, 22(7):2331–2339, 2008.

61. T. Hoffmann and L. Dougan. Single molecule force spectroscopy using polyproteins. *Chemical Society Reviews*, 41(14):4781–4796, 2012.

62. B. Cappella and G. Dietler. Force-distance curves by atomic force microscopy. *Surface Science Reports*, 34(1–3):1–104, 1999.

63. K. Mitsui, M. Hara, and A. Ikai. Mechanical unfolding of a2-macroglobulin with atomic force microscope. *FEBS Letters*, 385:29–33, 1996.

64. M. Rief. Single molecule force spectroscopy on polysaccharides by atomic force microscopy. *Science*, 275(5304):1295–1297, 1997.

65. R. Perez-Jimenez, S. Garcia-Manyes, S. R. K. Ainavarapu, and J. M. Fernandez. Mechanical unfolding pathways of the enhanced yellow fluorescent protein revealed by single molecule force spectroscopy. *The Journal of Biological Chemistry*, 281(52):40010–40014, 2006.

66. S. M. Morsi, A. Pakzad, A. Amin, R. S. Yassar, and P. A. Heiden. Chemical and nano-mechanical analysis of rice husk modified by ATRP-grafted oligomer. *Journal of Colloid and Interface Science*, 360(2):377–385, 2011.

67. S. B. Kaemmer. *Application Note #133: Introduction to Bruker's ScanAsyst and PeakForce Tapping AFM Technology*. Technical report, Bruker Nano Surfaces Division, Santa Barbara, CA, 2011.

68. D. Alsteens, V. Dupres, S. Yunus, J. P. Latgé, J. J. Heinisch, and Y. F. Dufrêne. High-resolution imaging of chemical and biological sites on living cells using peak force tapping atomic force microscopy. *Langmuir*, 28(49):16738–16744, 2012.

69. P. Trtik, J. Kaufmann, and U. Volz. On the use of peak-force tapping atomic force microscopy for quantification of the local elastic modulus in hardened cement paste. *Cement and Concrete Research*, 42(1):215–221, 2012.

70. B. Pittenger, A. Slade, A. Berquand, P. Milani, A. Boudaoud, O. Hamant, and M. Radmacher. *Application Note #141: Toward Quantitative Nanomechanical Measurements on Live Cells with PeakForce QNM.* Technical report, Bruker Nano Surfaces Division, Santa Barbara, CA, 2013.

71. H. G. Hansma. Surface biology of DNA by atomic force microscopy. *Annual Review of Physical Chemistry*, 52:71–92, 2001.

72. E. L. Florin, V. T. Moy, and H. E. Gaub. Adhesion forces between individual ligand-receptor pairs. *Science*, 264(5157):415–417, 1994.

73. V. T. Moy, E. L. Florin, and H. E. Gaub. Intermolecular forces and energies between ligands and receptors. *Science*, 266(5183):257–259, 1994.

74. T. Ando, N. Kodera, E. Takai, D. Maruyama, K. Saito, and A. Toda. A high-speed atomic force microscope for studying biological macromolecules. *Proceedings of the National Academy of Sciences of the United States of America*, 98(22):12468–12472, 2001.

75. D. Yamamoto, T. Uchihashi, N. Kodera, H. Yamashita, S. Nishikori, T. Ogura, M. Shibata, and T. Ando. High-speed atomic force microscopy techniques for observing dynamic biomolecular processes. *Methods in Enzymology*, 475(10):541–564, 2010.

76. T. Ando, T. Uchihashi, and N. Kodera. High-speed atomic force microscopy. *Japanese Journal of Applied Physics*, 51(8):08KA02, 2012.

77. T. Ando, N. Kodera, Y. Naito, T. Kinoshita, K. Furuta, and Y. Y. Toyoshima. A high-speed atomic force microscope for studying biological macromolecules in action. *ChemPhysChem*, 4(11):1196–1202, 2003.

78. T. Ando, N. Kodera, T. Uchihashi, A. Miyagi, R. Nakakita, H. Yamashita, and K. Matada. High-speed atomic force microscopy for capturing dynamic behavior of protein molecules at work. *e-Journal of Surface Science and Nanotechnology*, 3:384–392, 2005.

79. T. Uchihashi, N. Kodera, and T. Ando. Guide to video recording of structure dynamics and dynamic processes of proteins by high-speed atomic force microscopy. *Nature Protocols*, 7(6):1193–1206, 2012.

80. S. R. Manalis, S. C. Minne, A. Atalar, and C. F. Quate. Interdigital cantilevers for atomic force microscopy. *Applied Physics Letters*, 69(25):3944, 1996.

81. T. Sulchek, S. C. Minne, J. D. Adams, D. A. Fletcher, A. Atalar, C. F. Quate, and D. M. Adderton. Dual integrated actuators for extended range high speed atomic force microscopy. *Applied Physics Letters*, 75(11):1637, 1999.

82. T. Sulchek, G. G. Yaralioglu, C. F. Quate, and S. C. Minne. Characterization and optimization of scan speed for tapping-mode atomic force microscopy. *Review of Scientific Instruments*, 73(8):2928, 2002.

83. J. D. Adams, L. Manning, B. Rogers, M. Jones, and S. C. Minne. Self-sensing tapping mode atomic force microscopy. *Sensors and Actuators A: Physical*, 121(1):262–266, 2005.

84. B. Rogers, L. Manning, T. Sulchek, and J. D. Adams. Improving tapping mode atomic force microscopy with piezoelectric cantilevers. *Ultramicroscopy*, 100(3–4):267–276, 2004.

85. B. Rogers, T. Sulchek, K. Murray, D. York, M. Jones, L. Manning, S. Malekos et al. High speed tapping mode atomic force microscopy in liquid using an insulated piezo-electric cantilever. *Review of Scientific Instruments*, 74(11):4683, 2003.

86. A. D. L. Humphris, M. J. Miles, and J. K. Hobbs. A mechanical microscope: High-speed atomic force microscopy. *Applied Physics Letters*, 86(3):034106, 2005.

87. J. K. Hobbs, C. Vasilev, and A. D. L. Humphris. Real time observation of crystallization in polyethylene oxide with video rate atomic force microscopy. *Polymer*, 46(23):10226–10236, 2005.

88. D. A. Walters, J. P. Cleveland, N. H. Thomson, P. K. Hansma, M. A. Wendman, G. Gurley, and V. Elings. Short cantilevers for atomic force microscopy. *Review of Scientific Instruments*, 67(10):3583, 1996.

89. M. B. Viani, T. E. Schaffer, A. Chand, M. Rief, H. E. Gaub, and P. K. Hansma. Small cantilevers for force spectroscopy of single molecules. *Journal of Applied Physics*, 86(4):2258, 1999.

90. M. B. Viani, T. E. Schaffer, G. T. Paloczi, L. I. Pietrasanta, B. L. Smith, J. B. Thompson, M. Richter et al. Fast imaging and fast force spectroscopy of single biopolymers with a new atomic force microscope designed for small cantilevers. *Review of Scientific Instruments*, 70(11):4300, 1999.

91. T. Fukuma, Y. Okazaki, N. Kodera, T. Uchihashi, and T. Ando. High resonance frequency force microscope scanner using inertia balance support. *Applied Physics Letters*, 92(24):243119, 2008.

92. N. Kodera, H. Yamashita, and T. Ando. Active damping of the scanner for high-speed atomic force microscopy. *Review of Scientific Instruments*, 76(5):053708, 2005.

93. N. Kodera, M. Sakashita, and T. Ando. Dynamic proportional-integral-differential controller for high-speed atomic force microscopy. *Review of Scientific Instruments*, 77(8):083704, 2006.

94. G. E. Fantner, P. Hegarty, J. H. Kindt, G. Schitter, G. A. G. Cidade, and P. K. Hansma. Data acquisition system for high speed atomic force microscopy. *Review of Scientific Instruments*, 76(2):026118, 2005.

95. G. E. Fantner, G. Schitter, J. H. Kindt, T. Ivanov, K. Ivanova, R. Patel, N. Holten-Andersen et al. Components for high speed atomic force microscopy. *Ultramicroscopy*, 106(8–9):881–887, 2006.

96. P. K. Hansma, G. Schitter, G. E. Fantner, and C. Prater. Applied physics. High-speed atomic force microscopy. *Science*, 314(5799):601–602, 2006.

97. Y. L. Lyubchenko and L. S. Shlyakhtenko. AFM for analysis of structure and dynamics of DNA and protein-DNA complexes. *Methods*, 47(3):206–213, 2009.

98. N. H. Thomson, M. Fritz, M. Radmacher, J. P. Cleveland, C. F. Schmidt, and P. K. Hansma. Protein tracking and detection of protein motion using atomic force microscopy. *Biophysical Journal*, 70(5):2421–2431, 1996.

99. O. Pietrement, D. Pastre, S. Fusil, J. Jeusset, M.-O. David, F. Landousy, L. Hamon, A. Zozime, and E. Le Cam. Reversible binding of DNA on NiCl2-treated mica by varying the ionic strength. *Langmuir*, 19(15):2536–2539, 2003.

100. G. R. Abel, E. A. Josephs, N. Luong, and T. Ye. A switchable surface enables visualization of single DNA hybridization events with atomic force microscopy. *Journal of the American Chemical Society*, 135:6399–6402, 2013.

101. E. A. Josephs and T. Ye. A single-molecule view of conformational switching of DNA tethered to a gold electrode. *Journal of the American Chemical Society*, 134(24):10021–10030, 2012.

102. E. A. Josephs and T. Ye. Electric-field dependent conformations of single DNA molecules on a model biosensor surface. *Nano Letters*, 12(10):5255–5261, 2012.

103. M. Endo and H. Sugiyama. Single-molecule imaging of dynamic motions of biomolecules in DNA origami nanostructures using high-speed atomic force microscopy. *Accounts of Chemical Research*, 47(6):1645–1653, 2014.

104. A. J. Lee, M. Szymonik, J. K. Hobbs, and C. Wälti. Tuning the translational freedom of DNA for high speed AFM. *Nano Research*, 8(6):1811–1821, 2014.

105. S. Kasas, N. H. Thomson, B. L. Smith, H. G. Hansma, X. Zhu, M. Guthold, C. Bustamante, E. T. Kool, M. Kashlev, and P. K. Hansma. *Escherichia coli* RNA polymerase activity observed using atomic force microscopy. *Biochemistry*, 36(3):461–468, 1997.

106. M. Guthold, M. Bezanilla, D. A. Erie, B. Jenkins, H. G. Hansma, and C. Bustamante. Following the assembly of RNA polymerase-DNA complexes in aqueous solutions with the scanning force microscope. *Proceedings of the National Academy of Sciences of the United States of America*, 91(26):12927–12931, 1994.

107. D. A. Erie, G. L. Yang, H. C. Schultz, and C. Bustamante. DNA bending by Cro protein in specific and nonspecific complexes: Implications for protein site recognition and specificity. *Science*, 266(5190):1562–1566, 1994.

108. J. van Noort, F. Orsini, A. Eker, C. Wyman, B. de Grooth, and J. Greve. DNA bending by photolyase in specific and non-specific complexes studied by atomic force microscopy. *Nucleic Acids Research*, 27(19):3875–3880, 1999.

109. S. H. Leuba, G. Yang, C. Robert, B. Samori, K. van Holde, J. Zlatanova, and C. Bustamante. Three-dimensional structure of extended chromatin fibers as revealed by tapping-mode scanning force microscopy. *Proceedings of the National Academy of Sciences of the United States of America*, 91(24):11621–11625, 1994.

110. I. Sorel, O. Piétrement, L. Hamon, S. Baconnais, E. Le Cam, and D. Pastré. The EcoRI-DNA complex as a model for investigating protein-DNA interactions by atomic force microscopy. *Biochemistry*, 45(49):14675–14682, 2006.

111. N. Kodera, D. Yamamoto, R. Ishikawa, and T. Ando. Video imaging of walking myosin V by high-speed atomic force microscopy. *Nature*, 468(7320):72–76, 2010.

112. M. Shibata, H. Yamashita, T. Uchihashi, H. Kandori, and T. Ando. High-speed atomic force microscopy shows dynamic molecular processes in photoactivated bacteriorhodopsin. *Nature Nanotechnology*, 5(3):208–212, 2010.

113. T. Uchihashi, R. Iino, T. Ando, and H. Noji. High-speed atomic force microscopy reveals rotary catalysis of rotorless F1-ATPase. *Science*, 333(6043):755–759, 2011.

114. A. Rajendran, M. Endo, K. Hidaka, and H. Sugiyama. Direct and single-molecule visualization of the solution-state structures of G-hairpin and G-triplex intermediates. *Angewandte Chemie (International Edition)*, 126(16):4191–4196, 2014.

115. A. Rajendran, M. Endo, K. Hidaka, P. L. T. Tran, J.-L. Mergny, and H. Sugiyama. Controlling the stoichiometry and strand polarity of a tetramolecular G-quadruplex structure by using a DNA origami frame. *Nucleic Acids Research*, 41(18):8738–8747, 2013.

116. Y. Sannohe, M. Endo, Y. Katsuda, K. Hidaka, and H. Sugiyama. Visualization of dynamic conformational switching of the G-quadruplex in a DNA nanostructure. *Journal of the American Chemical Society*, 132(46):16311–16313, 2010.

117. M. Endo, M. Inoue, Y. Suzuki, C. Masui, H. Morinaga, K. Hidaka, and H. Sugiyama. Regulation of B-Z conformational transition and complex formation with a Z-form binding protein by introduction of constraint to double-stranded DNA by using a DNA nanoscaffold. *Chemistry*, 19(50):16887–16890, 2013.

118. Y. Suzuki, M. Endo, C. Cañas, S. Ayora, J. C. Alonso, H. Sugiyama, and K. Takeyasu. Direct analysis of Holliday junction resolving enzyme in a DNA origami nanostructure. *Nucleic Acids Research*, 42:1–8, 2014.

119. M. Endo, K. Tatsumi, K. Terushima, Y. Katsuda, K. Hidaka, Y. Harada, and H. Sugiyama. Direct visualization of the movement of a single T7 RNA polymerase and transcription on a DNA nanostructure. *Angewandte Chemie (International Edition)*, 51(35):8778–8782, 2012.

120. Y. Suzuki, M. Endo, Y. Katsuda, K. Ou, K. Hidaka, and H. Sugiyama. DNA origami based visualization system for studying site-specific recombination events. *Journal of the American Chemical Society*, 136(1):211–218, 2014.

121. S. Xu, S. Miller, P. E. Laibinis, and G. Y. Liu. Fabrication of nanometer scale patterns within self-assembled monolayers by nanografting. *Langmuir*, 15(21):7244–7251, 1999.

122. P. V. Schwartz. Meniscus force nanografting: Nanoscopic patterning of DNA. *Langmuir*, 17(19):5971–5977, 2001.

123. K. Salaita, Y. Wang, J. Fragala, R. A. Vega, C. Liu, and C. A. Mirkin. Massively parallel dip-pen nanolithography with 55,000-pen two-dimensional arrays. *Angewandte Chemie (International Edition)*, 45(43):7220–7223, 2006.

124. S. Liu, R. Maoz, and J. Sagiv. Template guided self-assembly of *Au*55 clusters on nano-lithographically defined monolayer patterns. *Nano Letters*, 2(10):1055–1060, 2002.

125. E. B. Cooper, S. R. Manalis, H. Fang, H. Dai, K. Matsumoto, S. C. Minne, T. Hunt, and C. F. Quate. Terabit-per-square-inch data storage with the atomic force microscope. *Applied Physics Letters*, 75(22):3566, 1999.

126. P. Vettiger, J. Brugger, M. Despont, U. Drechsler, U. Dürig, W. Häberle, M. Lutwyche et al. Ultrahigh density, high-data-rate NEMS-based AFM data storage system. *Microelectronic Engineering*, 46(1):11–17, 1999.

127. H. Shin, K. M. Lee, W. K. Moon, J. U. Jeon, G. Lim, Y. E. Pak, J. H. Park, and K. H. Yoon. An application of polarized domains in ferroelectric thin films using scanning probe microscope. *IEEE Transactions on Ultrasonics, Ferroelectrics, and Frequency Control*, 47(4):801–807, 2000.

128. B. Bhushan and K. J. Kwak. Platinum-coated probes sliding at up to 100 mm s⁻¹ against coated silicon wafers for AFM probe-based recording technology. *Nanotechnology*, 18(34):345504, 2007.

129. B. Bhushan, K. J. Kwak, and M. Palacio. Nanotribology and nanomechanics of AFM probe-based data recording technology. *Journal of Physics: Condensed Matter*, 20(36):365207, 2008.

130. H. J. Mamin and D. Rugar. Thermomechanical writing with an atomic force micro-scope tip. *Applied Physics Letters*, 61(8):1003, 1992.

131. H. Kado and T. Tohda. Nanometer-scale recording on chalcogenide films with an atomic force microscope. *Applied Physics Letters*, 66(22):2961, 1995.

132. S. Gidon, O. Lemonnier, B. Rolland, O. Bichet, C. Dressler, and Y. Samson. Electrical probe storage using Joule heating in phase change media. *Applied Physics Letters*, 85(26):6392, 2004.

133. C. D. Wright, M. Armand, and M. M. Aziz. Terabit-per-square-inch data storage using phase-change media and scanning electrical nanoprobes. *IEEE Transactions on Nanotechnology*, 5(1):50–61, 2006.

134. T. Ohkubo, J. Kishigami, K. Yanagisawa, and R. Kaneko. Submicron magnetizing and its detection based on the point magnetic recording concept. *IEEE Transactions on Magnetics*, 27(6):5286–5288, 1991.

135. J. Nakamura, M. Miyamoto, S. Hosaka, and H. Koyanagi. High-density thermomag-netic recording method using a scanning tunneling microscope. *Journal of Applied Physics*, 77(2):779, 1995.

136. L. Zhang and J. A. Bain. Dependence of thermomagnetic mark size on applied STM voltage in Co-Pt multilayers. *IEEE Transactions on Magnetics*, 38(5):1895–1897, 2002.

137. R. C. Barrett. High-speed, large-scale imaging with the atomic force microscope. *Journal of Vacuum Science & Technology B: Microelectronics and Nanometer Structures*, 9(2):302, 1991.

138. K. Franke, J. Besold, W. Haessler, and C. Seegebarth. Modification and detection of domains on ferroelectric PZT films by scanning force microscopy. *Surface Science Letters*, 302(1–2):L283–L288, 1994.

139. E. Betzig, J. K. Trautman, R. Wolfe, E. M. Gyorgy, P. L. Finn, M. H. Kryder, and C. H. Chang. Near-field magneto-optics and high density data storage. *Applied Physics Letters*, 61(2):142–144, 1992.

140. R. Maoz, E. Frydman, S. R. Cohen, and J. Sagiv. Constructive nanolithography: Site-defined silver self-assembly on nanoelectrochemically patterned monolayer templates. *Advanced Materials*, 12(6):424–429, 2000.

141. S. Hoeppener, R. Maoz, S. R. Cohen, L. Chi, H. Fuchs, and J. Sagiv. Metal nanoparticles, nanowires, and contact electrodes self-assembled on patterned monolayer templates—A bottom-up chemical approach. *Advanced Materials*, 14(15):1036–1041, 2002.

142. R. M. Stockle, Y. D. Suh, V. Deckert, and R. Zenobi. Nanoscale chemical analysis by tip-enhanced Raman spectroscopy. *Chemical Physics Letters*, 318(1–3):131–136, 2000.
143. B. S. Yeo, J. Stadler, T. Schmid, R. Zenobi, and W. Zhang. Tip-enhanced Raman spectroscopy—Its status, challenges and future directions. *Chemical Physics Letters*, 472(1–3):1–13, 2009.
144. S. Sun, K. S. L. Chong, and G. J. Leggett. Nanoscale molecular patterns fabricated by using scanning near-field optical lithography. *Journal of the American Chemical Society*, 124(11):2414–2415, 2002.
145. S. Sun, M. Montague, K. Critchley, M. S. Chen, W. J. Dressick, S. D. Evans, and G. J. Leggett. Fabrication of biological nanostructures by scanning near-field photolithography of chloro-methylphenylsiloxane monolayers. *Nano Letters*, 6(1):29–33, 2006.
146. S. Sun and G. J. Leggett. Micrometer and nanometer scale photopatterning of self-assembled monolayers of phosphonic acids on aluminum oxide. *Nano Letters*, 7(12):3753–3758, 2007.
147. E. Ul-Haq, Z. Liu, Y. Zhang, S. A. A. Ahmad, L. S. Wong, S. P. Armes, J. K. Hobbs et al. Parallel scanning near-field photolithography: The Snomipede. *Nano Letters*, 10(11):4375–4380, 2010.
148. Z. Liu, E. Ul-Haq, J. K. Hobbs, G. J. Leggett, Y. Zhang, J. M. R. Weaver, and C. J. Roberts. Parallel scanning near-field photolithography in liquid: The Snomipede. *Microelectronic Engineering*, 88:2109–2112, 2011.
149. M. F. Crommie, C. P. Lutz, and D. M. Eigler. Confinement of electrons to quantum corrals on a metal surface. *Science*, 262(5131):218–220, 1993.

# 4 Environmental Fate and Effects of Nanomaterials in Aquatic Freshwater Environments

*Arno C. Gutleb, Sébastien Cambier, Teresa Fernandes, Anastasia Georgantzopoulou, Thomas A.J. Kuhlbusch, Iseult Lynch, Ailbhe Macken, Kahina Mehennaoui, Ruth Moeller, Carmen Nickel, W. Peijnenburg, and Tomasso Serchi*

## CONTENTS

**ABSTRACT** Due to the promise of groundbreaking innovations for many technical applications via nanotechnologies, the potential toxicity of nanomaterials (NMs) is of high societal and scientific interest at present. However, NMs require new or adapted testing methods, as a result of key differences from molecular chemicals related to their surface reactivity. In particular, the dose reaching a cell or an organism needs thorough evaluation and the exact localization of NMs is of equally high importance. Both depend on the interaction of

NMs with biological molecules that adhere to their surface and alter behaviour, dissolution of ions, uptake and localization. The current state of specific quantitative structure–activity relationships for NMs and of grouping of such materials is conceptualized. Finally, the legal frameworks existing in the European Union, in particular the REACH regulation, but also the regulatory situation in United States, and how these are adapting to accommodate the specificities of NPs/NMs for environmental safety assessment are presented.

## 4.1  INTRODUCTION

The potential toxicity of nanomaterials (NMs) is of high societal and scientific interest, due to the promise of groundbreaking innovations for many technical applications.

However, there are many pitfalls for the novice in NM toxicity testing as not all techniques, general knowledge and standard procedures that have been developed and validated over decades to test chemicals can be applied exactly to nanoparticles (NPs; all dimensions between 1 and 100 nm; nanoscale) and nanostructures (NS; at least one dimension in the nanoscale). Applying standard tests to the assessment of hazard of NPs or NSs, without considering the possibility of any artifacts due to their particulate form and behavior, or lack of awareness of the specific features of NPs and NSs that may lead to the need for adjustments of toxicity tests, could lead to misevaluation of the toxicity potential due to false-positive and/or false-negative results which could occur unexpectedly. As a result of any nanoparticulate peculiarities, primarily related to their large surface area-to-volume ratio and consequent high surface energy, toxicity can often not be related to the actual size, mass, or surface area of the single NPs, NSs, or their agglomerates/aggregates (Auffan et al., 2009; Lankoff et al., 2012). It is generally accepted that for these particulate materials, toxicity is greatly influenced by some inherent and not yet fully understood the properties of the particles, such as ions dissolving from the particle surface or molecules adhering to the surface. The latter process might induce interference with the uptake of NPs/NSs by an organism and therefore the effective concentration, which may have implications for any observed toxicity. This chapter describes these specificities including agglomeration/ aggregation and the fate in the environment and the interaction with molecules present in the environment (nanobiological interactions and biological transformation). Furthermore, processes involved in uptake and tissue distribution or particle interaction with legacy chemicals and the potential for leaching of ions from NMs and their influence on NP/NS toxicity in aquatic organisms will be described.

Commonly used aquatic model organisms and protocols for NP/NS testing will be described and the fact that there is a discrepancy in relevance of results from organisms exposed to NPs in artificially pristine laboratory conditions compared to field conditions will be addressed. Another important aspect that will be discussed in this chapter is the NP dose at the relevant site within an organism and available techniques to visualize and (semi)quantify the distribution and localization of particles within the organism. The current state of NP/NS-specific quantitative structure–activity relationships (QSAR) and of grouping of NPs will also be described. Finally, the legal frameworks existing in the European Union, the Registration, Evaluation,

Authorisation and Restriction of Chemicals (REACH) regulation in particular, but also the regulatory situation in the United States, and how they are adapting to accommodate the specificities of NPs/NSs for safety assessment will be discussed.

## 4.2  NPs ENTERING THE ENVIRONMENT AND THEIR AGGLOMERATION AND FATE

With increased use of NPs in consumer products and tools, the likelihood of their release into environmental compartments during their life cycle also increases, as has been confirmed by some recent findings (Benn and Westerhoff, 2008; Kaegi et al., 2008, 2010). Which environmental compartments are affected, and to what extent, by the use of NPs depends on the application and usage of products that incorporate NPs and ultimately the environmental concentration of any potentially hazardous NP/NS.

Current applications of NPs/NSs in applications such as cosmetics, paints, or textiles can lead to direct release into surface water (Kaegi et al., 2008, 2010) as well as transport of the released NPs into the wastewater system and ultimately to sewage treatment plants (Gottschalk et al., 2009). These processes have been studied in various laboratory and field investigations, indicating that the highest fraction of the added NPs (80%–95%) is adsorbed to the sewage sludge (Gartiser et al., 2014; Kaegi et al., 2011, 2013; Kiser et al., 2009; Limbach et al., 2008; Westerhoff et al., 2011). However, the fraction of NPs that are not adsorbed to the sludge may ultimately leave the sewage plant and enter surface waters, and possibly lead to a risk to aquatic or benthic organisms as well as drinking water resources (Gartiser et al., 2014).

The detection of NPs in the environment as a prerequisite to understand fluxes and fate is far from being a routine analysis, and challenges include the low expected mass concentrations, the presence of natural and incidental (arising from, for example, combustion, natural or man-made) particles of similar size and elemental composition, and the agglomeration processes occurring and the lack of certified reference materials (Duester et al., 2014). At the Joint Research Centre (JRC), a repository of representative NMs was established[*] and these NMs have been prepared under Good Laboratory Practice (GLP) and can serve as international benchmarks.

It has to be noted that the NPs entering the environment can be the pristine, unaltered NPs as produced in industry but also processed NPs as well as environmentally altered NPs. They are mostly either still embedded in the product matrix or highly agglomerated after the release, but under some circumstances also the release of single nano-objects may occur (Nowack and Bucheli, 2007). Furthermore, alterations of the NPs after their release into the environment may occur affecting their environmental behavior as well as the hazard potential and this needs to be considered in future studies (Al-Kattan et al., 2014; Bondarenko et al., 2013).

One common alteration of released NPs, similar to naturally occurring colloids, is their tendency to agglomerate, for example, with other suspended particulate matter, settling and deposition in sediments, all limiting their mobility. These mechanisms

---

[*] http://ihcp.jrc.ec.europa.eu/our_activities/nanotechnology/nanomaterials-repository

lead ultimately to relatively low NP concentrations in the water phase and correspondingly higher concentrations possibly accumulating in sediments (Baalousha, 2009; Handy et al., 2012; Klaine et al., 2008; Von der Kammer et al., 2010). Agglomeration processes also directly affect the bioavailability of the NPs and determine which organisms may be most exposed to them (Duester et al., 2014).

Because agglomeration is of such importance in environmental waters, the parameters influencing this process are of high relevance to predict NP behavior. Parameters are related to the NP itself (structure, composition, surface chemistry, or reactivity) and the corresponding environmental conditions such as pH, ionic strength, type, and concentration of natural organic matter (NOM) (French et al., 2009; Guzman et al., 2006). A detailed description of agglomeration processes and the influencing parameters is given by, for example, Cornelis et al. (2014) and Von der Kammer et al. (2010). Briefly, two basic mechanisms determine the agglomeration behavior: steric and electrostatic mechanisms. Whereas steric stabilization is affected by adsorbed molecules at the surface of the NP, which physically hinder NP clustering, electrostatic stabilization is determined by repulsive interactions of the electric double layer. The Derjaguin, Landau, Verwey, and Overbeek (DLVO) theory can be used as starting point to describe the stability of a suspension or NP dispersion in environmental media (Derjaguin and Landau, 1941; Verwey and Overbeek, 1948) and considers the electrostatic repulsion and van der Waals adsorption.

Other environmental chemical processes such as oxidation and reduction, sulfidization, dissolution, adsorption, change of surface functionalization, and biotransformation can occur, and they will also influence the agglomeration process. The kinetics of these processes are dependent on the abiotic and biotic conditions of the media as well as the physical–chemical properties of the NM itself. A detailed description can be found in the works of Klaine et al. (2008), Lowry et al. (2012), Nickel et al. (2013), and Nowack et al. (2012).

The above-described mechanisms and processes are not only of relevance to understand the environmental behavior and fate of NMs but also for their impact on the environment (Muth-Köhne et al., 2013). For example, Ag NPs exerted increased toxicity to zebrafish (*Danio rerio*) embryos as the materials underwent a wastewater treatment process compared to the pristine NPs. This indicates that environmental processes may also influence the potential hazard of NMs. However, the process or factor that was responsible for these observations has not been identified, but likely it has to do with the interaction of the NPs and organic molecules present in complex environmental matrices. Further studies have investigated the bioaccessibility of Ag and Ag NPs in soil with aging and varying organic matter content (Coutris et al., 2012). This study showed that for some forms of Ag the NPs can be more bioaccessible than Ag ions, which may have implications for the risk assessment (RA) of Ag NPs and whether their RA is covered by using existing frameworks for RA of metals to predicted metal NP hazard in the environment. However, exposure models for NMs are still in their infancy. But basic approaches exist to predict the environmental behavior and fate for freshwater environment. A detailed description can be found in Ballousha et al. (2015).

## 4.3   NANOBIO INTERACTION—NANOENVIRONMENTAL INTERACTION

The high surface area-to-volume ratio of NPs, and their consequent high surface energy, drives NPs to acquire a coating of macromolecules in order to reduce or pacify the surface energy (Monopoli et al., 2011). Numerous studies have shown that the composition of NMs and the biofluid (e.g., human serum, cell culture media containing dispersants such as sodium dodecyl sulfate (SDS) or albumin) influences the identity of the biomolecules in the so-called NP corona, and evidence is emerging that the composition of the corona can be correlated with NP uptake by cells (Walkey et al., 2014; Fleischer and Payne, 2014; Lynch et al., 2007; Walczyk et al., 2010).

The first report of an environmentally relevant protein corona emerged in 2013, confirming that the concept of the protein corona is equally applicable to ecotoxicological investigations in soil and aquatic species as to toxicological ones. Thus, the composition of the protein corona of Ag NPs exposed to coelomic fluid of the earthworm *Eisenia fetida* was compared to that formed from fetal bovine serum (FBS), and the uptake of the corona-coated NPs was found to be higher for the coelomic fluid-coated particles (despite being tested using *E. fetida* macrophages which should preferentially uptake nonnative coronas [opsonization*]) than for the nonnative FBS-coated particles (Hayashi et al., 2013). Interestingly, over time the FBS-coated Ag NPs competitively acquired a biological identity native to the *E. fetida* cells *in situ* suggesting secretion of proteins by *E. fetida* in response to the presence/uptake of NPs (Hayashi et al., 2013). Such processes have also been observed in cell culture for toxicity studies, using uncoated $SiO_2$ NPs that within 1 h had acquired a cellular corona (Lesniak et al., 2012), and for polystyrene NPs incubated initially in serum and subsequently in cytosolic fluid (to mimic particle cellular uptake), which shed some of the serum proteins and replaced them with cytosolic ones (Lundqvist et al., 2011). So-called conditioning effects have also been reported for gold NPs, whereby the dynamic composition of the extracellular environment evolves as cells deplete and secrete biomolecules, resulting in altered protein coronas and NP–cell interactions (Albanese et al., 2014). Such effects are also highly likely to occur in the environment where a number of species produce so-called exudates or exudomes of secreted biomolecules that can potentially bind to NPs (Lynch et al., 2014a), with important examples of secreted biomolecules being exopolysaccharides (EPS) in biofilms (Sutherland, 2001) and fish exudate which plays an important role in *Daphnia* life cycle (Maszczyk and Bartosiewicz, 2012). More relevant to environmental systems, however, is the interaction of NPs–NSs with environmental constituents, which may lead to a true *environmental corona*.

In aquatic environments, the most abundant macromolecules, and thus the ones that most likely to come into contact with and bind to NPs, are aquatic humic substances, which account for 30%–50% of the organic carbon in water (Thurman et al.,

---

* The rendering of bacteria and other foreign substances subject to phagocytosis via addition of an opsonin (i.e., a complement moiety).

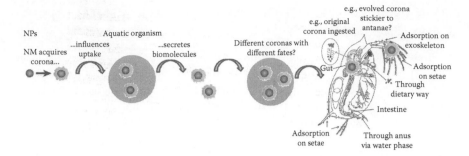

**FIGURE 4.1** Schematic of impact of NP biomolecule/environmental macromolecule corona, its evolution, and its potential impact on NP uptake and localization. The initially adsorbed NPs whose corona dictates their route and rate of entry influence the organism to secrete biomolecules, which alter the corona of the NPs further and may impact on further uptake and effects.

1982). The aquatic fulvic acid (FA) fraction contains substances with molecular weights ranging from 500 to 2000 and is monodisperse, whereas the aquatic humic acid (HA) fraction contains substances with molecular weights ranging from 1000 to greater than 10,000 and is generally polydisperse. The stabilization of NPs by humic substances has been extensively studied, with both HA and FA having been shown to stabilize a range of different NPs, including Ag NPs, carbon nanotubes, quantum dots (QDs), and metal oxides (see Table 1 in the work of Lynch et al. [2014b] which provides a comparison of the factors affecting HA or FA interactions with NMs). However, studies of competitive binding of HA and FA to NPs, and the dynamics of exchange of humic substances by other environmental biomolecules such as EPSs or organismal exudates, have yet to be performed. An initial step toward addressing this problem used quartz crystal microgravimetry with dissipation (QCM-D) monitoring to understand the relative impact of HA and FA fraction of the NOM on the stability and mobility of silver NPs (Furman et al., 2013), although significantly more work is needed to bring such approaches and knowledge to the degree of maturity available being already achieved for NP–human protein interactions described earlier. The understanding of these processes will also contribute to the refinement and more realistic toxicity testing (Figure 4.1).

## 4.4 TOXICITY TESTING IN FRESHWATER SPECIES

Industrial products and waste tend to end up in aquatic systems, which are considered as a major and final reservoir for many environmental contaminants including NPs. The large amount and nonregulated use of NPs leads to a growing concern about their potential uptake by and toxic effects toward aquatic biota (Canesi et al., 2012; Klaine et al., 2008; Moore, 2006). Toxic effects of NPs were observed at concentrations ranging from μg/l to mg/l either in the water column or in sediments (Kahru and Dubourguier, 2010).

When assessing the toxicity of NPs to freshwater organisms, several approaches can be taken. As previously mentioned, there is a need to have standardized acceptable methods for the toxicological assessment of NPs for regulatory purposes such as REACH. In this case, there needs to be a toolbox of relevant, validated, and approved methods (e.g., the Organisation for Economic Co-operation and Development [OECD], the International Organization for Standardization [ISO], etc.) for toxicological assessment and characterization for the hazard assessment of the NP in question. However, there is also a need for an environmentally relevant assessment to properly assess the mechanistic effects associated with the NP in a particular freshwater environment. For a more comprehensive RA, both aspects need to be investigated and there needs to be an assessment of the likely pathways for the NP to enter the freshwater environment (as previously discussed). Other important questions to be considered in a realistic RA of NPs in the freshwater environment include assessing if the NPs are likely to enter the food web (bioaccumulation, biomagnification) and what are the long-term consequences (chronic effects).

Algae, crustaceans, worms, bivalves, and fish have been used as model species in recent years to study NP/NS toxicity. Invertebrate species, for which standardized protocols exist (e.g., OECD TG 201 and 210) such as daphnids and worms, are widely used for toxicological assessment of chemical substances, including NPs/NSs, in aquatic systems (Farré et al., 2009; Rosenkranz et al., 2009). Given their importance as a major component of freshwater assemblages, in which they represent up to 95% of animals and so play an important ecological role, particularly given their role as a major trophic link in aquatic food chains (Baun et al., 2008), their study as potential targets of NPs is very important too.

Even though classical bioassays with mortality or growth inhibition as endpoints give valuable information about the general toxicity of NPs (and have relevance from a regulatory point of view), they do not provide information about the mechanisms of action toward aquatic species. This, of course, is not an issue exclusive to NMs and applies equally to the assessment of hazard of any chemical substances. Traditionally, in environmental regulatory testing, the focus is on the application of standard hazard tests with no requirements for more exploratory studies. As a consequence, the potential mechanisms underlying any hazard effects will go undetected. Considerations have to be taken when translating this approach to the assessment of hazard of NMs, given their status as potential emerging contaminants, for which mechanistic pathways are still largely unknown. An important aspect is the fact that exposures at environmentally relevant concentrations very often do not lead to mortality, which may raise the question of their potential hazard. Existing biomarkers widely used in the assessment of hazard of standard chemical substances can also be used to identify potential pathways via which NPs may affect an organism's physiology (Klaper et al., 2009). As for standard chemicals, a variety of biomarkers can serve as an *early warning* for effects such as for possible reproductive (embryo hatchability and growth) and population impacts (Musee et al., 2010; Zhu et al., 2010) even and especially at sublethal concentrations. Of particular interest would be the use of biomarkers in the detection of pathways which are particularly affected by exposures to materials in the nanoscale and particularly, the development and testing of any novel biomarkers of exposure or effect that are not relevant or applicable in exposures to conventional chemicals.

Exposure time has also been identified as an important parameter in the evaluation of NM hazard and this can also be different from exposures to conventional chemicals. Extending the acute *Daphnia magna* test to 72 h, from the 24 h and 48 h timepoints usually recommended by the OECD guideline 202, showed a significant increase in nTiO$_2$ toxicity, thus clearly highlighting the importance of the exposure duration as an important factor in nano-ecotoxicology (Zhu et al., 2010), although this observation cannot be extrapolated to all NMs, conditions and species.

Following on from these previous points, it can be highlighted that chronic endpoints at sublethal (potentially environmentally relevant) concentrations are of far greater interest in the case of NMs and their effects on the freshwater community. Many regulatory guidelines for chronic exposures exist for freshwater organisms (e.g., *D. magna* [OECD 211], fish [OECD 210]). Several morphological malformations as well as effects on reproduction during chronic tests with *D. magna* exposed to CuO NPs and microparticles were observed (De Rossetto et al., 2014). These chronic sublethal effects were observed at lower levels than the acute studies. However, all studies were still conducted at relatively high levels (mg/l concentrations). This study also highlighted the importance of having comparative toxicological studies with different forms of a similar substance (e.g., CuO, NPs, and CuO microparticles) with identical chemical abstracts service (CAS) numbers, but very different toxicological effects on the same organism.

NP toxicity studies carried out at sublethal levels on vertebrates are often performed on freshwater fish. The gills, gut, liver, and brain are the main affected organs for toxic effects of some manufactured NPs such as oxidative stress, cell pathologies, ion regulation, and vascular injuries (Farré et al., 2009; Handy et al., 2012).

Another important aspect is the potential for trophic transfer of NMs. This has investigated bivalves and algae (Renault et al., 2008) and also for zebrafish (Auffan et al., 2012). Bioaccumulation, biomagnification, and trophic transfers are all important, and are often overlooked, aspects of NP toxicity particularly because these processes may affect species which are pivotal in food chains (e.g., some crustaceans), being the link between primary producers and predators, which may constitute important economic species or may be of specific conservation value (Auffan et al., 2012). Studies focusing on these endpoints have shown varying results, depending on the species used to assess bioaccumulation, biomagnifications, and trophic transfer, and the NP investigated (Holbrook et al., 2008; Zhu et al., 2010). Zhu et al. (2010) found that there was a higher body burden of TiO$_2$ NP in the dietary exposure group compared to the aqueous exposure in a study with *D. magna*. This highlights the importance of diet as a potential route of exposure of NPs to higher organisms. Possible risks may need to be examined on a case-by-case scenario incorporating known physicochemical properties (of the bulk and NP of concern]) and existing data on the relevant bulk compound related to the NP of interest. In addition, the functionalization of NPs has been shown to influence the uptake and depuration behavior in organisms (e.g., *D. magna* and Zn NPs [Skjolding et al., 2014]), which is another important consideration in understanding the bioaccumulation, biomagnification, and trophic transfer of NPs and highlights the need to treat these investigations on a case-by-case scenario.

At present, the focus of environmental toxicity assessment of NMs with freshwater organisms needs to consider some important knowledge gaps in order to create a new and applicable framework to assess the environmental risks of manufactured

NMs and NPs. In relation to ecotoxicity testing, the specific gaps include the use of appropriate controls and reference materials (discussed later), and appropriate standardized dispersion protocols in order to allow for the comparison of data generated in different labs and allow for mutual acceptance of this data at the OECD level.

## 4.5 CONCENTRATION–RESPONSE CURVE AND PROBLEM OF DOSING

Toxicity testing of chemicals is usually performed by adding increasing amounts of a chemical to the exposure medium and this applies for both *in vitro* experiments and ecotoxicological experiments. The composition of the exposure medium in terms of nutrients and macrochemistry is selected such that the medium does not impose any initial stress on the organisms tested. In any case, the composition of the medium will always affect the chemistry (including fate and behavior) of the chemical substance being tested, and therefore, it is important that the information on the medium and the method used to mix with the chemicals tested is clearly specified. This is important for all chemical substances but of particular relevance to NMs given that they are particulate matter in the nanoscale whose fate in the medium is particularly dependent on its composition and the approach used in the suspension. Differences in media composition can have significant effects on NP/NS dispersion stability and surface charge/chemistry, and thus affect exposure conditions. This does affect the NP/NS dose at which organisms are exposed and will affect comparability across studies, as well as making accurate sequential dilutions. Conventional concentration response tests are also based upon a detailed understanding of the fate of the chemicals tested and on the maintenance of a homogeneous test concentration across the study which facilitates continuous exposure of the organisms tested. This has also proven challenging for NPs/NSs which as they agglomerate tend also to sediment. Within conventional test systems, and if needed (e.g., in the case of readily degradable or volatile chemicals), the test design can be modified by, for instance, the use of a flow-through system in which the concentration of the test chemical is kept constant. This can be adapted to NM exposures but could lead to artifacts and blockage of tubes. Finally, for conventional chemicals, the actual test concentration is determined and the effects are commonly expressed in terms of the actually dosed mass concentration.

In case of assessing concentration–response curves of NPs/NSs for specific test organisms, the common approach of adding increasing amounts of a test compound to a medium in which the performance of the test organism is optimized is in principle applicable. However, in practical terms, a few issues tend to occur. In sum, these problems are related to differing fate properties of NPs compared to the fate of nonparticulate chemicals, and to the general issue of the effective dose metrics of NPs. The key issues are summarized as follows:

1. The basic problem with regard to fate assessment of NPs is that the impact of the composition of the medium on the kinetics and the thermodynamics of the processes determining the fate of nanoscale particles is not well understood. Thus, it is currently not possible to fully forecast the fate of a given NP in a given test system, making it very difficult to predict the fate of NMs in

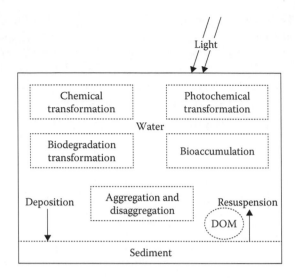

**FIGURE 4.2** Overview of the main processes affecting the fate of NPs in an aquatic test system. DOM, dissolved organic matter.

a realistic environmental setting. A general overview of the main processes determining the fate of particles in water is depicted in Figure 4.2. In common practice, especially aggregation and disaggregation and subsequent sedimentation are critical processes determining the fate of NMs in aquatic systems. Although the basic model formulations for assessing these processes have been available for several decades (with the most common theory being the DLVO theory [Derjaguin and Landau, 1941; Verwey and Overbeek 1948]), it is not yet possible to fully quantify these processes. In this respect, it should be noted that aggregation is in part due to homoaggregation (i.e., the aggregation of particles of similar composition), as well as heteroaggregation (aggregation of particles of different composition). This implies that at higher particle concentrations relatively more aggregates are likely to be formed (the rate of this process will also depend on the particles and medium), in turn likely to result in reduced effective exposure of biota to nanoscale single particles at higher particle concentrations.

2. With regard to the issue of dose metrics, the general basic observation is that, in contrast to nonparticulate chemicals, the effects of NPs are not always proportional to the mass dosed. Thus, particles of the same chemical composition but of, for instance, different size, different morphology, or different surface charge have been shown to induce a different response at the same exposure concentration when the exposure concentration is expressed in terms of mass dosed (Cohen et al., 2014). As an illustration, a typical example is schematically depicted in Figure 4.3. In this specific example, concentration–response curves are shown as obtained when exposing a test system (e.g., cell or organism) to chemically similar particles of different size, at increasing concentrations.

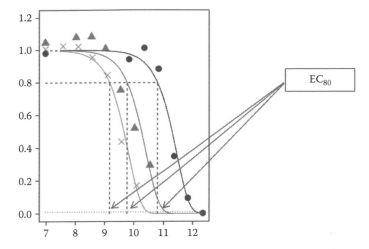

**FIGURE 4.3** Schematic example of concentration–response curves obtained for particles of similar chemical composition but of different sizes. The mass of the particles dosed is given on the *x*-axis, whereas the scaled effects are plotted on the *y*-axis. The 80% response levels are marked to indicate the dissimilarity of responses. $EC_{80}$, effective concentration (concentration at which 80% of the studied effect is observed in relation to a control, within the specified experimental time).

## 4.6 ION VERSUS PARTICLE FORMS OF A CHEMICAL

Depending in part on their intrinsic composition, NPs released into aqueous medium have a variety of potential fates, including being partially or totally converted into the ionic form dependent on their elemental composition, complexing with other ions, molecules, or molecular groups, agglomerating or remaining in the NP form. In some case, it has been postulated that the observed toxic effects of metal and metal oxide NPs can be attributed to the released metal ions rather than to the NPs themselves (e.g., in the case of ZnO NPs) (Brun et al., 2014; Mwaanga et al., 2014). Although titanium and gold NPs are relatively stable in the aquatic environment (García-Negrete et al., 2013; Schmidt and Vogelsberger, 2006), silver-, zinc-, and copper-based NPs are relative soluble (Adams et al., 2006; Johnston et al., 2010). Silver NPs are described as very toxic materials in the aquatic environment, and it has been suggested that this toxicity is primarily based on released aqueous free silver ions ($Ag^+$). This release of $Ag^+$ and their toxicity was described to be strongly linked to the water chemistry but also linked to the properties of Ag NPs (shape, size, coating) (Sharma et al., 2014). For instance, by increasing the water conductivity, the silver release also increased, although when increasing the chloride ($Cl^-$) concentration the $Ag^+$ toxicity decreased due to the medium-dependent speciation of $AgCl_x^{(x-1)-}$ (Chambers et al., 2014). This link between dissolution and toxicity of Ag NPs has not been universally proved (e.g., Mallevre et al., 2014). Studies on zinc NPs have also linked their toxicity in aquatic systems to the release of free ions ($Zn^{2+}$) resulting in an increase of oxidative stress (Brun et al., 2014; Mwaanga et al., 2014). This release of $Zn^{2+}$ occurs as soon as the NPs are dispersed in the medium, and this release of ions is strongly influenced by the pH, the presence of natural organic compounds, and the concentration of NPs (Miao et al., 2010). However, the toxicity of NPs is not only and directly linked to ion release as different responses were observed for ZnO NPs and the corresponding ions although this is still under debate (Xu et al., 2013). Copper NP toxicity is also described to be linked to the release of copper ($Cu^{2+}$) ions in water (Heinlaan et al., 2008). Cu NPs may show distinct effects compared to copper ions. For instance, Cu NPs caused a significant inhibition of the antipredator behavior of juvenile rainbow trout compared to those exposed to a similar quantity of free $Cu^{2+}$ ions by decreasing its ability to detect the alarm substance (Sovová et al., 2014). Altogether, these different data lead to the conclusion that the toxicity of NPs in an aqueous environment is strongly linked to their behavior and fate, which includes solubility of certain metal NPs.

## 4.7 TISSUE DISTRIBUTION, CONCENTRATION, AND LOCALIZATION

An understanding of the bioavailability and bioaccumulation of NPs in exposed organisms is essential for assessing potential environmental risks and for the meaningful interpretation of data from toxicity studies (Kruszewski et al., 2011).

A number of experiments using several organisms as models for uptake have shown that NPs are internalized and distributed across organs and tissues, and that the surface charge of NPs as well as the presence of organic matter can influence uptake and toxicity. *Artemia salina* (brine shrimp) were not able to excrete the ingested NPs and $TiO_2$ aggregates were observed in their gut (Ates et al., 2013). The size as well as the synthesis method affected the uptake of Ag NPs by *D. magna* and the effects could not be attributed solely to ions present (Georgantzopoulou et al., 2013). Polystyrene beads of 20 and 1000 nm size were taken up by *D. magna* and localized in the gastrointestinal tract, and a translocation in lipid storage droplets was observed. Furthermore, the 1000 nm beads were faster and more effectively cleared compared to the 20 nm beads (Rosenkranz et al., 2009). CdSe/ZnS quantum dots (QDs) were found in the gastrointestinal tract and brood chamber of daphnids and higher amounts of COOH-coated QDs were internalized compared to the $NH_2$- and PEG-coated ones (Feswick et al., 2013).

Zn was found to be distributed across organs (eyes, brain, yolk sack, spinal cord) in zebrafish embryos exposed to both ZnO NPs and free Zn derived from $ZnCl_2$, and the distribution showed similar patterns, indicating that effects are due to free Zn rather than the particulate form (Brun et al., 2014). Silver was mostly distributed in the liver of Japanese medaka (*Oryzias latipes*) exposed to polyvinylpyrrolidone (PVP), citrate-coated Ag NPs, or $AgNO_3$ irrespective of the form or surface coating showing a slow elimination rate with the Ag levels remaining unaltered after depuration (Jung et al., 2014). BSA-coated Ag NPs were localized in the skin, heart, and brain of zebrafish embryos (*Danio rerio*) (Asharani et al., 2008) and Ag NPs were found to be attached to the surface of exposed fathead minnow (*Pimephales promelas*) embryos, and aggregates were engulfed in vacuoles inside the embryos (Laban et al., 2010). Ag uptake was observed in the liver, intestine, and gallbladder of carp (*Cyprius carpio*) with a higher content being measured in the case of nanosized than microsized Ag (Gaiser et al., 2012).

Uptake of Ag occurred for all Ag forms (ionic, citrate- or HA-capped Ag NPs) by the freshwater snail *Lymnaea stagnalis* after either aqueous or dietary exposure although Ag elimination was much slower in the latter case during the depuration period (Croteau et al., 2011). Waterborne exposure of the estuarine snail *Peringia ulvae* to dissolved and citrate-capped Ag NPs led to increased Ag tissue levels although the dissolved Ag was twice more bioavailable (Khan et al., 2012). Fe and Cd were found to be rapidly accumulated and localized mostly in the digestive gland of the mussels *Mytilus galloprovincialis* exposed to polyethylene glycol (PEG)-functionalized $Fe_2O_3$ NPs and cadmium selenide QDs, though in much lower levels in the latter case, and only a decrease of 10% was observed after 72 h of depuration for the PEG-$Fe_2O_3$ NPs (Hull et al., 2013). BSA-coated Au NPs were internalized by the clam *Corbicula fluminea* and localized in the digestive gland with no signs of translocation in other organs (Hull et al., 2011).

The extent of bioaccumulation of $CeO_2$ NPs in the nematode *Caenorhabditis elegans* was dependent on the surface charge and the presence of HA in the exposure media. Positively charged dextran-coated $CeO_2$ showed greater accumulation and toxicity in *C. elegans* than the neutral or negatively charged NPs, although bioaccumulation decreased with increasing concentration of HA (Collin et al., 2014). It has been recently shown that Au NPs can be taken up by aquatic plants. The uptake and bioavailability were plant species dependent and were influenced by the presence of dissolved organic carbon (DOC). Lower gold levels in the plant tissues were observed in the presence of DOC (Glenn and Klaine, 2013).

In conclusion, the bioavailability and extent of uptake, distribution, and translocation can be influenced by several factors, for example, NP size, morphology, coating, surface charge, organism, or the presence of organic carbon in the exposure media, as well as the general composition of the media. It is not clear yet though how the material itself, its elemental composition, the coating, and the media composition (and their influence on the nature and composition of the macromolecules that bind to the NPs), individually and jointly, affect the uptake and distribution. Additional data, and application of statistical approaches such as principal components analysis, will support in teasing out these issues and highlight key drivers and interrelated properties of NMs that are linked to observed toxicity outcomes (Lynch et al., 2014b).

Analytical techniques such as inductively coupled plasma mass spectrometry (ICP-MS) or single particle ICP-MS have been used to study the tissue or in the case of small species whole organism concentration of certain elements (Franze and Engelhard, 2014; Georgantzopoulou et al., 2013; Lee et al., 2014b; Von der Kammer et al., 2012). However, such destructive techniques do not necessarily separate between different ionic or organic forms of a certain element, and in the case of small organisms (*D. magna* and other freshwater crustaceans, fish embryos, etc.), no tissue specific information is generally obtained. Distribution and localization of NPs within an organism are often studied using transmission electron microscopy (TEM) (Gliga et al., 2014; Kumbiçak et al., 2014; Moret et al., 2014). Currently the highest resolution and in parallel information on the level of the isotope distribution can only be acquired by the combination of secondary ion mass spectrometry (SIMS) with various probes such as time-of-flight SIMS (TOF-SIMS) (Lee et al.,

2014a) or the NanoSIMS50 with a lateral resolution of 50 nm (Audinot et al., 2013). The later probe showed accumulation of smaller sized Ag NPs localized in intestinal cells of exposed *D. magna* and additional accumulation of Ag in specific locations, seemingly developing oocytes (Georgantzopoulou et al., 2013).

## 4.8    TROJAN HORSE CONCEPT AND LEGACY COMPOUNDS

The Trojan horse concept has two manifestations in nanosafety of relevance to aquatic nanotoxicity: one in which the NP acts as a carrier of another pollutant or contaminant, binding it to the surface and carrying it to a cellular location that it would otherwise not reach (e.g., Baun et al., 2008), and another whereby the components of the NP itself are the hidden entity, as in the case of dissolving NPs such as Ag or ZnO, whereby the NP form can reach locations (e.g., cellular lysosomes) that the ionic form would not, and as it dissolves produces much higher local concentrations of the dissolved ionic species than could occur via equilibrium-based diffusion (Sabella et al., 2014). In the first scenario, it is the potential for enhanced bioavailability of so-called legacy industrial pollutants, such as heavy metals (e.g., cadmium [Chen et al., 2012]), lead and mercury (Bindler et al., 2011), and polychlorinated biphenyls (PCBs[*]) (Montaño et al., 2013), which is the primary concern (Zhang et al., 2012), although there is also the potential for mixture effects such as antagonistic or synergistic effects, a completely untouched area at present (Wang et al., 2014a), as well as for binding of PCBs to NPs to affect their degradation, for example, by enhancing or reducing the efficiency of enzyme-mediated phase I biotransformation (Lu et al., 2013b), which is another under-studied research area. However, the ability of NPs to bind PCBs also offers enormous potential for bioremediation (De Windt et al., 2005; Velzeboer et al., 2014) and biosensing (Tang et al., 2014) of PCB contamination, and as such safety concerns should be addressed in parallel with the development of environmental applications. Another potentially important co-pollutant class are the endocrine disruptors, considered as an emerging pollutant class associated with altered reproductive function in males and females and developmental issues.[†] Here also, there is potential for binding of endocrine-disrupting molecules to NPs, although there is very limited literature on this at present (Wang et al., 2014b), or indeed for NPs themselves to act as endocrine disruptors including developmental effects (Lu et al., 2013a; Iavicoli et al., 2013 and references therein). There is also the potential for nano-based sensors and nano-enhanced catalysis of degradation of endocrine-disrupting chemicals (Chalasani and Vasudevan, 2013).

## 4.9    LACK OF QSAR MODELS

Computational modeling has emerged over the past decade as a reliable tool to estimate the underpinning parameters that control properties and effects of chemical substances on the basis of (quantitative) structure–activity relationship (Q)SAR, and

---

[*] A mixture of 209 possible manmade congeners with different number and localization of chlorine on a biphenyl backbone that were produced and used worldwide from 1929 until the late 1970s, when they were finally banned.

[†] http://www.who.int/ceh/risks/cehemerging2/en/.

read across from the vast amount of available data. Combined with powerful data-mining tools, these computational models offer a rapid way of filling data gaps due to the lack or limited availability of experimental data on new substances. The *in silico* models are now routinely used by researchers, industry, and regulators to estimate physicochemical properties, human health and ecotoxicological effects, and environmental behavior and fate of a wide range of chemical substances.

However, although *in silico* modeling approaches have advanced well for conventional chemicals, a relationship between the various physicochemical properties and toxicological effects of NPs that allows development of reliable models has not yet been established. From the limited available data on NPs, a few parameters, such as size, surface area, and surface characteristics, have so far been envisaged to be important in relation to effect assessment of NPs (Chaudhry et al., 2010; OECD, 2010a). However, there are most likely to be other yet-unidentified parameters that play an important role in driving the properties and effects of NPs. For example, there is emerging evidence to suggest that NPs, especially in the lower nanometer range, are able to cross biological membrane barriers and may reach different organs which are otherwise protected against the entry of larger particulate materials. Exposure to some NPs has also been shown to lead to adverse health effects, for example, due to an increased generation of oxyradicals. However, it is also known that NPs are not always more harmful compared to their corresponding larger forms and, as indicated above, even to NPs of the same source, chemical composition, and size. This indicates that there are parameters, other than size and surface area, which play an important role in determining the effects and interactions of NPs in biological systems. For example, surface coatings are likely to be very important in this regard (Walkey et al., 2014). Surface coatings provide a perfect case for applying a modeling approach to identify the key parameters that drive the properties and toxicological effects of NPs. A recent example of this approach, where a structural similarity led the experts to suspect a possible harmful effect on the basis of the so called fiber paradigm, is the case of multiwalled carbon nanotube and asbestos fibers (Poland et al., 2008), although so far there are no examples from environmental studies where the *fiber paradigm* has been found applicable.

The few attempts that have recently been made to develop preliminary models of NP properties and effects have not only indicated the tantalizing possibility that a QSAR approach may indeed be feasible and useful in predicting the physicochemical properties and effects of NPs, but also revealed the challenges facing QSAR modeling of NM toxicity and areas that need research. Until now, it is unknown to what extent specific characteristics of NMs or combinations of these characteristics are indicative for the toxic effects of particles. The question still to be answered is which physicochemical descriptors constitute a full description of an NP for QSAR modeling purposes, and should their measurement or estimation be standardized?

## 4.10   GROUPING

The classical grouping concept implies considering closely related materials as a group, or category, rather than as individual materials. In a category approach to hazard assessment, not every material needs to be tested for every single endpoint.

Instead, for a given category would allow for the estimation of the hazard of untested endpoints. Use of information from structurally related materials to predict the endpoint information for another material is called read-across. Grouping is thus the process of assigning chemical substances to a *group* or *category* of similar chemicals. The basis for grouping may be provided by, for instance, similar or regular patterns of physicochemical, biological, exposure, biokinetic, or environmental fate properties (see, for instance, Sayes et al. [2013]). For conventional organic chemicals, these similarities are often due to common functional groups or to a constant pattern of a given property across the group. Grouping of chemicals is, among others, important as it is one of the first steps (definition of the applicability domain) in developing predictive models for fate and effect endpoints, like QSAR models. Related to grouping is the concept of read-across in which the grouping concept is applied to predict endpoint-specific information for a target substance by using data for the same endpoint from other substances. Therefore, the need to test every substance for every endpoint is avoided. During the analog approach of read-across, a very limited number of substances are compared, whereas the structural or category approaches that are common features of grouping imply comparing larger numbers of substances (for further information, see European Chemicals Agency [ECHA] Guidance on QSAR and grouping of chemicals).*

Compared to the grouping of conventional chemicals, the grouping of NMs is a more complex challenge. So far, no single physical or chemical material property—be it surface, volume, or reactive oxygen species generation—has perfectly correlated with the observed biological effects for various types of NPs. More than a dozen physicochemical properties of NMs have been identified that could potentially contribute to hazardous interactions, including chemical composition, size, shape, aspect ratio, surface charge, redox activity, dissolution, crystallinity, surface coatings, and the state of agglomeration or dispersion (Thomas et al., 2011). Some of these characteristics also influence biokinetic processes, and thus the fraction of NPs that reach the target site, whereas potentially other—or another set of—material properties influence the biological effects. This may confound the correlation between one single material property and an apical effect. Complexity of NPs thus requires developing a more comprehensive grouping allowing identification of interlined or covarying properties and quantification of contributions of multiple physicochemical endpoints to a reduced set of principle components that can be linked with toxicity, such as intrinsic NM properties and extrinsic properties such as interactions and inherent composition, as described by Lynch et al. (2014b). At the same time, this complexity offers the opportunity to include more information: NP grouping should not be restricted to the determination of NS–activity relationships, but takes into account the entire source-to-adverse outcome pathway, including changing of physicochemical characteristics of an NP. Therefore, pillars of NP grouping should include the physicochemical characteristics of an NP as well as the uptake, biodistribution, and biopersistence (biokinetics) of an NP in an organism and the physicochemical characteristics of an NP inside the organism. Finally, early biological effects need to be considered when grouping NPs.

---

* http://www.echa.europa.eu/documents/10162/13632/information_requirements_r6_en.pdf.

## 4.11    CONTROL MATERIALS

The focus of efforts in the development of control or reference materials to date has been on development of reference NMs for physicochemical characteristics, typically in pristine environments, which bear no relevance to the state/characteristics of the particles in *real* environments, such as in rivers, lakes, wastewater treatment plants, sewage sludges, sediments, and soil (Stefaniak et al., 2013). As far back as 2010, the NanoImpactNet project identified a need for test materials for ecotoxicology, and several potential particles were identified, including $TiO_2$ NPs, polystyrene beads labeled with fluorescent dyes, and Ag NPs (Stone et al., 2010). Stefaniak et al. (2013) identified some key challenges to the development of reference NMs, including the need for the synthesis route(s) to be sufficiently robust to facilitate the required level of reproducibility, and the need for appropriate metrology to perform precise and reproducible measurements of the relevant physical, chemical, and/or biological phenomena. They suggested that realistic ambitions for development of reference or control NMs would include (1) suitable materials (including positive and negative controls) to develop harmonized protocols for *in vitro* and *in vivo* toxicity testing and elucidate mechanisms of toxicity resulting from nanoscale properties and (2) materials to verify instrument or method performance and operator or laboratory proficiency, thereby improving consistency in interpreting exposure and toxicity data (Hole et al., 2013; Stefaniak et al., 2013). Development of multiparametric reference NMs for nanotoxicity testing, based, for example, on silica NPs, with three traceable properties, namely, composition, size, and surface area, has been suggested, along with more advanced hybrid materials to facilitate also tracing and tracking of the NPs (Orts-Gil et al., 2013). The ability to trace the materials is especially important in the environment where there can be high background levels due to naturally occurring particles. Stable-isotope enrichment or labeling of NPs, and the development of such approaches into reference NMs or standards, is thus an important advance for environmental toxicity assessment in aquatic environments. Valsami-Jones and colleagues have produced stable-isotope-enriched NPs of Ag, ZnO, CuO, and $Ce_2O_3$, which have been used in environmental fate studies in a range of aquatic and sediment species (e.g., Croteau et al., 2014; Khan et al., 2013; Larner et al., 2012). These offer significant promise as control or reference materials for ecotoxicity assessment.

## 4.12    REGULATORY ASPECTS OF NMs

The risk assessment (RA) of chemicals is considered by regulators to take risk management decisions aiming to avoid negative impact of chemicals on human health and the environment. There is, meanwhile, consensus that the classical RA paradigm as applied to traditional chemicals is generally also appropriate to assess the safety of NMs, although certain aspects that need evaluation still require further research, and may need methodological adaptations to specificities of NMs (EFSA, 2009; FAO/WHO, 2010; OECD, 2010b; OECD Communication; SCENIHR, 2009).

For NMs, regulatory risk management is a developing process. Although countries do not have and do not currently develop specific legislation to manage risk to human health and environmental safety, existing frameworks providing sufficient risk management options for chemicals, which may also be applied to NMs, need adaptations. In the European Union, amendments to certain sector-specific jurisdictions for placing on the market chemical products have already been made. Examples include the legislation for Biocidal Products (Regulation [EC] No. 528/2012), Cosmetics (Regulation [EC] No. 1223/2009) and for Food Contact Materials (Regulation [EC] No. 10/2011). In addition, national reporting and notification regimes specific for NMs have been implemented like the decree (Ministere de l'écologie, du développement durable, des transports et du logement, 2012) and a European product register is in discussion (European Commission, website a).

In the European Economic Area, the REACH (Regulation [EC] No. 1907/2006) and the classification, labelling and packaging (CLP) (Regulation [EC] No. 1272/2008) are the horizontal frameworks for assessing and managing risks of chemical substances on their own, in mixtures or in articles when manufactured, placed on the market and used. These regulations apply in addition to sector-specific legislation, and although they do not include yet specific requirements for the safety assessment of NMs, these materials meet the regulations' substance definition (ECHA, website a). The European Commission concluded that the REACH Regulation forms the best framework for risk management of NMs occurring as substances or in mixtures, but specific requirements have been proven necessary (European Commission, 2012). Safety assessment in the REACH registration procedure still hampers for reasons such as outstanding implementation of a legally binding definition of NM (as recommended by the European Commission [2011] but being currently in review), nano-specific provisions for substance identification, and adaptation of tonnage-dependent information requirements. Starting from an annual production volume above 1 ton/year manufacturers and importers have to register substances and submit certain data on (eco)toxicity, environmental fate (in particular according to OECD testing methods), and an RA in a tiered approach dependent on volume and hazard. Guidance documents for REACH registration and chemical safety assessment have been complemented by the ECHA with specific recommendations for NMs in the year 2012 (ECHA, website b). The adaptation of REACH information requirements to NM safety assessment is currently in law-making process (European Commission, website b; Matrix Insight Ltd., 2014).

For the United States, regulation of NMs is also a work in progress. No law has been or is being developed specifically for NMs and a statutory definition has also not been implemented in existing regulations. Existing regulation includes in particular the Toxic Substance Control Act (TSCA), and further specific regulations such as the Federal Insecticide, Fungicide and Rodenticide Act (FIFRA), the Federal Food, Drug and Cosmetic Act, and further acts on occupational health and safety, consumer, and environmental protection apply (OECD, 2010c). The U.S. Environmental Protection Agency (EPA)'s management of NMs under TSCA has evolved from a voluntary approach to a regulatory approach (EPA, 2011). The TSCA requires notification of an intended use to the EPA before commercialization. This information-gathering authority, however, includes no minimum toxicity and environmental fate data, thus due to lack of data RA of NM relying much on category and computational approaches (although information and risk management measures may be requested case-by-case risk or exposure based). Shortcomings of the regulation of NMs under the TSCA such as the lack of upfront testing have been reviewed by the

EPA and new regulatory rules on reporting and record keeping, including production volume and processes, exposure, release, and available health and safety data, have been proposed for NMs (Federal Register, 2015; U.S. EPA, website [a,b]; U.S. EPA, 2011). Moreover, the United States and Canada participate in the Regulatory Cooperation Council (RCC) Nano Initiative aiming in approximation of regulatory processes with respect to nanoscale materials (Government of Canada, website on nanotechnology working plan). Further information on regulatory regimes for different OECD regions (Canada, Australia, the United States, and European Union), including applicable legislative frameworks, planned amendments, NM definitions, and regulatory challenges, is available with the OECD report of the questionnaire on regulatory regimes for manufactured NMs 2010–2011 (OECD, 2014).

## 4.13   RESEARCH NEEDS

In the last years, an exponential increase in publications dealing with effects of NMs can be observed and this vast amount of reports was termed *Babylonian plethora of studies* (Krug, 2014).

Overall there is a need for research on many aspects of NMs and the fact that more and more materials appear with multiple shapes, sizes, surface chemistries, and so on asks for a comprehensive and intelligent testing approach based on sound understanding of general principles of toxicology and under consideration of NM special features (e.g., Scott-Fordsman et al., 2014; Stone et al., 2014).

There is a need for standard materials and procedures to address many of the questions all the way from the environmental fate to the distribution and effect pathways within an organism.

Furthermore, the mechanisms involved in the translocation of NPs to other organs in the organism following the initial uptake need to be studied in detail.

Other important issues comprise the following:

1. Quantification of agglomeration/aggregation in dependency of the composition of the medium
2. Environmental corona formation and its impact on ecotoxicity
3. Extension of the concepts of grouping and read-across toward NPs
4. Development of NP-specific descriptors suitable as a basis for deriving (quantitative) NP property relationships
5. Development of suited dose metrics for expressing adverse effects
6. Development of NP-specific RA, preferably based on NP-specific metrics to substitute for mass as in the case of *conventional* chemicals
7. Consideration of aged NPs rather than testing solely pristine materials

If we do not follow and respect these rules covering the design of toxicological investigations, and we continue without the harmonization of experimental processes through, for example, the use of SOPs or other standardized protocols, then future support programs, whether national or international, are doomed to failure too, and their results will only contribute further to the Babylonian plethora of low-value results that exists today.

**Krug, 2014, p. 12318**

## ACKNOWLEDGMENTS

ACG, ALM, CN, TK, and TS thank the COST Action ES1205 ENTER for the kind support and exchange of ideas and discussions. ACG, TS, WP, TK, TFF, and IL thank FP7 FUTURENANONEEDS (Grant Agreement No. 604602) for financial support. The contribution of ACG, TS, and SC was partly performed within the framework of the "Small Particles—Environmental Behaviour and Toxicity of Nanomaterials and Particulate Matter" project. AG contributed within the Fonds National de la Recherche (FNR) funded "Core2010" program NANEAU II (C10/SR/799842). KH was supported by the Fonds National de la Recherche (FNR) Luxembourg through the PhD Grant NANOGAM [9229040].

## REFERENCES

Adams, L.K., Lyon, D.Y., McIntosh, A., Alvarez, P.J. 2006. Comparative toxicity of nano-scale $TiO_2$, $SiO_2$ and ZnO water suspensions. *Water Sci. Technol.* 54, 327–334.

Albanese, A., Walkey, C.D., Olsen, J.B., Guo, H., Emili, A., Chan, W.C. 2014. Secreted bio-molecules alter the biological identity and cellular interactions of nanoparticles. *ACS Nano* 8, 5515–5526.

Al-Kattan, A., Wichser, A., Zuin, S., Arroyo, Y., Golanski, L., Ulrich, A., Nowack, B. 2014. Behaviour of $TiO_2$ released from nano-$TiO_2$-containing paint and comparison to pristine nano-$TiO_2$. *Environ. Sci. Technol.* 48, 6710–6718.

Asharani, P.V., Lian Wu, Y., Gong, Z., Valiyaveettil, S. 2008. Toxicity of silver nanoparticles in zebrafish models. *Nanotechnology* 19, 255102.

Ates, M., Daniels, J., Arslan, Z., Farah, I.O. 2013. Effects of aqueous suspensions of titanium dioxide nanoparticles on *Artemia salina*: Assessment of nanoparticle aggregation, accu-mulation, and toxicity. *Environ. Monit. Assess.* 185, 3339–3348.

Audinot, J.-N., Georgantzopoulou, A., Piret, J.-P., Gutleb, A.C., Dowsett, D., Migeon, H.-N., Hoffmann, L. 2013. Identification and localization of nanoparticles in tissues by mass spectrometry. *Surf. Interface Anal.* 45, 230–233.

Auffan, M., Rose, J., Wiesner, M.R., Bottero, J.Y. 2009. Chemical stability of metallic nanoparticles: A parameter controlling their potential cellular toxicity *in vitro*. *Environ. Poll.* 157, 1127–1133.

Auffan, M., Santaella, C., Thiéry, A., Paillès, C., Rose, J., Achouak, W., Thill, A. et al. 2012. Ecotoxicity of inorganic nanoparticles: From unicellular organisms to invertebrates. In B. Bhushan (ed), *Encyclopedia of Nanotechnology*. Springer, PP. 623–636.

Baalousha, M. 2009. Aggregation and disaggregation of iron oxide nanomaterials: Influence of particle concentration, pH and natural organic matter. *Sci. Total Environ.* 407, 2093–2101.

Baun, A., Hartmann, N.B., Grieger, K., Kusk, K.O. 2008. Ecotoxicity of engineered nanopar-ticles to aquatic invertebrates: A brief review and recommendations for future toxicity testing. *Ecotoxicology* 17, 387–395.

Benn, T.M., Westerhoff, P. 2008. Nanoparticle silver released into water from commercially available sock fabrics. *Environ. Sci. Tech.* 42, 4133–4139.

Bindler, R., Rydberg, J., Renberg, I. 2011. Establishing natural sediment reference condi-tions for metals and the legacy of long-range and local pollution on lakes in Europe. *J. Paleolimnol.* 45, 519–531.

Bondarenko, O., Juganson, K., Ivask, A., Kasmets, K., Mortimer, M., Kahru, A. 2013. Toxicity of Ag, CuO and ZnO nanoparticles to selected environmentally relevant test organisms and mammalian cells *in vitro*: A critical review. *Arch. Toxicol.* 7, 1181–1200.

Brun, N.R., Lenz, M., Wehrli, B., Fent, K. 2014. Comparative effects of zinc oxide nanoparticles and dissolved zinc on zebrafish embryos and eleuthero-embryos: Importance of zinc ions. *Sci. Total Environ.* 476–477, 657–666.

Canesi, L., Ciacci, C., Fabbri, R., Marcomini, A., Projana, G., Gallo, G. 2012. Bivalve molluscs as a unique target group for nanoparticle toxicity. *Marine Environ. Res.* 76, 16–21.

Chalasani, R., Vasudevan, S. 2013. Cyclodextrin-functionalized $Fe_3O_4@TiO_2$: Reusable, magnetic nanoparticles for photocatalytic degradation of endocrine-disrupting chemicals in water supplies. *ACS Nano* 7, 4093–4104.

Chambers, B.A., Afrooz, A.R., Bae, S., Aich, N., Katz, L., Saleh, N.B., Kirisits, M.J. 2014. Effects of chloride and ionic strength on physical morphology, dissolution, and bacterial toxicity of silver nanoparticles. *Environ. Sci. Technol.* 48, 761–769.

Chaudhry, Q., Bouwmeester, H., Hertel, R.F. 2010. The current risk assessment paradigm in relation to regulation of nanotechnologies. In G.A. Hodge, D.M. Bowman, and A.D. Maynard (eds), *International Handbook on Regulating Nanotechnologies*. Cheltenham: Edward Elgar, pp. 124–143.

Chen, Q., Yin, D., Zhu, S., Hu, X. 2012. Adsorption of cadmium(II) on humic acid coated titanium dioxide. 2012. *J. Colloid Interface Sci.* 367, 241–248.

Cohen, J.M., Teeguarden, J.G., Demokritou, P. 2014. An integrated approach for the *in vitro* dosimetry of engineered nanomaterials. *Part. Fibre Toxicol.* 11, 20.

Collin, B., Oostveen, E., Tsyusko, O.V., Unrine, J.M. 2014. Influence of natural organic matter and surface charge on the toxicity and bioaccumulation of functionalized ceria nanoparticles in *Caenorhabditis elegans*. *Environ. Sci. Technol.* 48, 1280–1289.

Cornelis, G., Hund-Rinke, K., Kuhlbusch, T.A.J., Van den Brink, N., Nickel, C. 2014. Fate and bioavailability of engineered nanoparticles in soils: A review. *Crit. Rev. Environ. Sci. Technol.* (in press).

Coutris, C., Joner, E.J., Oughton, D.H. 2012. Aging and soil organic matter content affect the fate of silver naoparticles in the environment. *Sci. Total Environ.* 420, 327–333.

Croteau, M.N., Misra, S.K., Luoma, S.N., Valsami-Jones, E. 2011. Silver bioaccumulation dynamics in a freshwater invertebrate after aqueous and dietary exposures to nanosized and ionic. *Agric. Environ. Sci. Technol.* 45, 6600–6607.

Croteau, M.N., Misra, S.K., Luoma, S.N., Valsami-Jones, E. 2014. Bioaccumulation and toxicity of CuO nanoparticles by a freshwater invertebrate after waterborne and dietborne exposures. *Environ. Sci. Technol.* 48, 10929–10937.

Derjaguin, B.V., Landau, L.D. 1941. Theory of the stability of strongly charged lyophobic sols and of the adhesion of strongly charged particles in solutions of electrolytes. *Acta Physicochim. U.R.S.S.* 14, 633–662.

De Rossetto, A.L., Melegari, S.P., Ouriques, L.C., Matias, W.G. 2014. Comparative evaluation of acute and chronic toxicities of CuO nanoparticles and bulk using *Daphnia magna* and *Vibrio fischeri*. *Sci. Total Environ.* 490, 807–814.

De Windt, W., Aelterman, P., Verstraete, W. 2005. Bioreductive deposition of palladium (0) nanoparticles on *Shewanella oneidensis* with catalytic activity towards reductive dechlorination of polychlorinated biphenyls. *Environ. Microbiol.* 7, 314–325.

Duester, L., Burkhardt, M., Gutleb, A.C., Kaegi, R., Macken, A., Meermann, B., von der Kammer, F. 2014. Towards a comprehensive and realistic risk evaluation of engineered nanomaterials (ENMs) in the urban water cycle. *Front. Chem.* 2, 39.

ECHA (website a). *Nanomaterials*. http://www.echa.europa.eu/regulations/nanomaterials (consulted on 26.08.2014).

ECHA (website b). *Guidance on Information Requirements and Chemical Safety Assessment*. http://www.echa.europa.eu/guidance-documents/guidance-on-information-requirements-and-chemical-safety-assessment (consulted on 11.08.2015).

EFSA. 2009. Scientific opinion of the scientific committee on a request from the European commission on the potential risks arising from nanoscience and nanotechnologies on food and feed safety. *The EFSA J.* 958, 1–39. http://www.efsa.europa.eu/sites/default/ files/scientific_output/files/main_documents/sc_op_ej958_nano_en%2C3.pdf (consulted on 11.08.2015).

European Commission. 2011. Commission Recommendation of 18 October 2011 on the definition of nanomaterial (2011/696/EU), Official Journal of the European Union, L 275/38, EN, October 2011. http://eur-lex.europa.eu/legal-content/EN/TXT/PDF/?uri= CELEX:32011H0696&from=EN (consulted on 26.08.2014).

European Commission. 2012. Communication from the Commission to the European Parliament, the Council and the European Economic and Social Committee: Second Regulatory Review on Nanomaterials. COM(2012)572 final. http://www.europarl.europa.eu/registre/docs_autres_ institutions/commission_europeenne/com/2012/0572/COM_COM%282012%290572_ EN.pdf (consulted on 26.08.2014).

European Commission (website a). Public consultation on transparency measures for nanomaterials on the market. http://ec.europa.eu/enterprise/sectors/chemicals/reach/nanomaterials/public-consultation_en.htm (consulted on 26.08.2014).

European Commission (website b). Consultation on the modification of the REACH Annexes on Nanomaterials. http://ec.europa.eu/environment/consultations/nanomaterials_2013_ en.htm (consulted on 26.08.2014).

FAO/WHO. 2010. FAO/WHO expert meeting on the application of nanotechnologies in the food and agriculture sectors: Potential food safety implications, 2010, Meeting report. http:// whqlibdoc.who.int/publications/2010/9789241563932_eng.pdf (consulted on 26.08.2014).

Farré, M., Gajda-Schrantz, K., Kantiani, L., Carceló, D. 2009. Ecotoxicity and analysis of nanomaterials in the aquatic environment. *Anal. Bioanal. Chem.* 393, 81–95.

Federal Register (The Daily Journal of the United States Government). 2015. *Chemical Substances When Manufactured or Processed as Nanoscale Materials; TSCA Reporting and Recordkeeping Requirement.* A Proposed rule by the Environmental Protection Agency on 04/06/2015. https://www.federalregister.gov/articles/2015/04/06/2015-07497/ chemical-substances-when-manufactured-or-processed-as-nanoscale-materials-tsca-reporting-and (consulted on 11.08.2015).

Feswick, A., Griffitt, R.J., Siebein, K., Barber, D.S. 2013. Uptake, retention and internalization of quantum dots in Daphnia is influenced by particle surface functionalization. *Aquat. Toxicol.* 130–131, 210–218.

Fleischer, C.C., Payne, C.K. 2014. Nanoparticle-cell interactions: Molecular structure of the protein corona and cellular outcomes. *Acc. Chem. Res.* 47, 2651–2659.

Franze, B., Engelhard, C. 2014. Fast separation, characterization, and speciation of gold and silver nanoparticles and their ionic counterparts with micellar electrokinetic chromatography coupled to ICP-MS. *Anal. Chem.* 86, 5713–5720.

French, R.A., Jacobson, A.R., Kim, B., Isley, S.L., Penn, R.L., Baveye, P.C. 2009. Influence of ionic strength, pH, and cation valence on aggregation kinetic of titanium dioxide nanoparticles. *Environ. Sci. Technol.* 43, 1354–1359.

Furman, O., Usenko, S., Lau, B.L.T. 2013. Relative importance of the humic and fulvic fractions of natural organic matter in the aggregation and deposition of silver nanoparticles. *Environ. Sci. Technol.* 47, 1349–1356.

Gaiser, B.K., Fernandes, T.F., Jepson, M.A., Lead, J.R., Tyler, C.R., Baalousha, M., Biswas, A. et al. 2012. Interspecies comparisons on the uptake and toxicity of silver and cerium dioxide nanoparticles. *Environ. Toxicol. Chem.* 31, 144–154.

Garcia-Negrete, C.A., Blasco, J., Volland, M., Rojas, T.C., Hampel, M., Lapresta-Fernandez, A., Jimenez de Haro, M.C., Soto, M., Fernandez, A. 2013. Behaviour of Au-citrate nanoparticles in seawater and accumulation in bivalves at environmentally relevant concentrations. *Environ. Poll.* 174, 134–141.

Gartiser, S., Nickel, C., Stintz, M., Damme, S., Schaeffer, A., Erdinger, L., Kuhlbusch, T.A.J. 2014. Behaviour of nanoscale titanium dioxide in laboratory wastewater treatment plants according to OECD 303 A. *Chemosphere* 104, 197–204.

Georgantzopoulou, A., Balachandran, Y.L., Rosenkranz, P., Dusinska, M., Lankoff, A., Wojewodzka, M., Kruszewski, M. et al. 2013. Ag nanoparticles: Size- and surface-dependent effects on model aquatic organisms and uptake evaluation with NanoSIMS. *Nanotoxicology* 7, 1168–1178.

Glenn, J.B., Klaine, S.J. 2013. Abiotic and biotic factors that influence the bioavailability of gold nanoparticles to aquatic macrophytes. *Environ. Sci. Technol.* 47, 10223–10230.

Gliga, A.R., Skoglund, S., Wallinder, I.O., Fadeel, B., Karlsson, H.L. 2014. Size-dependent cytotoxicity of silver nanoparticles in human lung cells: The role of cellular uptake, agglomeration and Ag release. *Part. Fibre. Toxicol* 11, 11.

Gottschalk, F., Sonderer, T., Scholz, R.W., Nowack, B. 2009. Modeled environmental concentrations of engineered nanomaterials ($TiO_2$, ZnO, Ag, CNT, fullerenes) for different regions. *Environ. Sci. Tech.* 43, 9216–9222.

Guzman, K.A.D., Finnegan, M.P., Banfield, J.F. 2006. Influence of surface potential on aggregation and transport of titania nanoparticles. *Environ. Sci. Technol.* 40, 7688–7693.

Handy, R.D., Cornelis, G., Fernandes, T., Tsyusko, O., Decho, A., Sabo-Attwood, T., Metcalfe, C. et al. 2012. Ecotoxicity test methods for engineered nanomaterials: Practical experiences and recommendations from the bench. *Environ. Toxicol. Chem.* 31, 15–31.

Hayashi, Y., Miclaus, T., Scavenius, C., Kwiatkowska, K., Sobota, A., Engelmann, P., Scott-Fordsmand, J.J., Enghild, J.J., Sutherland, D.S. 2013. Species differences take shape at nanoparticles: Protein corona made of the native repertoire assists cellular interaction. *Environ. Sci. Technol.* 47, 14367–14375.

Heinlaan, M., Ivask, A., Blinova, I., Dubourguier, H.C., Kahru, A. 2008. Toxicity of nanosized and bulk ZnO, CuO and $TiO_2$ to bacteria *Vibrio fischeri* and crustaceans *Daphnia magna* and *Thamnocephalus platyurus*. *Chemosphere* 71, 1308–1316.

Holbrook, R.D., Murphy, K.E., Morrow, J.B., Cole, K.D. 2008. Trophic transfer of nanoparticles in a simplified invertebrate food web. *Nat. Nanotechnol.*, 3, 352–355.

Hole, P., Sillence, K., Hannell, C., Maguire, C.M., Roesslein, M., Suarez, G., Capracotta, S. et al. 2013. Interlaboratory comparison of size measurements on nanoparticles using Nanoparticle Tracking Analysis (NTA). *J. Nanopart. Res.* 15, 2101.

Hull, M.S., Chaurand, P., Rose, J., Auffan, M., Bottero, J.-Y., Jones, J.C., Schultz, I.R., Vikesland, P.J. 2011. Filter-feeding bivalves store and biodeposit colloidally stable gold nanoparticles. *Environ. Sci. Technol.* 45, 6592–6599.

Hull, M.S., Vikesland, P.J., Schultz, I.R. 2013. Uptake and retention of metallic nanoparticles in the Mediterranean mussel (*Mytilus galloprovincialis*). *Aquat. Toxicol.* 140–141, 89–97.

Iavicoli, I., Fontana, L., Leso, V., Bergamaschi, A. 2013. The effects of nanomaterials as endocrine disruptors. *Int. J. Mol. Sci.* 14, 16732–16801.

Johnston, H.J., Hutchison, G., Christensen, F.M., Peters, S., Hankin, S., Stone, V. 2010. A review of the *in vivo* and *in vitro* toxicity of silver and gold particulates: Particles attributes and biological mechanisms responsible for the observed toxicity. *Crit. Rev. Toxicol.* 40, 328–346.

Jung, Y.-J., Kim, K.-T., Kim, J.Y., Yang, S.-Y., Lee, B.-G., Kim, S.D. 2014. Bioconcentration and distribution of silver nanoparticles in Japanese medaka (*Oryzias latipes*). *J. Hazard. Mater.* 267, 206–213.

Kaegi, R., Sinnet, B., Zuleeg, S., Hagendorfer, H., Mueller, E., Vonbank, R., Boller, M., Burkhardt, M. 2010. Release of silver nanoparticles from outdoor facades. *Environ. Pollut.* 158, 2900–2905.

Kaegi, R., Ulrich, A., Sinnet, B., Vonbank, R., Wichser, A., Zuleeg, S., Simmler, H. et al. 2008. Synthetic TiO(2) nanoparticle emission from exterior facades into the aquatic environment. *Environ. Pollut.* 156, 233–239.

Kaegi, R., Voegelin, A., Ort, C., Sinnet, B., Thalmann, B., Krismer, J., Hagendorfer, H., Elumelu, M., Mueller, E. 2013. Fate and transformation of silver nanoparticles in urban waste water systems. *Water Res.* 47, 3866–3877.

Kaegi, R., Voegelin, A., Sinnet, B., Zuleeg, S., Hagendorfer, H., Burkhardt, M., Siegrist, H. 2011. Behavior of metallic silver nanoparticles in a pilot wastewater treatment plant. *Environ. Sci. Tech.* 45, 3902–3908.

Kahru, A., Dubourguier, H.C. 2010. From ecotoxicology to nanoecotoxicology. *Toxicology* 269, 105–119.

Khan, F.R., Laycock, A., Dybowska, A., Larner, F., Smith, B.D., Rainbow, P.S., Luoma, S.N., Rehkämper, M., Valsami-Jones, E. 2013. Stable isotope tracer to determine uptake and efflux dynamics of ZnO nano- and bulk particles and dissolved Zn to an estuarine snail. *Environ. Sci. Technol.* 47, 8532–8539.

Khan, F.R., Misra, S.K., García-Alonso, J., Smith, B.D., Strekopytov, S., Rainbow, P.S., Luoma, S.N., Valsami-Jones, E. 2012. Bioaccumulation dynamics and modeling in an estuarine invertebrate following aqueous exposure to nanosized and dissolved silver. *Environ. Sci. Technol.* 46, 7621–7628.

Kiser, M.A., Westerhoff, P., Benn, T., Wang, Y., Perez-Rivera, J., Hristovski, K. 2009. Titanium nanomaterial removal and release from wastewater treatment plants. *Environ. Sci. Technol.* 43, 6757–6763.

Klaine, S.J., Alvarez, P.J.J., Batley, G.E., Fernandes, T.F., Handy, R.D., Lyon, D.Y., Mahendra, S., McLaughlin, M.J., Lead, J.R. 2008. Nanomaterials in the environment: Behavior, fate, bioavailability, and effects. *Environ. Tox. Chem.* 27, 1825–1851.

Klaper, R., Crago, J., Barr, J., Arndt, D., Setyowati, K., Chen, J. 2009. Toxicity biomarker expression in daphnids exposed to manufactured nanoparticles: Changes in toxicity with functionalization. *Environ. Pollut.* 157, 152–156.

Krug, H.F. 2014. Nanosafety research—Are we on the right track? *Angew. Chem. Int. Edit.* 53, 12304–12319.

Kruszewski, M., Brzoska, K., Brunborg, G., Asare, N., Dobrzynska, M., Dusinska, M., Fjellsbø, L. et al. 2011. Toxicity of silver nanomaterials on higher eukaryotes. *Adv. Mol. Toxicol.* 5, 179–218.

Kumbiçak, U., Cavas, T., Cinkiliç, N., Kumbiçak, Z., Vatan, O., Yilmaz, D. 2014. Evaluation of *in vitro* cytotoxicity and genotoxicity of copper-zinc alloy nanoparticles in human lung epithelial cells. *Food. Chem. Toxicol.* 73, 105–112.

Laban, G., Nies, L.F., Turco, R.F., Bickham, J.W., Sepúlveda, M.S. 2010. The effects of silver nanoparticles on fathead minnow (*Pimephales promelas*) embryos. *Ecotoxicology* 19, 185–195.

Lankoff, A., Sandberg, W.J., Wegierek-Ciuk, A., Lisowska, H., Refsnes, M., Sartowska, B., Schwarze, P.E., Meczynska-Wielgosz, S., Wojewodzka, M., Kruszewski, M. 2012. The effect of agglomeration state of silver and titanium dioxide nanoparticles on cellular response of HepG2, A549, and THP-1 cells. *Toxicol. Lett.* 208, 197–213.

Larner, F., Dogra, Y., Dybowska, A., Fabrega, J., Stolpe, B., Bridgestock, L.J., Goodhead, R. et al. 2012. Tracing bioavailability of ZnO nanoparticles using stable isotope labeling. *Environ. Sci. Technol.* 46, 12137–12145.

Lee, P.L., Chen, B.C., Golavelli, G., Yin, S.Y., Lei, S.H., Jhang, C.L. Lee, W.R., Ling, Y.C. 2014a. Development and validation of TOF-SIMS and CLSM imaging method for cytotoxicity study of ZnO nanoparticles in HaCaT cells. *J. Hazard. Mater.* 277, 3–12.

Lee, S., Bi, X., Reed, R.B., Ranville, J.F., Herckes, P., Westerhoff, P. 2014b. Nanoparticle size detection limits by single particle ICP-MS for 40 elements. *Environ. Sci. Technol.* 48, 10291–10300.

Lesniak, A., Fenaroli, F., Monopoli, M.P., Åberg, C., Dawson, K.A., Salvati, A. 2012. Effects of the presence or absence of a protein corona on silica nanoparticle uptake and impact on cells. *ACS Nano* 6, 5845–5857.

Limbach, L.K., Bereiter, R., Mueller, E., Krebs, R., Gaelli, R., Stark, W.J. 2008. Removal of oxide nanoparticles in a model wastewater treatment plant: Influence of agglomeration and surfactants on clearing efficiency. *Environ. Sci. Technol.* 42, 5828–5833.

Lowry, G.V., Gregory, K.B., Apte, A.C., Lead, J.R. 2012. Transformations of nanomaterials in the environment. *Environ. Sci. Technol.* 46, 6893–6899.

Lu, X., Liu, Y., Kong, X., Lobie, P.E., Chen, C., Zhu, T. 2013a. Nanotoxicity: A growing need for study in the endocrine system. *Small* 9, 1654–1671.

Lu, Z., Ma, G., Veinot, J.G., Wong, C.S. 2013b. Disruption of biomolecule function by nanoparticles: How do gold nanoparticles affect phase I biotransformation of persistent organic pollutants? *Chemosphere* 93, 123–132.

Lundqvist, M., Stigler, J., Cedervall, T., Berggård, T., Flanagan, M.B., Lynch, I., Elia, G., Dawson, K. 2011. The evolution of the protein corona around nanoparticles: A test study. *ACS Nano* 5, 7503–7509.

Lynch, I., Cedervall, T., Lundqvist, M., Cabaleiro-Lago, C., Linse, S., Dawson, K.A. 2007. The nanoparticle-protein complex as a biological entity; a complex fluids and surface science challenge for the 21st century. *Adv. Colloid. Interface Sci.* 134–135, 167–174.

Lynch, I., Dawson, K.A., Lead, J.R., Valsami-Jones, E. 2014a. Nanoscience and the environment. In J.R. Lead and E. Valsami-Jones (eds), *Series: Frontiers of Nanoscience*, Volume 7. Richard E. Plamer. Amsterdam, the Netherlands: Elsevier.

Lynch, I., Weiss, C., Valsami-Jones, E. 2014b. A strategy for grouping of nanomaterials based on key physico-chemical descriptors as a basis for safer-by-design NMs. *Nano Today* 9, 266–270.

Mallevre, F., Fernandes, T.F., Aspray, T.J. 2014. Silver, zinc oxide and titanium dioxide nanoparticle ecotoxicity to bioluminescent *Pseudomonas putida* in laboratory medium and artificial wastewater. *Environ. Poll.* 195, 218–225.

Maszczyk, P., Bartosiewicz, M. 2012. Threat or treat: The role of fish exudates in the growth and life history of Daphnia. *Ecosphere* 3: art91.

Matrix Insight Ltd. 2014. Request for services in the context of the FC ENTR/2008/006, lot 3: A study to support the impact assessment of regulatory options for nanomaterials in the framework of REACH. http://ec.europa.eu/enterprise/sectors/chemicals/reach/nanomaterials/index_en.htm (consulted on 26.08.2014).

Miao, A.J., Zhang, X.Y., Luo, Z., Chen, C.S., Chin, C.S., Santschi, P.H., Quigg, A. 2010. Zinc oxide-engineered nanoparticles: Dissolution and toxicity to marine phytoplankton. *Environ. Toxicol. Chem.* 29, 2814–2822.

Ministère de l'écologie, du développement durable, des transports et du logement. 2012. Décret n 2002-232 du 17 février 2012 relatif à la déclaration annuelle des substances à l'état nanoparticulaire pris en application de l'article L. 523-4 du code de l'environnement. *J. officiel République Française (JORF)* n 0043 du 19 février 2012 page 2863 texte n 4. http://www.legifrance.gouv.fr/affichTexte.do?cidTexte=JORFTEXT000025377246&categorieLien=id (consulted on 26.08.2014).

Monopoli, M.P., Walczyk, D., Campbell, A., Elia, G., Lynch, I., Bombelli, F.B., Dawson, K.A. 2011. Physical-chemical aspects of protein corona: Relevance to in vitro and in vivo biological impacts of nanoparticles. *J. Am. Chem. Soc.* 133, 2525–2534.

Montaño, M., Gutleb, A.C., Murk, A.J. 2013. Persistent toxic burdens of halogenated phenolic compounds in humans and wildlife. *Environ. Sci. Technol.* 47, 6071–6081.

Moore, M.N. 2006. Do nanoparticles present ecotoxicological risks for the health of the aquatic environment? *Environ. Int.* 32, 967–976.

Moret, F., Selvestrel, F., Lubian, E., Mognato, M., Celotti, L., Mancin, F., Reddi, E. 2014. PEGylation of ORMOSIL nanoparticle differently modulates the in vitro toxicity toward human lung cells. *Arch. Toxicol.* 89, 607–620.

Musee, N., Oberholster, P.J., Sikhwivhilu, L., Botha, A.M. 2010. The effects of engineered nanoparticles on survival, reproduction, and behaviour of freshwater snail, *Physa acuta* (Draparnaud, 1805). *Chemosphere* 81, 1196–1203.

Muth-Köhne, E., Sonnack, L., Schlich, K., Hischen, F., Baumgartner, W., Hund-Rinke, K., Schafers, C., Fenske, M. 2013. The toxicity of silver nanoparticles to zebrafish embryos increases through sewage treatment processes. *Ecotoxicology* 22, 1264–1277.

Mwaanga, P., Carraway, E.R., Schlautman, M.A. 2014. Preferential sorption of some natural organic matter fractions to titanium dioxide nanoparticles: Influence of pH and ionic strength. *Environ. Monit. Assess.* 186, 8833–8844.

Nickel, C., Hellack, B., Nogowski, A., Babick, F., Stintz, M., Maes, H., Schäffer, A., Kuhlbusch, T. *Mobility, Fate and Behaviour of TiO₂ Nanomaterials*, Final Report (UFOPlan) FKZ 3710 65 414, Federal Ministry for the Environment, Nature Conservation and Nuclear Safety, 2013. UBA Bericht 76/2013 Hrsg.: Umweltbundesamt, Förderkennzeichen 3710 65 414, UBA-FB 001741/E, p. 194.

Nowack, B., Bucheli, T.D. 2007. Occurrence, behavior and effects of nanoparticles in the environment. *Environ. Poll.* 150, 5–22.

Nowack, B., Ranville, J.F., Diamond, S., Gallego-Urrea, J.A., Metcalfe, C., Rose, J., Horne, N., Koelmans, A.A., Klaine, S.J. 2012. Potential scenarios for nanomaterial release and subsequent alteration in the environment, critical review. *Environ. Toxicol. Chem.* 31, 50–59.

OECD. 2004. *Guidelines for the Testing of Chemicals. Section 2: Effects on Biotic Systems.* Test No. 202: *Daphnia* sp. acute immobilisation test. Paris, France: Organisation for Economic Co-operation and Development.

OECD. 2010a. *OECD Environment, Health and Safety Publications Series on the Safety of Manufactured Nanomaterials: Guidance Manual for the Testing of Manufactured Nanomaterials: OECD's Sponsorship Programme*; First Revision, June 2, 2010, ENV/JM/MONO(2009)20/REV.

OECD. 2010b. Report of the workshop on risk assessment of manufactured nanomaterials in a regulatory context, ENV/JM/MONO(2010)10. http://www.oecd.org/officialdocuments/publicdisplaydocumentpdf/?cote=env/jm/mono%282010%2910&doclanguage=en (consulted on 26.08.2014).

OECD. 2010c. Report of the questionnaire on regulatory regimes for manufactured nanomaterials, ENV/JM/MONO(2010)12. http://www.oecd.org/officialdocuments/publicdisplaydocumentpdf/?doclanguage=en&cote=env/jm/mono%282010%2912 (consulted on 26.08.2014).

OECD. 2011. *Guidelines for the Testing of Chemicals. Section 2: Effects on Biotic Systems.* Test No. 201: Freshwater alga and cyanobacteria, growth inhibition test. Paris, France: Organisation for Economic Co-operation and Development.

OECD. 2014. *Report of the Questionnaire on Regulatory Regimes for Manufactured Nanomaterials 2010–2011*, ENV/JM/MONO(2014)28. http://www.oecd.org/officialdocuments/publicdisplaydocumentpdf/?cote=env/jm/mono%282014%2928&doclanguage=en (consulted on 06.11.2014).

OECD Communication. 2012. *Six Years of OECD Work on the Safety of Manufactured Nanomaterials: Achievements and Future Opportunities.* http://www.oecd.org/env/ehs/nanosafety/Nano%20Brochure%20Sept%202012%20for%20Website%20%20(2).pdf (consulted on 26.08.2014).

Orts-Gil, G., Natte, K., Österle, W. 2013. Multi-parametric reference nanomaterials for toxicology: State of the art, future challenges and potential candidates. *RSC Adv.* 3, 18202–18215.

Poland, C.A., Duffin, R., Kinloch, I., Maynard, A., Wallace, W.A.H., Seaton, A., Stone, V., Brown, S., MacNee, W., Donaldson, K. Carbon nanotubes introduced into the abdominal cavity of mice show asbestos-like pathogenicity in a pilot study. *Nat. Nanotechnol.* 3, 423–428.

Renault, S., Baudrimont, M., Mesmer-Dudons, N., Gonzalez, P., Mornet, S., Brisson, A. 2008. Impacts of gold nanoparticle exposure on two freshwater species: A phytoplanktonic alga (*Scenedesmus subspicatus*) and a benthic bivalve (*Corbicula fluminea*). *Gold Bull.* 41, 116–126.

Rosenkranz, P., Chaudhry, Q., Stone, V., Fernandes, T.F. 2009. A comparison of nanoparticle and fine particle uptake by *Daphnia magna. Environ. Toxicol. Chem.* 28, 2142–2149.

Sabella, S., Carney, R.P., Brunetti, V., Malvindi, M.A., Al-Juffali, N., Vecchio, G., Janes, S.M. et al. 2014. A general mechanism for intracellular toxicity of metal-containing nanoparticles. *Nanoscale* 6, 7052–7061.

Sayes, C.M., Smith, P.A., Ivanov, I.V. 2013. A framework for grouping nanoparticles based on their measurable characteristics. *J. Nanomed.* 8 (Suppl. 1), 45–56.

SCENIHR. 2009. *Risk Assessment of Products of Nanotechnologies.* http://ec.europa.eu/health/ph_risk/committees/04_scenihr/docs/scenihr_o_023.pdf (consulted on 26.08.2014).

Schmidt, J., Vogelsberger, W. 2006. Dissolution kinetics of titanium dioxide nanoparticles: observation of an unusual kinetic size effect. *J. Phys. Chem. B.* 110, 3955–3963.

Scott-Fordsmand, J.J., Pozzi-Mucelli, S., Tran, L., Aschberger, K., Sabella, S., Vogel, U., Poland, C. et al. 2014. A unified framework for nanosafety is needed. *Nano Today.* 9, 546–549.

Sharma, V.K., Siskova, K.M., Zboril, R., Gardea-Torresdey, J.L. 2014. Organic-coated silver nanoparticles in biological and environmental conditions: fate, stability and toxicity. *Adv. Colloid. Interface Sci.* 204, 15–34.

Skjolding, L.M., Winther-Nielsen, M., Baun, A. 2014. Trophic transfer of differently functionalised zinc oxide nanoparticles from crustaceans (*Daphnia magna*) to zebrafish (*Danio rerio*). *Aquat. Toxicol.* 157, 101–108.

Sovová, T., Boyle, D., Sloman, K.A., Vanegas Pérez, C., Handy, R.D. 2014. Impaired behavioural response to alarm substance in rainbow trout exposed to copper nanoparticles. *Aquat. Toxicol.* 152, 195–204.

Stefaniak, A.B., Hackley, V.A., Roebben, G., Ehara, K., Hankin, S., Postek, M.T., Lynch, I., Fu, W.E., Linsinger, T.P., Thünemann, A.F. 2013. Nanoscale reference materials for environmental, health and safety measurements: Needs, gaps and opportunities. *Nanotoxicology* 7, 1325–1337.

Stone, V., Nowack, B., Baun, A., van den Brink, N., von der Kammer, F., Dusinska, M., Handy, R. et al. 2010. Nanomaterials for environmental studies: Classification, reference material issues, and strategies for physico-chemical characterization. *Sci. Total Environ.* 408, 1745–1754.

Stone, V., Pozzi-Mucelli, S., Tran, L., Aschberger, K., Sabella, S., Vogel, U., Poland, C. et al. 2014. ITS-NANO—Prioritising nanosafety research to develop a stakeholder driven intelligent testing strategy. *Part. Fibre Toxicol.* 11, 9.

Sutherland, I. 2001. Biofilm exopolysaccharides: A strong and sticky framework. *Microbiology* 147, 3–9.

Tang, H., Meng, G., Huang, Q., Zhu, C., Huang, Z., Li, Z., Zhang, Z., Zhang, Y. 2014. Urchin-like Au-nanoparticles@Ag-nanohemisphere arrays as active SERS-substrates for recognition of PCBs. *SC Adv.* 4, 19654–19657.

Thomas, C.R., George, S., Horst, A.M., Ji, Z., Miller, R.J., Peralta-Videa, J.R., Xia, T. et al. 2011. Nanomaterials in the environment: From materials to high-throughput screening to organisms. *ACS Nano* 5, 13–20.

Thurman, E.M., Wershaw, R.L., Malcolm, R.L., Pinckney, D.J. 1982. Molecular size of aquatic humic substances. *Organ. Geochem.* 4, 27–35.

U.S. EPA. December 2011. *EPA Needs to Manage Nanomaterial Risks More Effectively*, Report No. 12-P-0162. http://www.epa.gov/oig/reports/2012/20121229-12-P-0162.pdf (consulted on 26.08.2014).

U.S. EPA (website a). 2015. *Regulatory Development and Retrospective Review Tracker: Nanoscale Materials; Chemical Substances When Manufactured, Imported, or Processed as Nanoscale Materials; Reporting and Recordkeeping Requirements.* http://yosemite.epa.gov/opei/rulegate.nsf/byRIN/2070-AJ54 (consulted on 26.08.2014).

U.S. EPA (website b). 2015. *Control of Nanoscale Material under the Toxic Substances Control Act.* http://www.epa.gov/oppt/nano/ (consulted on 10.08.2015).

Velzeboer, I., Kwadijk, C.J., Koelmans, A.A. 2014. Strong sorption of PCBs to nanoplastics, microplastics, carbon nanotubes, and fullerenes. *Environ. Sci. Technol.* 48, 4869–4876.

Verwey, E.J.W., Overbeek, J.Th.G. 1948. *Theory of the Stability of Lyophobic Colloids.* Amsterdam, the Netherlands: Elsevier.

Von der Kammer, F., Ferguson, P.L., Holden, P.A., Masion, A., Rogers, K.R., Klaine, S.J., Koelmans, A.A., Horne, N., Unrine, J.M. 2012. Analysis of engineered nanomaterials in complex matrices (environment and biota): General considerations and conceptual case studies. *Environ. Toxicol. Chem.* 31, 32–49.

Von Der Kammer, F., Ottofuelling, S., Hofmann, T. 2010. Assessment of the physico-chemical behavior of titanium dioxide nanoparticles in aquatic environments using multi-dimensional parameter testing. *Environ. Poll.* 158, 3472–3481.

Walczyk, D., Bombelli, F.B., Monopoli, M.P., Lynch, I., Dawson, K.A. 2010. What the cell "sees" in bionanoscience. *J. Am. Chem. Soc.* 132, 5761–5768.

Walkey, C.D., Olsen, J.B., Song, F., Liu, R., Guo, H., Olsen, D.W., Cohen, Y., Emili, A., Chan, W.C. 2014. Protein corona fingerprinting predicts the cellular interaction of gold and silver nanoparticles. *ACS Nano.* 8, 2439–2455.

Wang, D., Gao, Y., Lin, Z., Yao, Z., Zhang, W. 2014a. The joint effects on *Photobacterium phosphoreum* of metal oxide nanoparticles and their most likely coexisting chemicals in the environment. *Aquat Toxicol.* 154, 200–206.

Wang, Q., Chen, Q., Zhou, P., Li, W., Wang, J., Huang, C., Wang, X., Lin, K., Zhou, B. 2014b. Bioconcentration and metabolism of BDE-209 in the presence of titanium dioxide nanoparticles and impact on the thyroid endocrine system and neuronal development in zebrafish larvae. *Nanotoxicology* 8, 196–207.

Westerhoff, P., Song, G., Hristovski, K., Kiser, M.A. 2011. Occurrence and removal of titanium at full scale wastewater treatment plants: Implications for $TiO_2$ nanomaterials. *J. Environ. Monit.* 13, 1195–1203.

Xu, M., Li, J., Hanagata, N., Su, H., Chen, H., Fujita, D. 2013. Challenge to assess the toxic contribution of metal cation released from nanomaterials for nanotoxicology—the case of ZnO nanoparticles. *Nanoscale* 7, 4763–4769.

Zhang, B., Chen, L., Choi, J.J., Hennig, B., Toborek, M. 2012. Cerebrovascular toxicity of PCB153 is enhanced by binding to silica nanoparticles. *J. Neuroimm. Pharmacol.* 7, 991–1001.

Zhu, X., Wnag, J., Zhang, X., Chang, Y., Chen, Y. 2010. Trophic transfer of $TiO_2$ nanoparticles from daphnia to zebrafish in a simplified freshwater food chain. *Chemosphere* 79, 928–933.

# 5 Magnetron-Sputtered Hard Nanostructured TiAlN Coatings

## Strategic Approach toward Potential Improvement

*Vishal Khetan, Nathalie Valle, Marie Paule Delplancke, and Patrick Choquet*

## CONTENTS

**ABSTRACT**   Magnetron sputtered hard nanostructured TiAlN has gained high importance in the field of protective tribological coatings. Nevertheless, its use regarding high-temperature ($\geq 800°C$) applications such as dry high-speed machining still remains a challenge. There are several strategies, which have been used to improve these coatings in terms of higher oxidation resistance, higher fracture toughness, or better tribological behavior at high temperatures.

Most significant among them are adjustment of process parameters during deposition and addition of new elements. Specially, addition of elements such as Si, C, V, Cr, Ta, or Y has shown a significant beneficial impact on these properties. Addition of new element also leads to development of TiAlN-based hard nanocomposite coatings. For a better performance of these coatings, an in-depth understanding of their structure and their correlation to oxidation and wear mechanisms over a wider range of temperatures is needed. This chapter focuses on elaborating the significance of these strategies, which would aid in further development of these coatings keeping magnetron sputtering as a technique of choice for their deposition.

## 5.1   INTRODUCTION

Hard coatings have found prime importance in the cutting tool, aerospace, and automobile industry [1–3]. Coating enhances component performance, reliability, and service life, and permits lighter with more compact designs for a number of machines. Reduced energy consumption and the use of environmentally benign products in smaller quantities are further advantages that come into play in the building of machines and equipment just as they can in engine and vehicle making. Components of machine tools, textile machines, injection molding equipment for plastics, and equipment for food processing now come with hard coatings as a standard feature commercially.

The applications require high hot hardness and high thermal stability of the coating material. In the current scenario, temperature domain of 800°C–900°C, where the chemical wear due to the formation of soft oxides with poor adherence dominates over the abrasive wear, is of great interest. It is why the research on hard oxidation- and wear-resistant coatings play an important role to hence protect systems from high-temperature degradation.

Common hard coatings investigated for this application in the scientific area are carbide–nitride of chromium and titanium. One of these compounds investigated and used widely in industry is TiAlN [1–3]. Additional alloying elements such as Ta [4–11] have shown improving trends of these properties. Deeper understanding of the wear and oxidation mechanisms of these coatings is needed to develop strategies for the improvement of their performance and will be discussed.

This chapter should help researchers to further understand TiAlN-based hard coatings with properties such as the following:

- High hardness
- High toughness
- High-temperature behavior and thermal stability of individual phases
- High-temperature wear and oxidation resistance

This chapter summarizes the state of art for TiAlN coatings keeping physical vapor deposition techniques as a source to realize them. To summarize, this chapter focuses on the development of TiAlN-based coatings and properties exhibited by them by modulating their chemistry and deposition parameters involved to realize these coatings, and describes the following:

- The establishment of the relations between the deposition parameters and the coating microstructure, hardness, oxidation, and tribology of TiAlN coatings
- The influence of alloying elements taking example of tantalum as an alloying element
- The understanding of the oxidation and the wear behavior for these coatings

## 5.2   TERNARY NITRIDES: TiAlN COATINGS

TiAlN is the most investigated compound in the field of TiN-based hard coatings and has gained high importance in the field of wear- and oxidation-resistant coatings [1,12–27]. With addition of Al in TiN coatings, mechanical, oxidation, and tribological behavior of TiN coatings can be tuned significantly. As Al has an important role to play in modulating the individual properties of the TiN coatings, in Sections 5.2.1 through 5.2.4 the individual properties exhibited by TiAlN will be detailed and discussed individually.

### 5.2.1   STRUCTURE

In TiAlN coatings with cubic NaCl (c) structure, aluminum substitutes for titanium in the TiN structure. The deposition parameters and chemistry (aluminum concentration) can influence the structure of TiAlN coatings significantly [1,12–28]. It has been demonstrated that a change in deposition parameters such as substrate bias, substrate temperature, or deposition angle imposed no major effect on the composition and phase formation of the TiAlN coatings, but had significant influence on the development of their microstructure and surface morphology. The most important parameters being Al concentration, substrate bias voltage, and substrate temperature will be discussed in this section.

#### 5.2.1.1   Influence of Al Concentration

The Al content within the $Ti_{1-x}Al_xN$ coatings plays an important role in tuning the structure of TiN coatings [1,16,19,20]. Figure 5.1a shows X-ray diffraction (XRD) patterns from $Ti_{1-x}Al_xN$ films with $x = 0, 0.1, 0.3, 0.5, 0.6, 0.7, 0.9, 1$, respectively, deposited using arc ion plating [19]. The XRD pattern of pure TiN reflected (111) and (200) peaks with a NaCl structure. For films with $x = 0.6$, the peaks gradually shifted to higher diffraction angle in proportion to the $x$ value, indicating that the lattice parameter decreased with the addition of Al. It has been reported that the change in lattice parameter was attributed to the substitution of Al atoms with Ti atoms in TiN, keeping the NaCl structure [23]. The lattice parameter decreased from ~4.23Å for $x = 0$, that is, pure TiN, to 4.17Å for $x = 0.6$. For Al contents exceeding the maximum solubility ($x_{max} \sim 0.7$, depending on the deposition conditions used) in the cubic phase, a mixed cubic NaCl and wurtzite ZnS (w) structure is formed [21]. An example for this transformation comes from Ohnuma work. XRD profiles obtained in his work are displayed in Figure 5.1b. The figure shows that $Ti_{0.6}Al_{0.4}N$ film had a B1 structure, $Ti_{0.42}Al_{0.58}N$ had a B4 + B1 structure, and $Ti_{0.3}Al_{0.7}N$ had a B4 structure [21].

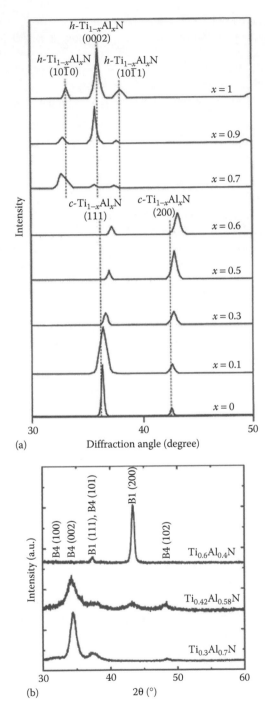

**FIGURE 5.1** XRD patterns of the TiAlN coatings varying Al concentration obtained by (a) Kiumra, A. et al. (Data from Kiumra, A. et al., *Surf. Coat. Technol.*, 120–121, 438–441, 1999.) and (b) Hörling, A. et al. (Data from Ohnuma, H. et al., *Surf. Coat. Technol.*, 177–178, 623–626, 2004.)

### 5.2.1.2    Influence of Substrate Bias Voltage

Substrate bias voltage influences the coating properties as it helps to enhance the ion bombardment on growing films by controlling the kinetic energy and the behaviors of ions on growing films [22–23]. Figure 5.2 presents an example for this behavior. It shows the XRD patterns as a function of the substrate bias voltage after the deposition of TiAlN coatings [22]. These diffraction patterns show orientations in crystallographic planes {111} and {200} corresponding to the TiAlN coatings. The film was strongly {111} textured at lower bias voltage ($V_s = 40$ V). The lower adatom mobility at lower negative substrate bias voltage also favors the {111} orientation because the highest number of atoms per unit area can be incorporated at low-energy sites. With increase in negative substrate bias voltage, adatom mobility increases and mass transfer from {111} texture to less compact and imperfect crystallites corresponding to {200} is expected to take place [22]. Increasing negative bias voltage increases {200}-to-{111} ratio in TiAlN coatings [22,23]. Increase in the ion bombardment energy leads to decrease in the intensity of the {111} peak as the Al atoms are incorporated to the films. This observation is consistent with the increase in deformation energy, promoted by the ions bombardment, resulting in higher defect concentrations.

### 5.2.1.3    Influence of Substrate Temperature

Substrate temperature could significantly influence the structure of TiAlN coatings. Researchers have studied the influence of temperature on TiAlN coatings when deposited using magnetron sputtering [23–25].

Wuhrer and Yeung studied the effect of substrate temperature (120°C, 240°C, and 360°C) on the microstructure and property development of TiAlN coatings

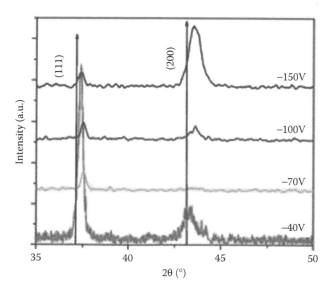

**FIGURE 5.2** XRD patterns of the TiAlN coatings varying substrate bias voltage. (Data from Devia, D.M. et al., *Appl. Surf. Sci.*, 257, 6181–6185, 2011.)

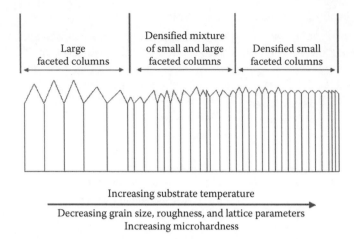

**FIGURE 5.3** Schematic representation of two-dimensional structure zone model showing the effect of substrate temperature on the microstructure and property development of (Ti,Al)N coatings deposited with a 30° magnetron configuration. (Data from Wuhrer, R. and Yeung, W. Y., *J. Mater. Sci.*, 37, 1993–2004, 2002.)

deposited with a 30° magnetron configuration [23]. They determined that as the substrate temperature increased, the coating structure was densified with development of fine grain size and reduced surface roughness. The reason behind this change in microstructure and property enhancement is attributed to an increased translational kinetic energy of the depositing atoms and a greater thermal energy provided to the substrate and the coating material with increasing substrate temperature. Figure 5.3 illustrates a summary of the structural development of the coatings deposited at 120°C–360°C.

Further, Shetty et al. investigated TiAlN coatings deposited at different substrate temperatures ranging from room temperature to 650°C [25]. The θ–2θ scans of TiAlN films deposited are shown in Figure 5.4. A single-phase cubic B1 NaCl-type structure was observed at all temperatures. As the substrate temperature was increased, the coatings displayed an (111) out-of-plane orientation. Also, increasing the substrate temperature leads to a growth with most dense planes of the coating parallel to the substrate surface. Because the (111) plane is the densest in the TiAlN crystal hence with increasing substrate temperature, an (111) out-of-plane orientation was observed. Similar kind of result was also observed by Wuhrer and Yeung [23]. Shetty et al. [25] also demonstrated that with increasing substrate temperature coatings become homogenous, thus decreasing the influence of flux angle.

### 5.2.2 HARDNESS AND ELASTIC MODULUS

After the structure of TiAlN coatings being discussed, the next step toward the evaluation of these coatings is to study their mechanical properties (hardness,

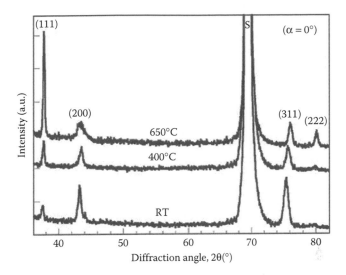

**FIGURE 5.4** Bragg–Brentano XRD patterns of TiAlN coating deposited with $\alpha = 0°$ (deposition angle) with a substrate bias of $-25$ V. Deposition temperatures are indicated on the plots. The symbol (S) corresponds to the substrate. (Data from Shetty, A. R., et al., *Thin Solid Films*, 519, 4262–4270, 2011.)

elastic modulus, and intrinsic stress). Hardness can be defined as the ability of a coating to resist plastic deformation, generally measured using indentation technique. Hardness may also refer to resistance to scratching, abrasion, cutting, or penetration. An elastic modulus is a quantity that evaluates a coating's resistance to being deformed elastically (i.e., nonpermanently) when a force is applied to it. The elastic modulus of a coating is defined as the slope of its stress–strain curve in the elastic deformation region. Both these properties of the coating strongly depend on the microstructure and stresses developed within the coating. Both these parameters are controlled by the crystallographic structure and intrinsic defects within the coating, which lead to elastic constants demonstrated by the coatings. The elastic constants in turn lead to the overall hardness and elastic modulus displayed by the coating.

The variation of deposition parameters and chemistry leads to the variation in microstructure and stresses developed within the coating. Hence, hardness and elastic modulus of TiAlN coatings are influenced by Ti/Al ratio and deposition parameters such as substrate bias voltage, substrate temperature, and deposition angle [1]. Many researchers have studied these variations and they will be described in Sections 5.2.2.1 through 5.2.2.3.

### 5.2.2.1 Influence of Al Concentration

The Al content strongly influences the hardness and Young's modulus of TiAlN coatings. Figure 5.5 demonstrates that the hardness and Young's modulus of TiAlN coatings increase with increasing Al content in the coating [29–31]. Hardness reaches its maximum value at an Al concentration of 50%. This concentration can

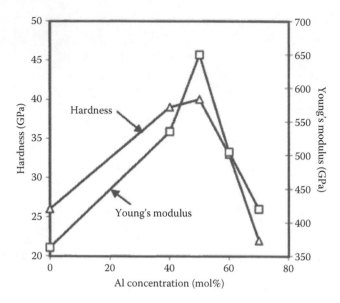

**FIGURE 5.5** Hardness and Young's modulus as a function of the Al concentration in the TiAlN films. (Data from Zhou, M. et al., *Thin Solid Films*, 339, 203–208, 1999.)

vary with change in deposition techniques and deposition conditions. Both hardness and Young's modulus values decreased drastically beyond 50% Al content due to the formation of hexagonal wurtzite phase as displayed in Figure 5.5. Formation of hexagonal AlN phase deteriorates the mechanical and tribological properties of TiAlN coatings [31].

Further, Kutschej et al. investigated four different $Ti_{1-x}Al_xN$ coatings with different Al/(Al + Ti) atomic ratios (= 0.54, 0.67, 0.69, and 0.76 in the coating) deposited on high-speed steel with constant deposition parameters using magnetron sputtering [31]. These coatings were tested for their hardness. The results demonstrated that coatings with a dominant face-centered cubic (fcc) phase (i.e., $x = 0.54$ and 0.67) were much harder than the dual-phase and single-phase hexagonal close-packed (hcp) coatings. The coating with $x = 0.54$ showed a hardness value of 33 GPa whereas the coating with $x = 0.76$ showed a hardness value of only 19 GPa.

Along with Al content, hardness is also associated with the amount of residual stress in the film, which is modified due to Al content and the deposition parameters involved for a particular deposition technique. Though compressive residual stresses are desirable to some extent in retarding the crack propagation and in improving the fracture toughness of the coating, too high of a compressive stress causes poor adhesion of the film to the substrate [30]. The TiAlN coatings have been demonstrated to have a lower internal stress in the range of 4.5–5.2 GPa combined with a microhardness as high as 32.4 GPa (3300 HK) [30]. Therefore, TiAlN coating is preferred for wear-resistant applications over a TiN coating considering the mechanical properties exhibited by them.

### 5.2.2.2 Influence of Substrate Bias Voltage

Although the composition of the coatings only varied slightly under the different bias voltages investigated, a densified structure with finer grain size and improved surface morphology developed as the bias voltage increased. A hardness enhancement of the coatings from 1500 HV (0 V) to 2300 HV (200 V) was achieved by Wuhrer et al. [23]. This strength enhancement is believed to be related to the densified coating structure and improved surface morphology. In previous studies of magnetron sputtering process, Messier et al. [32] reported that increasing the negative bias could result in a transition of the zone 1 to zone T microstructure of Thornton's model [33] of sputtered coatings. As a result, as the substrate bias increased, the open porous columnar structure of the $Ti_{0.5}Al_{0.5}N$ coatings was suppressed with a substantial hardness enhancement.

### 5.2.2.3 Influence of Substrate Temperature

It was thought interesting to study the role of substrate temperature on the hardness and Young's modulus of the TiAlN coatings as it influences the structure of the coating significantly. Wuhrer et al. [23] reported that when the substrate temperature was increased from 400°C to 480°C, the coating microstructure became denser, which improved the lattice and strengthened the interface. The microhardness values displayed in Figure 5.6 show a similar structure development in spite of the coatings that were produced at lower substrate temperatures. Despite the absence of major changes in the deposition rate and composition, the microhardness of the coatings increased from 1600 to 2200 HV as the substrate temperature increased from 120°C to 360°C.

Shetty et al. [25] reported similar trends of hardness and Young's modulus of films deposited at a bias of −25 V for different substrate temperatures. The hardness and Young's modulus show an increasing trend with increase in substrate temperature. The observations aid us to postulate that as the substrate temperature is increased, the surface diffusion is enhanced, resulting in improved hardness compared to those at room temperature.

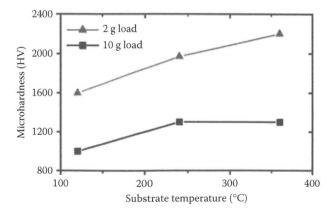

**FIGURE 5.6** Microhardness (HV) and Young's modulus of TiAlN coatings as a function of temperature. (Data from Wuhrer, R. and Yeung, W. Y., *J. Mater. Sci.*, 37, 1993–2004, 2002.)

## 5.2.3   Oxidation Resistance of TiAlN Coatings

Titanium nitride film has been applied as a hard coating on cutting tools, but it is known to be easily oxidized at high temperatures (above 500°C). TiAlN coatings exhibit higher oxidation resistance compared with TiN coatings. TiN oxidizes rapidly at temperatures higher than 600°C. With addition of Al in TiN coatings, their oxidation resistance could be enhanced up to 800°C [18,34,35].

Researchers have investigated the oxidation kinetics of TiAlN coatings over the past two decades. McIntyre et al. [18] and Ichimura and Kawana [34] independently studied oxidation mechanisms occurring in TiN and TiAlN coatings over the temperature range from 750°C to 900°C. Both of them showed that inward diffusion of oxygen and outward diffusion of Al are processes controlling the mechanism of oxidation within this temperature range. McIntyre et al. [18] confirmed this process using iridium marker experiments.

Before discussing the role of Al in improving the oxidation resistance of TiN, theoretical understanding of oxidation kinetics of nitride coatings is very important. This would facilitate for better understanding of this section.

Oxidation process of nitride coatings can be divided into two stages. At the initial stage of oxidation, the mass gain $w$ usually follows a linear rate law with time $t$:

$$w = k_1 t$$

where $k_1$ is the linear rate constant.

The rate-limiting reaction step of this stage is the surface reaction of oxygen or the diffusion through the gas phase and can be designated as *Regime 1* [9]. As soon as a dense oxide scale is formed on top of the nitride coating, the mass gain rate is retarded and controlled by the diffusion of the reactants through this oxide scale. This bulk/grain boundary diffusion is based on Fick's diffusion law, and here the mass-gain normally follows a parabolic rate law (*Regime 2*):

$$w = \sqrt{\left(k_2 t\right)}$$

where $k_2$ is the parabolic rate constant [9,35].

For very high-temperature oxidation (>850°C), often a paralinear behavior is observed, where a predominant parabolic Regime 2 is accompanied by a linear rate law:

$$w = \sqrt{\left(k_2 t\right)} + c k_1 t$$

with a reduced (by the prefactor $c$; usually much smaller than 1) linear rate constant. This is the case if the formed oxide scale suffers damage, for example, due to cracks [9].

Matsui et al. reported the oxidation rates of the TiAlN films together with those of Ti and TiN coatings [36]. It was assumed that the oxidation follows a diffusion-limiting process based on the parabolic law. He demonstrated that the oxidation rate of TiN is equivalent to that of Ti. The oxidation rate decreased as the aluminum ion substituted for the titanium ion with increasing $x$ from 0.1 to 0.5 in TiAlN.

It is important now to bring in the thermodynamics of diffusion process taking place in TiAlN coatings for improved understanding of oxidation kinetics. The oxidation rate of TiAlN during the diffusion-limiting process is given by

$$D(T) = D_0 \exp\left(-\frac{Ea}{kT}\right)$$

where:

$D_0$ is a pre-exponential factor
$Ea$ is the activation energy of oxidation
$k$ is Boltzmann's constant
$T$ is the temperature

The activation energies of the diffusion of Ti and O in $TiO_2$ and the diffusion of Al and O in $Al_2O_3$ are shown obtained from literature available in Table 5.1. From the $E_a$ values, it can be seen that $Al_2O_3$ has a better barrier to oxygen diffusion than $TiO_2$. Accordingly, the activation energy of oxidation of the $Ti_{1-x}Al_xN$ coatings increases with increasing aluminum content. This shows that the rate-determining process of oxidation changes from oxygen diffusion in $TiO_2$ to that in $Al_2O_3$. Moreover, the fast diffusion of aluminum in $TiO_2$ results in a possible segregation of aluminum at the coating surface. Higher energy of activation for O diffusion in alumina makes the oxidation parabolic in nature by making further inner diffusion of O difficult. Also, it was observed that the activation energy of diffusion of Al within the alumina layer was comparatively high retarding further oxidation of coating through Al outer diffusion.

This provides a basis to reduce oxidation and diffusion wear at the surfaces (*oxidation and diffusion wear* involves a chemical reaction/movement of ions with the surface induced by the tribological [frictional] contact and yielding products that have a detrimental effect on the tribological processes). Joshi et al. [35] also investigated the oxidation behavior of TiAlN over a wider temperature range (600°C–877°C) and highlighted the existence of two different oxidation mechanisms prevailing during oxidation of these coatings by focusing on oxidation chemistry. They demonstrated via Auger electron spectroscopy depth profiles that inward diffusion of oxygen controlled the oxidation at <700°C, whereas both inward diffusion of O and outward diffusion of Al controlled the oxidation process at ≥800°C.

### TABLE 5.1
### Activation Energies of Diffusion

| Process | E (kJ/mol) | Reference |
|---|---|---|
| O diffusion in $Al_2O_3$ | 460 | [37] |
| O diffusion in $TiO_2$ | 251 | [38] |
| Ti diffusion in $TiO_2$ | 257 | [39] |
| Al diffusion in $Al_2O_3$ | 477 | [40] |

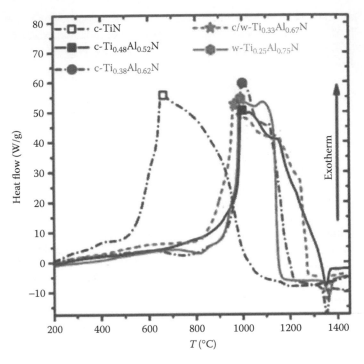

**FIGURE 5.7** DSC analysis in synthetic air of $Ti_{1-x}A_xN$ with Al contents of $x = 0, 0.52, 0.62,$ 0.67, and 0.75. (Data from Chen, L. et al., *Surf. Coat. Technol.*, 206, 2954–2960, 2012.)

Further, the structure of TiAlN coatings also influences the oxidation behavior of TiAlN coatings. Chen et al. [17] provides an illustration through dynamic differential scanning calorimetry (DSC) experiments of powdered $Ti_{1-x}Al_xN$ freestanding coating samples up to 1450°C in synthetic air as shown in Figure 5.7. All Al-containing coatings investigated were found to exhibit an onset of the pronounced exothermic peak due to oxidation at $\geq$800°C, hence, at least ~300°C above that for c-TiN. The DSC data also showed that the single-phase cubic coating with the highest Al content (c-$Ti_{0.38}Al_{0.62}N$) as well as the single-phase wurtzite coating (w-$Ti_{0.25}Al_{0.75}N$) exhibit the highest onset (~900°C) and peak temperature (~1000°C) for oxidation reaction. It could also be observed that the dual-phase (cubic and wurtzite) coating c/w-$Ti_{0.33}Al_{0.67}N$ showed a ~100°C lower onset temperature for the pronounced oxidation reaction compared to other Al-containing coatings investigated. The endothermic reaction at ~1350°C was independent of the chemical composition and the reason behind this was sintering processes of the powdered samples.

### 5.2.4 Tribological Properties of TiAlN Coatings

Before discussing tribological performance of TiAlN coatings, it is essential to define a tribological system. A tribological system consists of component surfaces that are in moving contact and thus become tribologically active (Figure 5.8). Friction and the resulting wear depend heavily on the composition and structure of the materials

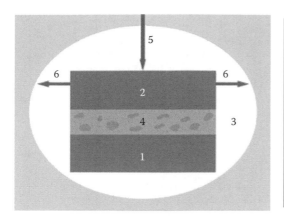

| | |
|---|---|
| | 1 Base object |
| | 2 Opposed body |
| | 3 Surrounding influences: |
| | Temperature |
| | Relative humidity |
| | Pressure |
| | 4 Intermediate materials: |
| | Oil |
| | Grease |
| | Water |
| | Particles |
| | Contaminants |
| | 5 Load |
| | 6 Motion |

**FIGURE 5.8**    Illustration of typical tribological system.

in the system. A further effect is often due to lubricants such as oil, grease, or water. Particles on the surfaces also affect wear. Other factors at work include the prevailing temperatures, the loads applied, and the loading modes such as sliding, rolling, oscillating, and pulsating. All these effects govern the tribological system and its behavior, and thus the extent, nature, and progress of wear. The fundamental connection is thus between friction and wear.

Oxidation resistance along with properties such as hot hardness, toughness, surface roughness, and thermal stability are important factors determining the tribological performance of a hard coating specifically at high temperatures (above 500°C) [41]. Ball-on-disc test is one of the methods used to measure the tribological properties of the coatings. Depending on the relative hardness, wear takes place either on the ball or on the coating surface. The wear resistance is expressed in terms of worn volume. The incorporation of Al into the fcc structure of TiN results in the formation of a metastable ternary solid solution TiAlN. This is well known to improve the cutting performance [42] caused by its high hot hardness due to age hardening [43] and excellent oxidation behavior up to temperatures of 700°C due to the formation of a protective Al-rich oxide layer at the coating surface [44].

Ball-on-disc experiments under dry sliding ambient conditions (room temperature) were performed by Vmcoille et al. [44] in order to compare the tribological behavior of different (Ti,X)N coatings. The coatings under consideration were TiN, TiAlN, TiNbN, and TiCN. Tribological properties were studied against corundum counterbody, as a function of coating composition and sliding speed. Experiments revealed that (Ti,X)N coatings wear via micro-abrasion caused by coating particles that are trapped in the contact. The nature of the wear debris accounted for the observed differences in coating wear. The steady state coefficient of frictions against corundum for TiN and (Ti,Al)N were in the range of 1–1.2 compared with a lower value of 0.1–0.2 for Ti(C,N). The coating wear volume (on the coating) as measured by profilometry was found to be markedly different for various types of coatings. In general, the wear decreased as the sliding speed increased after going through a maximum in the speed range of 0.05–0.10 m s$^{-1}$. At room temperature, the wear rate of (Ti,Al)N coatings increased with increasing

Al content in the coating. A coating containing 17% Al wore at a comparable rate with that of TiN, whereas coatings containing 50% Al wore 4–5 times more than TiN coatings. The wear of Ti(C,N) coatings was the lowest among others due to its low coefficient of friction.

Han et al. [45] performed ball-on-disc wear tests on TiAlN-coated M2 steels under lubricated conditions in the temperature range of 400°C–600°C. The coating persisted well during high-temperature wear up to 500°C under a load of 4.0 N. The major failure during wear cycling was observed to originate from local macroscopic arc droplets.

A further investigation on tribological performance of TiAlN coatings for various Al concentrations was made by Kutschej et al. [31] at room temperature, 500°C and 700°C using steel ball as a counterbody. As described in Section 5.2.1.1, increasing Al content leads to a change from a single-phase to a dual-phase structure. The single-phase fcc TiAlN coatings displayed lower friction coefficients compared to that of dual-phase and hcp films. In contrast to the friction coefficient, the negative wear rate of the dual-phase and hcp coatings with high Al contents were lower than that of coatings with a dominant fcc phase. This was explained through the sticking effects at room temperature between the coating and the ball being more pronounced at higher Al contents and a dual-phase TiAlN structure. It was speculated to be due to the formation of highly disturbed Al-rich column boundaries providing reactive elements which interfere with the steel ball inducing sticking effects.

In the work of Kutschej [31], $Ti_{0.33}Al_{0.67}N$ coating having a dominant fcc phase was considered as a reference coating to evaluate the effect of temperature on wear behavior of these coatings. At room temperature, the wear track was broad with abrasive wear and also some adhesive material as shown in Figure 5.9. At 500°C, the track width decreased and more transfer material was observed (Figure 5.9). Finally at 700°C, temperature caused the formation of a rather narrow wear track accompanied by pronounced transfer material. The small tracks were explained by the softening of the steel ball, which especially occurred at temperatures higher than 500°C. This was explained as follows: At the beginning of the experiment, when the ball comes in contact with the coating surface, plastic deformation of the ball takes place along with pronounced transfer from the ball to the film. The higher the temperature, the more material transfers at the beginning of the experiment could be observed. This protected the coating from further abrasion during the remaining experiment and the wear track itself remained narrow. This could possibly explain the lower friction coefficients because of sliding of the ball against transferred steel sticking on the coating surface, which was favored at higher temperatures. Also, oxidation of the transferred steel and the coating itself could have played a role in determining the tribological behavior at 700°C. Positive and negative wears also decreased with higher Al contents at 700°C due to the preferred formation of oxides that protect the coating surface against the counter body. The study demonstrated that low friction and high hardness are required and fcc TiAlN structure should be preferred because of the excellent mechanical properties, with an Al content as high as possible which favors good wear performance against stainless steel counterparts.

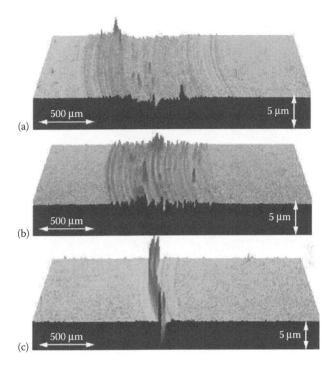

**FIGURE 5.9** Optical 3D images of wear tracks of the $Ti_{0.33}Al_{0.67}N$ coating after ball-on-disc tests at (a) 25°C, (b) 500°C, and (c) 700°C. (Data from Kutschej, K. et al., *Surf. Coat. Technol.*, 200, 2358–2365, 2005.)

## 5.3 INFLUENCE OF ADDITION OF ELEMENTS: EFFECT OF Ta ADDITION ON TiAlN COATINGS

One of the possible ways how to further tune the material properties is the concept of multicomponent alloying Cr, Y, and Ta to retard the decomposition process to higher thermal loads [4–11,15,46]. Tantalum is used as an example to illustrate the positive impact of an additional element to TiAlN coatings when added in adequate quantity in this chapter. Other elements such as Cr and Y show similar effect, but they are out of scope of this chapter. Ta addition to TiAlN enhances its hardness, thermal stability, and oxidation resistance further. TiAlTaN coatings have been deposited and investigated by both cathodic arc evaporation and magnetron sputtering techniques [4–11]. Computational studies have also been made on coating extensively due to its promising properties exhibited for cutting tool applications with its durability [46,47]. In Sections 5.3.1 and 5.3.2, key properties, namely, microstructure, oxidation, and wear behavior which are significantly varied TiAlN coatings once they are Ta alloyed, would be described.

### 5.3.1 MICROSTRUCTURAL EVALUATION

Microstructural studies made on the effect of Ta alloying into TiAlN hard coatings deposited using unbalanced magnetron sputtering [6]. Thin films were deposited

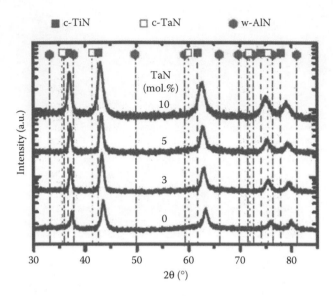

**FIGURE 5.10**   XRD pattern of the investigated as deposited single-phase cubic $(Ti_{1-x}Al_x)_{1-y}Ta_yN$ coatings, exhibiting a peak shift to lower diffraction angles with increasing TaN mole fractions [47].

on low-alloy steel substrates by unbalanced magnetron sputtering in a mixed Ar/ $N_2$ plasma discharge ($P_{N2}$ was 40% of the total working gas pressure of 0.5 Pa) from powder metallurgically produced Ti/Al/Ta compound targets (PLANSEE, 99.9% purity) of a constant Ti/Al ratio of 1:1 and increasing Ta content from 0 to 2, 4, and 8 at.%.

The diffraction patterns in Figure 5.10 demonstrated that all as deposited $(Ti_{1-x}Al_x)_{1-y}Ta_yN$ films crystallize in a single-phase cubic (B1) structure. An increasing mole fraction of TaN from 0% to 3% to 5% to 10% results in a shift in the diffraction peaks to lower angles, corresponding to an increase in lattice parameter from 4.148Å to 4.168Å to 4.176Å to 4.188Å. The obtained results were verified using *ab initio* calculations [47].

The structure was also studied as a function of substrate bias voltage by Pfeiler et al. [4] using cathodic arc evaporation. XRD studies on TiAlTaN demonstrated a change with increasing bias voltage from a dual-phase structure containing cubic and hexagonal phases to a single-phase cubic structure. It has been reported on improved mechanical and tribological properties of Ti–Al–Ta–N, which was connected to a promoting effect of Ta on fcc at the expense of the hcp) phases [5].

### 5.3.2   OXIDATION AND TRIBOLOGICAL BEHAVIOR

Pfeiler et al. [4] and Rachbauer et al. [6] independently demonstrated experimentally that AlTiTaN coatings have better oxidation resistance than TiAlN coatings within the temperature range of 800°C–950°C when deposited using cathodic arc

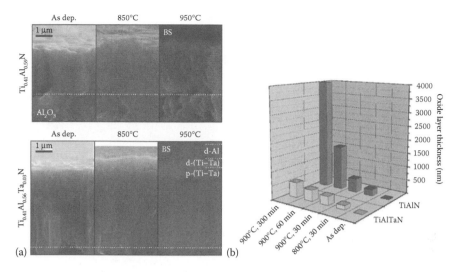

**FIGURE 5.11** (a) Fracture cross-sectional images of the $Ti_{0.41}Al_{0.59}N$ and $Ti_{0.41}Al_{0.56}Ta_{0.03}N$ films in the as-deposited (as dep.) state and after oxidation at 850°C (scanning electron microscope [SEM] images) and 950°C for 20 h. BS, backscatter electron SEM images. (Data from Rachbauer, R. et al., *Surf. Coat. Technol.*, 211, 98–103, 2012.) (b) Oxide layer thickness of TiAlN and Ti–Al–Ta–N after oxidation in ambient air at various oxidation parameters. (Data from Pfeiler, M. et al., *J. Vac. Sci. Technol. A*, 27, 554–560, 2009.)

evaporation and magnetron sputtering, respectively. Ta when added to TiAlN plays a significant role in improving the oxidation resistance of the coating. The cross-sectional images shown in Figure 5.11a and the oxide thickness evaluation after oxidation in various conditions illustrated by Pfeiler et al. in Figure 5.11b illustrate the same [4,6]. It has been suggested in the literature that the replacement of $Ti^{4+}$ with $Ta^{5+}$ in TiO2 formed at an elevated temperature (≥800°C) reduces the oxygen vacancies, which limits the inward diffusion of oxygen in the coating and provides better oxidation resistance. This in turn aids in decreasing the extent of further $TiO_2$ formation.

Recently, Khetan et al. have demonstrated the oxidation mechanism of AlTiTaN coatings by correlating structure and chemistry of oxides formed at different temperatures (700°C–900°C). An illustration of the oxidation mechanism in the form of a schematic demonstrating the different types of oxides resulting from air annealing of AlTiTaN coatings in the temperature range of 700°C–950°C is shown in Figure 5.12. Three kinds of oxides were observed depending on the air oxidation temperature. At 700°C, an amorphous AlTiTa oxide is observed. Within the temperature range of 750°C–850°C, an AlTiTa oxide with crystalline rutile grains is formed with an Al-rich top layer facilitated by outward diffusion of Al. At temperatures from 850°C to 950°C, the process of crystallization is fast and two different types of AlTiTa oxide layers can be observed (Al rich and TiTa rich) along with a pure alumina layer on top of the AlTiTa oxide layer [10].

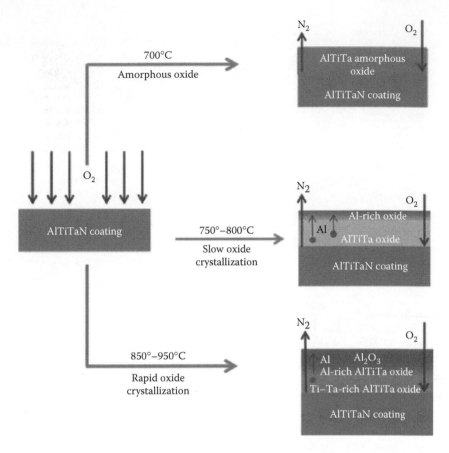

**FIGURE 5.12** Oxidation mechanism of AlTiTaN coatings when they are air annealed in the temperature range of 700°C–950°C. The evolution from the amorphous oxide layer to the crystalline oxide layer is represented by the dark to light contrast in the AlTiTa oxide layer. (Data from Khetan, V. et al., *ACS Appl. Mater. Interfaces*, 6, 4115–4125, 2014.)

The first study of tribological properties of TiAlTaN coatings was the study of the effect of bias voltage on tribological properties these coatings studied for temperatures up to 700°C [5]. It was observed that tribological behavior at room temperature is predominantly determined by the presence of moisture, which supports tribo-reactions that increase wear. In the absence of moisture at elevated temperatures, no loss of material was detected. It was observed that the wear performance of cathodic arc-evaporated AlTiTaN coatings was found to be similar to the wear resistance up to 700°C of TiAlN coatings. Wear investigation at 900°C revealed complete coating failure for TiAlN, whereas the TiAlTaN coating survived the temperature but was worn out completely in the wear track under dry sliding conditions. Therefore, the enhanced oxidation resistance did not provide the expected wear resistance of AlTiTaN coatings at temperatures above 700°C.

Nevertheless, a clear correlation between the structure of the oxide formed and the oxidation mechanism in TiAlN or TiAlTaN coatings has not been investigated.

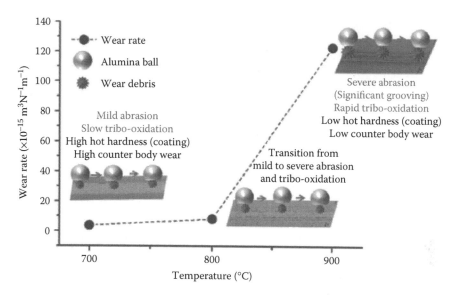

**FIGURE 5.13** Schematic representation of the different types of wear mechanisms observed on AlTiTaN coatings in the temperature range of 700°C–900°C. (Data from Khetan, V. et al., *ACS Appl. Mater. Interfaces*, 6, 15403–15411, 2014.)

Also, there is a need to understand the wear mechanisms of these TiAlTaN coatings and their correlation with oxidation resistance at temperatures above 700°C. Insight into the wear processes could aid later for the development of improved strategies of hard coatings concerning dry machining with working temperatures higher than 800°C. Establishing a correlation between the oxidation and wear behavior of these coatings can facilitate this objective.

Further Khetan et al. [11] investigated the wear mechanisms of AlTiTaN coatings at temperatures 700°C, 800°C, and 900°C to further understand the wear processes in these coatings (Figure 5.13). It was revealed that predominant abrasion and tribo-oxidation are the wear mechanisms for this type of coating, depending on the wear debris formed. At room temperature, it is abrasion leading to surface polishing. At 700°C and 800°C, slow tribo-oxidation and an amorphous oxide were observed, leading to reduce the wear rate compared to room temperature. At 900°C, significant increase in the wear rate was rationalized due to fast tribo-oxidation accompanied by grooving. This was the reason for failure of coating under tribo-contact at 900°C.

## 5.4 SUMMARY

Finally to summarize, with investigations discussed in this chapter, TiAlN-based coatings prove to be excellent candidate coatings whose properties can be tailored for a number of applications. By varying deposition parameters and chemistry illustrated in this chapter, we can tune their properties to our needs with limitations in term of high-temperature wear and oxidation resistance. Simultaneously, additional elements such as Ta further improve specific properties of TiAlN coatings such as oxidation resistance up to 900°C (improvement by 100°C). Investigations on wear behavior on these coatings are limited, but a scope to improve at temperatures above 850°C is still needed. Hence, it would be interesting to use the strategies discussed in this chapter to enhance both the wear and the oxidation resistance of the TiAlN coatings at high temperatures (>850°C). Tantalum was used as an example to illustrate the importance of addition of a new element to TiAlN coatings, but this is not a limit. Many other elements (Y, Cr, etc.) have been used to enhance the oxidation and the wear resistance of these coatings. Research in the direction of channeling the right permutation and combinations of these elements with adequate deposition parameters can lead to desired outcome explained in this chapter.

## REFERENCES

1. PalDey S.; Deevi S. C. Single layer and multilayer wear resistant coatings of (Ti,Al)N: A review. *Mater. Sci. Eng. A* **2003**, 342, 58–79.
2. Mayrhofer P. H.; Mitterer C.; Hultman L.; Clemens H. Microstructural design of hard coatings. *Prog. Mater. Sci.* **2006**, 51, 1032–1114.
3. Leyendecker T.; Lemmer O.; Esser S.; Ebberink J. The development of the PVD coating TiAlN as a commercial coating for cutting tools. *Surf. Coat. Technol.* **1991**, 48, 175–178.
4. Pfeiler M.; Scheu C.; Hutter H.; Schnöller J.; Michotte C.; Mitterer C.; Kathrein M. On the effect of Ta on improved oxidation resistance of Ti–Al–Ta–N coatings. *J. Vac. Sci. Technol. A*, **2009**, 27, 554–560.
5. Pfeiler M.; Fontalvo G. A.; Wagner J.; Kutschej K.; Penoy M.; Michotte C.; Mitterer C.; Kathrein M. Arc evaporation of Ti-Al-Ta-N coatings: The effect of bias voltage and Ta on high-temperature tribological properties. *Tribol. Lett.*, **2008**, 30, 91–97.
6. Rachbauer R.; Holec D.; Mayrhofer P. H. Increased thermal stability of TiAlN thin films by Ta alloying. *Surf. Coat. Technol.* **2012**, 211, 98–103.
7. Kathrein M.; Michotte C.; Polick P.; Coated tool, U.S. Patent 7,521,132 B2, April 21, **2009**.
8. Koller C. M.; Hollerweger R.; Sabitzer C.; Rachbauer R.; Kolozsvari S.; Paulitsch J.; Mayrhofer P. H. Thermal stability and oxidation resistance of arc evaporated TiAlN, TaAlN, TiAlTaN, and TiAlN/TaAlN coatings. *Surf. Coat. Technol.* **2014**, 259, 599–607.
9. Hollerweger R.; Reidl H.; Paulitsch J.; Arndt M.; Rachbauer R.; Polick P.; Primig S.; Mayrhofer P. H. Origin of high temperature oxidation resistance of Ti–Al–Ta–N coatings. *Surf. Coat. Technol.* **2014**, 257, 78–86.
10. Khetan V.; Valle N.; Duday D.; Michotte C.; Delplancke M. P.; Choquet P. Influence of temperature on oxidation mechanisms of fiber-textured AlTiTaN coatings. *ACS Appl. Mater. Interfaces* **2014**, 6, 4115–4125.
11. Khetan V.; Valle N.; Duday D.; Michotte C.; Mitterer C.; Delplancke M. P.; Choquet P. Temperature-dependent wear mechanisms for magnetron-sputtered AlTiTaN hard coatings. *ACS Appl. Mater. Interfaces* **2014**, 6, 15403–15411.
12. Cavaleiro A.; Trindade B.; Vieira M. T. The influence of the addition of a third element on the structure and mechanical properties of transition metal based nano-structured hard films: Part I—Nitrides. *Nanocomposite Thin Films.* J. M. Th. De Hosson and A. Cavaleiro (eds.). Kluwer, New York, **2006**, 261–314.
13. Chim Y. C.; Ding X. Z.; Zeng X. T.; Zhang S. Oxidation resistance of TiN, CrN, TiAlN and CrAlN coatings deposited by lateral rotating cathode arc. *Thin Solid Films* **2009**, 517, 4845–4849.
14. Kwasny W.; Mikuła J.; Dobrzański L. A. Fractal and multifractal characteristics of coatings deposited on pure oxide ceramics. *JAMME* **2006**, 17, 257–260.

15. Pfeiler M.; Zechner J.; Penoy M.; Michotte C.; Mitterer C.; Kathrein M. Improved oxidation resistance of TiAlN coatings by doping with Si or B. *Surf. Coat. Technol.* **2009**, 203, 3104–3110.
16. Makino Y.; Miyake S. Structural change and properties of pseudobinary nitrides containing AlN. *Trans. JWRI* **2001**, 30, 39–43.
17. Chen L.; Paulitsch J.; Du Y.; Mayrhofer P. H. Thermal stability and oxidation resistance of Ti–Al–N coatings. *Surf. Coat. Technol.* **2012**, 206, 2954–2960.
18. McIntyre D.; Greene J. E.; Hakansson G.; Sundgren J.-E.; Münz W.-D. Oxidation of metastable single phase polycrystalline Ti0.5Al0.5N films: Kinetics and mechanisms. *J. Appl. Phys.* **1990**, 67, 1542–1553.
19. Kimura A.; Hasegawa H.; Yamada K.; Suzuki T. Effects of Al content on hardness, lattice parameter and microstructure of $Ti_{1-x}Al_xN$ films. *Surf. Coat. Technol.* **1999**, 120–121, 438–441.
20. Hörling A.; Hultman L.; Odén M.; Sjölén J.; Karlsson L. Mechanical properties and machining performance of $Ti_{1-x}Al_xN$-coated cutting tools. *Surf. Coat. Technol.* **2005**, 191, 384–392.
21. Ohnuma H.; Nihira N.; Mitsuo A.; Toyoda K.; Kubota K.; Aizawa T. Effect of aluminum concentration on friction and wear properties of titanium aluminum nitride films. *Surf. Coat. Technol.* **2004**, 177–178, 623–626.
22. Devia D. M.; Restrepo-Parra E.; Arango P. J.; Tschiptschin A. P.; Velez J. M. TiAlN coatings deposited by triode magnetron sputtering varying the bias voltage. *Appl. Surf. Sci.* **2011**, 257, 6181–6185.
23. Wuhrer R.; Yeung W. Y. A study on the microstructure and property development of d.c. magnetron cosputtered ternary titanium aluminium nitride coatings: Part III—Effect of substrate bias voltage and temperature. *J. Mater. Sci.* **2002**, 37, 1993–2004.
24. Chen J. T.; Wang J.; Zhang F.; Zhang G. A.; Fan X. Y.; Wu Z. G.; Yan P. X. Characterization and temperature controlling property of TiAlN coatings deposited by reactive magnetron co-sputtering. *J. Alloys Compd.* **2009**, 472, 91–96.
25. Shetty A. R.; Karimi A.; Cantoni M. Effect of deposition angle on the structure and properties of pulsed DC magnetron sputtered TiAlN thin films. *Thin Solid Films* **2011**, 519, 4262–4270.
26. Suzuki T.; Huang D.; Ikuhara Y. Microstructures and grain boundaries of (Ti,Al)N films. *Surf. Coat. Technol.* **1998**, 107, 41–47.
27. Rauch J. Y.; Rousselot C.; Martin N. Structure and composition of $Ti_xAl_{1-x}N$ thin films sputter deposited using a composite metallic target. *Surf. Coat. Technol.* **2002**, 157, 138–143.
28. Lee S. H.; Lim J. W.; Lee D. K.; Han Y. H.; Lee J. J. A study on the characteristics of $(Ti_{1-x}Al_x)N$ coatings deposited by plasma-enhanced chemical vapour deposition after heat treatment. *Surf. Coat. Technol.* **2003**, 169, 371–374.
29. Zhou M.; Makino Y.; Nose M.; Nogi K. Phase transition and properties of Ti–Al–N thin films prepared by r.f.-plasma assisted magnetron sputtering. *Thin Solid Films* **1999**, 339, 203–208.
30. Derflinger V.; Brändle H.; Zimmermann H. New hard/lubricant coating for dry machining. *Surf. Coat. Technol.* **1999**, 113, 286–292.
31. Kutschej K.; Mayrhofer P. H.; Kathrein M.; Polcik P.; Tessadri R.; Mitterer C. Structure, mechanical and tribological properties of sputtered $Ti_{1-x}Al_xN$ coatings with $0.5{\leq}x{\leq}0.75$. *Surf. Coat. Technol.* **2005**, 200, 2358–2365.
32. Messier R.; Giri A.P.; Roy, R.A. Revised structure zone model for thin film physical structure. *J. Vac. Sci. Technol. A* **1984**, 2, 500–503.
33. Thornton J. A. High rate thick film growth. *Ann. Rev. Mater. Sci.* **1977**, 7, 239–260.
34. Ichimura, H.; Kawana, A. High-temperature oxidation of ion-plated TiN and TiAlN films. *J. Mater. Res.* **1993**, 8, 1093–1100.

35. Joshi, A.; Hu, H. S. Oxidation behavior of titanium-aluminium nitrides. *Surf. Coat. Technol.* **1995**, 76–77 (2), 499–507.

36. Matsui Y.; Hiratani M.; Nakamura Y.; Asano I.; Yano F. Formation and oxidation properties of ($Ti_{1-x}Al_x$) N thin films prepared by dc reactive sputtering. *J. Vac. Sci. Technol. A* **2002**, 20, 605–611.

37. Unnam J.; Shenoy R. N.; Clark R. K. Oxidation of commercial purity titanium. *Oxid. Met.* **1986**, 26, 231–252.

38. Oishi Y.; Kingery W. D. Oxygen diffusion in periclase crystals. *J. Chem. Phys.* **1960**, 33, 905–906.

39. Venkatu L.; Potet L. E. Diffusion of titanium of single crystal rutile. *Mater. Sci. Eng.* **1969/1970**, 5, 258–262.

40. Paladino A. E.; Kingery W. D. Aluminum ion diffusion in aluminum oxide. *J. Chem. Phys.* **1962**, 37, 957–962.

41. Findik F. Latest progress on tribological properties of industrial materials. *Mater. Des.* **2014**, 57, 218–244.

42. Knotek O.; Munz W.-D.; Leyendecker T. Industrial deposition of binary, ternary, and quaternary nitrides of titanium, zirconium, and aluminum. *J. Vac. Sci. Technol. A* **1987**, 5 (4), 2173.

43. Mayrhofer P. H.; Hfrling A.; Karlsson L.; Sjfle'n J.; Larsson T.; Mitterer C.; Hultman L. Self-organized nanostructures in the Ti–Al–N system. *Appl. Phys. Lett.* **2003**, 83 (10), 2049.

44. Vancoille E.; Celis J. P.; Roos J. R. Dry sliding wear of TiN based ternary PVD coatings. *Wear* **1993**, 165, 41–49.

45. Han J. G.; Yoon J. S.; Kim H. J.; Song K. High temperature wear resistance of (TiAl)N films synthesized by cathodic arc plasma deposition. *Surf. Coat. Technol.* **1996**, 86–87, 82–87.

46. Holec D.; Zhou L.; Rachbauer R.; Mayrhofer P. H. Alloying-related trends from first principles: An application to the Ti–Al–X–N system. *J. Appl. Phys.* **2013**, 113, 113510 (1–8).

47. Rachbauer R.; Holec D.; Mayrhofer P. H. Phase stability and decomposition products of Ti–Al–Ta–N thin films. *Appl. Phys. Lett.* **2010**, 97, 151901(1–3).

# 6 Functional Nanoceramics
## A Brief Review on Structure Property Evolutions of Advanced Functional Ceramics Processed Using Microwave and Conventional Techniques

*Santiranjan Shannigrahi and Mohit Sharma*

## CONTENTS

## 6.1 INTRODUCTION

On December 29, 1959, in an after-dinner speech at a conference in Pasadena, California, Richard Feynman famously quipped, "There's Plenty of Room at the Bottom." He thus presented a bold and enduring vision of a technological journey leading toward the atomic scale and toward the ultimate boundaries set by physical laws. The world has moved closer to what Feynman had visualized and the journey still continues. Scientists and engineers are primarily striving to create materials, gain fundamental insights into their properties, and use the nanoscale materials as components or building blocks to create novel structures or devices.[1,2] At nano length scales (1–100 nm), materials show unique properties and functions. However, in certain cases, the length scales for these novel properties may be under 1 nm (down to 0.1 nm for atomic and molecular manipulation) or over 100 nm (up to 300 nm in case of nanopolymers and nanocomposites).[3,4] Nanomaterials are also referred to as a convergent technology in which the boundaries separating discrete disciplines become blurred. Biochemists, materials scientists, electrical engineers, and molecular biologists may all be considered experts in the field if they are involved in the development of nanosized structures.[5,6]

Hence, the role of nanomaterials in modern technologies is becoming increasingly significant because of the feasibility and ease of adding new functions to the existing commercial products, apart from products made completely from nanomaterials through the bulk, which is relatively difficult. The expectations on nanomaterials are enormous, as their unique mechanical, optical, electrical, magnetic, thermal, and catalytical properties make them special ingredients for a number of applications. However, the market for nanotechnology over the years has not matched the initial hype, which was based on the expectation that the nanotechnology-based products will permeate through every industrial sector. Nevertheless, the market for nanotechnology products has grown significantly, especially in the consumer products area.[7,8]

Successful commercialization of nanomaterials is possible only when the material production and application development proceed in parallel with each other. Often, the material production is a challenging process, although it looks simple from the synthesis point of view. This is because the surface functionalities of nanoparticles have to be tailored keeping in mind applications of nanoparticles. For example, through surface modification, a number of properties of nanoparticles, such as dispersability in a suspension, compactability during subsequent consolidation, color in case of metal and semiconductor quantum dots, and compatibility with the matrix material in the case of nanocomposites, can be altered. Thus, in real practice, multiple synthesis techniques have to be evaluated for each application. Additional challenges in large-scale synthesis include consistency in product quality, cost of raw material and equipment, yield of the product, safety of the process, waste disposal, and environmental issues.[9,10]

When it comes to application development, many of the proven technologies based on nanomaterials are related to health products, including water filters, medical textiles, cosmetics, and drug delivery. Big companies have initiated programs on nanomaterials on their own or in collaboration with academic/R&D institutions. Many companies work on applications of nanomaterials for value addition to their products. Only a few institutes and industries are making efforts to develop scalable synthesis processes for mass production of nanomaterials. Although nanomaterial-based technology development is becoming very competitive, serious attention has to be paid to the safety and toxicology issues of nanomaterials during both synthesis and application development.[11,12]

In modern times, human life is heavily relying on various smart materials and systems, which can be considered as derivatives of nanomaterials and nanotechnology. A smart material is one that reacts to a change in its environment because of its intrinsic nature and not because of external electronics. The reaction may exhibit itself as a change in volume, a change in color, or a change in viscosity, and this may occur in response to a change in temperature, stress, electrical current, or magnetic field. The change is also completely reversible, and it usually occurs because the structure of the material (i.e., the way the atoms, molecules, or crystals in the materials are arranged) is changing. Therefore, the coating used on spectacle lenses to turn them into sunglasses on a sunny day is a smart material because it changes according to the level of ultraviolet (UV) light. Smart materials and smart structures are a new emerging materials system that combines contemporary materials science with information science. The smart system is composed of sensing, processing, actuating, feedback, self-diagnosing, and self-recovering subsystems. It uses the functional properties of advanced materials to achieve high performances with capabilities of recognition, discrimination, and adjustification in response to a change in its environment. Each component of this system must have functionality, and the entire system is integrated to perform a self-controlled smart action, similar to a living creature who can *think*, make judgment, and take actions. A smart system can be considered as a design philosophy that emphasizes predictivity, adaptivity, and receptivity. A smart system/structure is defined as a nonbiological physical structure with a definite purpose, means and imperative to achieve that purpose, and the pattern of functioning of a computer or turning machine.[13] Smart materials are a subset of the smart system, that is, smart structures at the microscopic or mesoscopic scale. Smart system is a nonbiological structure, which means that the system functions as a biological system too. Such a material will generally include at least one structural element, some means of sensing the environment and/or its own state, and some type of processing and adaptive control algorithm. The development of smart materials will undoubtedly be an essential task in many fields of science and technology, such as information science, microelectronics, computer science, medical treatment, life science, energy, transportation, and safety engineering and military technologies. Materials development in the future, therefore, should be directed toward creation of hyperfunctional materials that surpass even biological organs in some aspects. The current materials research is to develop various pathways that will lead the modern technology to the smart system.

Functional materials are considered as a group of smart materials that are distinctly different from structural materials. The physical and chemical properties of functional materials are sensitive to a variety of changes in the environment, such as temperature, pressure, electric field, magnetic field, optical wavelength, adsorbed gas molecules, and the pH value. These materials utilize their intrinsic properties to perform an intelligent action. Functional materials cover a broader range of materials than smart materials. Besides the materials belonging to the smart structure, any materials having functionality are attributed to functional materials, such as the ferroelectric $BaTiO_3$ (perovskite), the magnetic field sensor of $La_{1-x}Ca_xMnO_3$, high-frequency microwave (MW) applications of $Ni_{0.5}Zn_{0.5}Fe_2O_4$ (ferrite) surface acoustic wave sensor of $LiNbO_3$, liquid petroleum gas sensor of Pd-doped $SnO_2$, semiconductor light detectors (CdS, CdTe), high-temperature piezoelectric $Ta_2O_5$, fast-ion conductor $Y_2(Sn_yTi_{1-y})_2O_7$ (pyrochlore structure), the electric voltage-induced reversible coloring of $WO_3$, and high-temperature superconductors. Functional materials cover a wide range of organic and inorganic materials. This chapter focuses only on oxide functional materials. Preparations of complex oxides with functionality are a key challenge for materials development. Searching new routes to prepare materials and understanding the relationship between the structures and the properties are equally important. A key requirement in preparation of materials is to control the structural and compositional evolution for achieving superior properties. Nanocrystal engineered materials are a new trend of materials research, aiming to improve the performances of materials.

It is now obvious that in the majority of the functional materials, several different elements form a molecule, where the electronic structure of this cluster is very different from any of its original elemental configurations because of the transfer and/ or sharing of valence electrons among atoms. In general, only the valence electrons are most critical to bonding; the distribution and motion of valence electrons are usually described by the molecular orbitals. These valence states and molecular orbits are responsible for the functional properties of the molecule. The ligand field theory is designed to describe the molecular structure of an atom cluster. When different elements are combined to form a crystalline solid in which the atoms or atom groups (or molecules) are bonded together to form a three-dimensional structure with specified symmetries, the properties of the solid would depend on both the electronic structure of the atoms or atom groups and their spatial distribution. The molecular orbital theory and band structure theory are usually applied to elucidate the relationship between the structure and the properties. Based on the electron band structure, inorganic materials can be classified as conductors, semiconductors, and insulators. If a change is made in the crystal structure so that the band gap is reduced or eliminated, a transition from an insulator to a conductor is possible. Modification of a crystal structure can be performed by changing either the spatial distribution of atoms (such as bonding angles, bonding lengths, and symmetry of atom arrangement) or the chemical composition (such as from stoichiometry to nonstoichiometry). All these changes are referred to as structural evolution, which is closely related to the properties of the materials. Many functional properties of inorganic materials are determined by the elements with mixed valences in the structure unit, by which we mean that an element has two or more different valences while forming a compound. Valence mixture refers to a case in which several elements have different valences, but each one only has a single valence. In the periodic table of fundamental elements, 40 elements can form mixed valence compounds; transition d-block elements and lanthanide (Eu, Yb, Ce, Pr, Tb, etc.) are typical examples. Modern inorganic chemistry has shown that the oxidation state of any element can be modified under special conditions. Many oxide functional materials contain elements with mixed valences. This is a typical difference between functional materials and structural materials. The concept of mixed valence chemistry offers a pathway to design and synthesize new compounds with unique optical, electric, or magnetic properties. Research in functional materials in its broad sense always depends on the conception and synthesis of interesting novel mixed-valence compounds. The discovery of high-temperature superconductor compounds is a fascinating successful example of the mixed valence chemistry. We believe that exploring the possible structures of mixed valence compounds and their evolution behaviors may lead to many pathways to synthesize new functional materials.

## 6.2 SMART MATERIALS

Shape memory alloys (SMAs) are one of the most well-known types of smart material, and they have found extensive uses. A shape memory transformation was first observed in 1932 in an alloy made from gold and cadmium, and then later in brass in 1938. In 1962, an alloy of half titanium and half nickel was found to exhibit a significant shape memory effect, and nitinol (so named because it is made from nickel and titanium and its properties were discovered at the Naval Ordnance Laboratory, White Oak, Maryland) has become the most commonly used smart metal. Other SMAs include those based on copper (in particular CuZnAl), NiAl, and FeMnSi, but it should be noted that nitinol has by far the most superior properties. By changing how nitinol is processed, it can be trained or programmed to have one of three different properties. In the two-way shape memory effect, the material transforms between two different structures, one above and one below its transformation or memory temperature. At the memory temperature, the crystal structure of the material changes, resulting in a volume or shape change. Nitinol can be made with a transformation temperature anywhere between −100°C and +100°C, which makes it very versatile. Two-way shape memory metals can be used in a wide range of applications, many of which replace the traditional bimetallic strip. Coils of shape memory metal wire can be used as switches in temperature-controlled circuits, such as the switch that turns off a boiling kettle or a central heating thermostat. They can also be incorporated into window hinge systems to open and close them at a particular temperature.

*Chromic* materials change color with a change in one aspect of their environment, and there are many types.

*Electrochromic* materials literally change from transparent to opaque at the flick of a switch. Applying an electrical field to these materials causes a change in the structure and thus a change in color. *Photochromic* materials change color with a change in the level of UV light and have been widely used as coatings on spectacle lenses. Probably the most well-known color-changing materials are the *thermochromic* polymers, which change with variations in temperature. There are two types of thermochromic systems: those based on liquid crystals and those that rely on molecular rearrangement. In both cases, a change in the structure of the material occurs at a particular temperature giving rise to an apparent change in color. The change is reversible, so as the material cools down it changes color back to its original state. In liquid crystals, the change from colored to transparent takes place over a small temperature range (around 1°C) and arises as the crystals in the material change their orientation. Thermochromic materials have found a number of applications such as color-changing toothbrushes, baby spoons, which indicate whether food is too hot, and even kettles, which change color as the water is heated. The pigments can be incorporated into dyes for fabric to produce clothing that changes color with variations in temperature. Thermochromic inks can also be used for printing onto clothing and food packaging.

*Light-emitting polymers* are a relatively new group of materials that emit light when a voltage is applied. They are often called *organic light-emitting diodes* and the scope for their use is huge. Applications range from flexible lights for safety applications to roll-up televisions.

*Shape memory polymers* exhibit the same sort of behavior as their metal counterparts, in that they will show a shape change with a change in temperature. Uses of these materials include couplings and linings for pipes.

As with any other type of composite, *smart composites* are made by mixing different materials together. It can be ceramic polymer, ceramic metal, metal polymer, and so on. A new addition is quantum tunneling composite, which is made by adding very fine nickel powder to a polymer resin, and it is interesting because the electrical resistance of the material decreases as pressure is applied to it in an almost linear relationship. The change in resistance is not due to improved conductivity as the nickel particles are pushed into contact with each other; rather, it arises because of a quantum tunneling effect. Once the pressure is released, the resistance increases again. Although this is a very new material, there is a lot of scope for its use. Suggested applications include its use as a variable resistor, a switch for power tools, a switch for lighting, which can be laid under a carpet, an indicator in noncontact sports, and even as a pressure sensor in the fingertips of robotic arms and prosthetic limbs!

## 6.3 SMART FLUIDS

There are a number of different types of smart fluids, but in each case, the viscosity of the material changes in response to the specific stimulus. One of the most well-known smart fluids is the toy Silly Putty, which is a type of non-Newtonian fluid. The viscosity of Silly Putty is dependent on the rate at which it is deformed; the faster it is deformed, the more viscous it becomes. Silly Putty is actually a type of silicone compound called polyborosiloxane, which consists of long-chain molecules with bulky side groups. When the material is deformed slowly, the structure can flow, but when deformed rapidly, the structure locks together, and the material can become very brittle! Other types of smart fluids rely on a suspension of very fine, micron-sized particles in a carrier liquid such as glycerol or mineral oil. In an electro-rheological fluid, the viscosity increases in the presence of an electrical field. In a magneto-rheological fluid, the viscosity changes in the presence of a magnetic field. In both cases, the smart fluid changes from a liquid to a solid with the application of the relevant field. The small particles in the fluid align and are attracted to each other, resulting in a dramatic change in viscosity. The effect takes milliseconds to occur and is completely reversible by the removal of the field.

## 6.4 SMART CERAMICS

The piezoelectric effect was discovered in 1880 by Jacques and Pierre Curie, who conducted a number of experiments using quartz crystals. This probably makes piezoelectric materials the oldest type of smart material. These materials, which are mainly ceramics, have since found a number of uses. The piezoelectric effect and electrostriction are opposite phenomena, and in both, the shape change is associated with a change in the crystal structure of the material. Piezoelectric materials exhibit two crystalline forms: one form is ordered and relates to the polarization of the molecules and the other is a nonpolarized, disordered state. If a voltage is applied to the nonpolarized material, a shape change occurs, as the molecules reorganize to align in the electrical field. This is known as *electrostriction*. Conversely, an electrical field is generated if a mechanical force is applied to the material to change its shape. This is the piezoelectric effect. Although quartz has been in use for a very long period, the most commonly used piezoelectric ceramic today is lead zirconium titanate (PZT). The physical properties of PZT can be controlled by changing its chemistry and processing scheme. There are limitations associated with PZT; like all ceramics, it is brittle, giving rise to mechanical durability issues, and there are also problems associated with joining it with other components in a system. These materials have found a number of uses, including airbag actuators, earthquake detection systems, tiny linear motors, and damping systems. One exciting use of these materials is in flat panel speakers in which a voltage is used to make the material vibrate and produce sound waves. However, a majority of the lead-based ceramics contain more than 60% of lead. Because of the toxicity of lead, global trend is shifting toward the adoption of lead-free functional ceramics.

The functional materials are described from the mixed valences and stoichiometry points of view to understand the structural evolution and transformation of different materials systems. The mixed valence compounds are elucidated as the fundamental for performing unique functionality. There are numerous books describing the properties, preparations, electronic structures, and crystal structures of transition and rare earth metals and their oxides. One of the aims is to explore new routes for synthesizing functional materials from the fundamental structure building blocks.

Conventionally, there are two technical challenges in nanoceramics processing: high-temperature annealing and longer holding time. This stage is known as *sintering*. Nanoceramic sintering techniques are divided into four categories: pressureless sintering, pressure sintering, electrically assisted sintering, and other sintering-related techniques. Pressureless sintering has mainly evolved around modifying sintering schedules, improving nanoparticle-packing characteristics, and using additives to tailor the diffusion rates. Pressure sintering, which includes hot pressing, hot isostatic pressing, and sinter forging, can effectively achieve full densification for nanostructured ceramics, but microstructural inhomogeneity and sintered shape limitation are difficult to overcome. For electrically assisted sintering, many nanoceramics have been sintered to full density with spark plasma, even though the atomic diffusion process is not well understood; MW sintering can achieve fast heating and has ability to produce control grain growth.

*MW processing.* The application of MW energy to the processing of various materials, such as ceramics, metals, and composites, offers several advantages over conventional (CV) heating methods. Figure 6.1(a) and (b) shows the CV and MW furnace used for ceramic sintering. In CV furnace, the heating elements mostly used are silicon carbide (SiC) or molybdenum disilicide ($MoSi_2$) or superkanthal. During CV heating, materials get heated from outside to inside. This heat transfer requires time and depends on the materials. Moreover, heat treatment of thin films using CV furnace often results in delamination and cracks due to thermal mismatch between the substrate and the thin films materials. Moreover, properties of CV heat treated materials vary from furnace to furnace, which mostly arise due to the inhomogeneous thermal zone. In MW heating, the materials get heated from inside to outside. MWs are electromagnetic waves with wavelengths 1 mm to 1 m and corresponding frequencies between 300 MHz and 300 GHz. The most common frequency of 2.45 GHz is used for MW heating. These frequencies are chosen

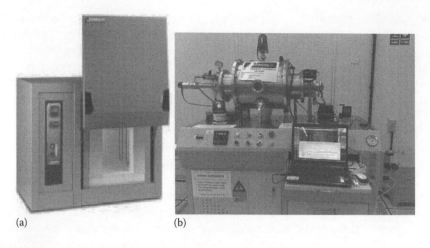

(a)                                    (b)

**FIGURE 6.1**    (a) CV furnace and (b) MW furnace.

for the MW heating based on two reasons: The first is that they are in one of the industrial, scientific, and medical (ISM) radio bands set aside for noncommunication purposes, and the second is that the penetration depth of the MWs is greater for these low frequencies. However, heating is not necessarily increased with decreasing frequency, as the internal field ($E$) can be low depending on the properties of the material. A frequency of 2.45 GHz is mostly used for household MW ovens, and 0.915 GHz is preferred for industrial/commercial MW ovens. MW furnaces with variable frequencies from 0.9 to 18 GHz have been developed for materials processing. MWs are coherent and polarized and can be transmitted, absorbed, or reflected depending on the material type.[14] MW heating is fundamentally different from the CV one in which thermal energy is delivered to the surface of the material by radiant and/or convection heating that is transferred to the bulk of the material via conduction. In contrast, MW energy is delivered directly to the material through molecular interaction with the electromagnetic field. MW heating is the transfer of electromagnetic energy to thermal energy and is energy conversion rather than heat transfer. Since MWs can penetrate the material and supply energy, heat can be generated throughout the volume of the material resulting in volumetric heating. Hence, it is possible to achieve rapid and uniform heating of thick materials. Therefore, the thermal gradient in the MW-processed material is the reverse of that in the material processed by CV heating. In CV heating, slow heating rates are selected to reduce steep thermal gradient leading to process-induced stresses. Thus, there is a balance between processing time and product quality. During MW processing, the potential exists to reduce processing time and enhance product quality, as MWs can transfer energy throughout the whole volume of the material. In this case, energy transfer occurring at a molecular level can have some additional advantages. When MW energy is in contact with materials having different dielectric properties, it will selectively couple with the higher loss tangent material. Therefore, MWs can be used for the selective heating of the materials.[15] One of the objectives of this chapter is to present a comparative analysis of both CV and MW processing of different materials. MW furnace consists of three major components: the source, the transmission lines, and the applicator, as shown in Figure 6.1(b). The source generates electromagnetic radiation, and the transmission lines deliver the electromagnetic energy from the source to the applicator, while the energy is either absorbed or reflected by the material in the applicator. Generally, the processing is performed within a metallic applicator (e.g., single-mode applicator, traveling wave applicator, and multimode applicator). The type of applicator used depends on the materials to be processed. The single-mode applicator and the traveling wave applicator are successful in processing materials of simple geometries. However, the multimode applicator has the capability to produce large and complex components. Therefore, multimode systems are used for industrial applications.[15]

*MW–material interaction.* MW energy is transferred to the material by interaction of the electromagnetic field at the molecular level. The dielectric properties determine the effect of the electromagnetic field on the material.[15] The interaction of MWs with a dielectric material results in translational motions of free or bound charges and rotation of the dipoles. The resistance of these induced motions due to inertial, elastic, and frictional forces causes loss resulting in volumetric heating.[14] The power absorbed per unit volume, $P$ (W/m$^3$), is expressed as[14]

$$P = \sigma |E|^2 = 2\pi f \varepsilon_0 \varepsilon'_r \tan \delta |E|^2 \tag{6.1}$$

where:
$E$ is the magnitude of the internal field (V/m)
$\sigma$ is the total effective conductivity (S/m)
$f$ is the frequency (GHz)
$\varepsilon_0$ is the permittivity of free space ($\varepsilon_0 = 8.86 \times 10^{-12}$ F/m)
$\varepsilon'_r$ is the relative dielectric constant
$\tan \delta$ is the loss tangent

Equation 6.1 demonstrates that the power absorbed varies linearly with the frequency, the relative dielectric constant, the loss tangent, and the square of the electric field. The penetration depth of MWs ($D$) at which the incident power is reduced by one half is expressed as[14]

$$D = \frac{3\lambda_0}{8.686\,\pi \tan \delta \left(\varepsilon'_r/\varepsilon_0\right)^{1/2}} \tag{6.2}$$

where $\lambda_0$ is the incident or free-space wavelength.

The relative dielectric constant and the loss tangent are the parameters that describe the behavior of a dielectric material under the influence of the MW field. During heating, the relative dielectric constant and the loss tangent change with temperature. One of the drawbacks is the thermal runaway, that is, when ceramics are processed in nonuniform electromagnetic fields, the local temperature varies within the material. If a local area reaches the critical temperature before the rest of the material, it begins to get heated more rapidly. Thus, the temperature of that local region begins to increase even more resulting in localized thermal runaway, leading to high enough stresses and material fracture.[15] Nevertheless, MW can still be a very useful and economic sintering process to achieve advanced ceramic.

### 6.4.1 MW-Processed Advanced Ceramics

In general, ceramics processing involves mostly two stages: calcination and sintering. Calcination is a comparatively low-temperature heat treatment process, where the unwanted carbon-based compounds are generally removed from the mixed component powders, and a nucleated form of ceramic develops. After the calcination, the nucleated ceramic powders are required to be consolidated into a desired shape and heat-treated at an elevated temperature, which is termed as sintering. At this stage, the ceramics generate a grain growth and densify. In functional ceramics, the functionality is mostly dipole centric; it may be the application of supercapacitor, pyroelectric sensor, ferroelectric random access memory, and so on. Therefore, the density closer to the theoretical value is more desired. However, at present, achieving high-density ceramics involves many challenges. Many different properties of ceramics depend on the final grain size of the ceramic body. In this regard, $BaTiO_3$ (BT) is being chosen, as it has several important functional properties, in addition to its lead-free nature. BT is stable under high temperatures in various applications capacitors. Especially, multilayer ceramic capacitors based on BT are one of the most important electronic components in surface-mounted electronic circuits.[14,16–18] Similarly, $NiZnFe_2O_4$ (NZF) is also one of the technologically important ferrites and is found very useful in high-frequency applications, as well as its MW absorption properties,[19] because of its high resistivity and low eddy current losses.[20–23] Moreover, BT and NZF are being studied as the multiferroic ceramic composites.[24,25] Although BT and NZF ceramics are well studied, they are still a topic of active research in order to further enhance or optimize their properties. Most of the useful properties of BT and NZF ceramics are governed by their grain size, so control of homogeneous grain growth at reduced temperatures will be quite useful for the scientific community and the industry.

This work attempts to compare the grain growth of commercial BT and NZF powders heat-treated by CV and MW techniques in the temperature range of 850°C–1000°C and time 0.5–2 h. We have also tested the effect of LiF as flux and its effect on the microstructures. Preparation of BT and NZF ceramics is carried out using MW technique simply by using the mixed oxides of the raw powders and consolidating them using cold isostatic pressure of 4, 6, and 8 tons followed by heating using MW at 1000°C for 15 min. Figure 6.2 depicts the crystalline quality of the MW-sintered ceramics.

*The heat-treated samples for their surface morphologies.* Figure 6.3 shows the surface morphologies of the various sintered BT samples. It is observed that the grains are coalesced more and increase with increasing pressure, which is

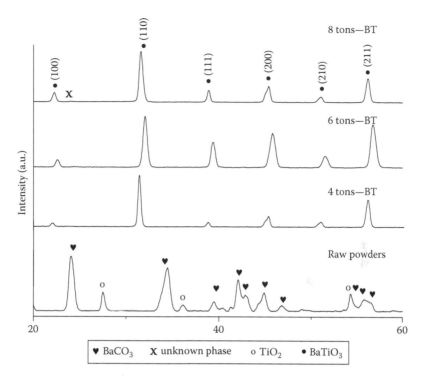

FIGURE 6.2   *XRD* patterns of BT from milled raw powders MW sintered at 1000°C for 15 min.

(a)                                                 (b)                                                 (c)

FIGURE 6.3   Surface morphologies of the BT ceramics prepared using different compressive forces (a) 4 tons, (b) 6 tons, and (c) 8 tons.

remarkable; of course, the threshold pressure needs to be established to produce highly dense ceramics, which may vary across materials systems. It is interesting to note that when a higher pressure is applied, big clusters of grains are surrounded by small nanosized particles. It is observed that the small nanosized particles coalesce to become big clusters.

The shrinkage of the sintered disc samples decreased with pressure. Table 6.1 depicts the linear shrinkage of the BT ceramic heat-treated using MW at 920°C for 10 min.

**TABLE 6.1**
**Linear Shrinkage of the BT Ceramics**

| Cold Consolidation Pressure (tons) | Dimensions (mm) | | | |
|---|---|---|---|---|
| | Diameter | | | |
| | Before Sintering ($D_b$) | After Sintering ($D_a$) | Linear Shrinkage (%) $D_b - D_a \times 100/D_b$ | Thickness |
| 4 | 10.23 | 8.4 | 17.88 | 1.79 |
| 6 | 10.24 | 8.49 | 17.09 | 1.55 |
| 8 | 10.24 | 8.66 | 15.43 | 1.6 |

Realizing the impact of the MW sintering experiments for BT ceramics, further study has been conducted to understand the mechanism of consolidation of the nanoceramic powders specifically for the BT and NZF ceramics. A thorough study of grain structure and electrical and magnetic properties of BT and NZF ceramics sintered using MW and CV techniques as the grain grows in polycrystalline ceramics during sintering process of powders is extremely important. In this aspect, *starting source powders* were taken as the nanoform of the individual ceramics, which were consolidated followed by heat-treating using various conditions. The crystalline quality of the sintered pellets was analyzed by XRD analysis and scanning electron microscope (SEM). An atomic force microscope (AFM) was used to statistically estimate the porosity in the sintered ceramics through surface topography analysis. A thin layer of silver paste was applied to the polished surfaces of the sintered BT and NZF samples for measuring their electrical properties. An impedance analyzer (HP, 4194A) and a ferroelectric testing system (*Radiant Technology, Inc.*, RT66A) were used to measure dielectric and ferroelectric properties, respectively. The magnetic properties of the NZF samples were measured using the physical property measurement system (PPMS; Quantum Design, San Diego, California).

The crystalline quality of all the BT and NZF ceramics sintered using MW and CV processes in the temperature range of 850°C–1000°C is analyzed. For the purposes of simplification and close comparison, the results are presented for the samples sintered at the temperature between 900°C and 1000°C for 2 h using CV process and 0.5 h using MW process, respectively, as these two types of recepies produced most reasonable results. Results for the specimens sintered at other temperatures are presented where applicable. Figures 6.4 and 6.5 depict the XRD spectra of the BT- and NZF-sintered specimens. No evidence of unwanted phase formation has been found, which indicates that there is no loss of stoichiometry in these ceramics, that is, they *remain phase pure single phase*.[14,26] However, sharper peaks observed for MW-sintered samples indicate bigger grain sizes compared to those observed for CV-sintered BT samples. All of the characterized peaks are marked with the standard Joint Committee on Powder Diffraction Standards card nos. 05-0626 and 019-0629, respectively, for BT and NZF ceramic specimens.

Figure 6.6 shows the shrinkage of the NZF ceramics disc samples before and after sintering. The average shrinkage for BT ceramics sintered using CV and MW methods is in the range of 5%–7%. Noticeable shrinkage of 9%–11% was observed

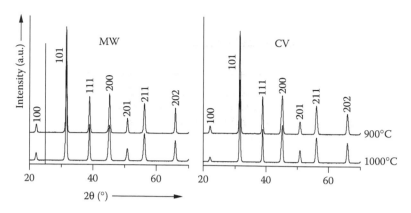

**FIGURE 6.4** XRD patterns of CV- and MW-sintered BT ceramics. (Data from Shannigrahi S.R. and Tan C.K.I., *Technologies*, 3, 47–57, 2015.)

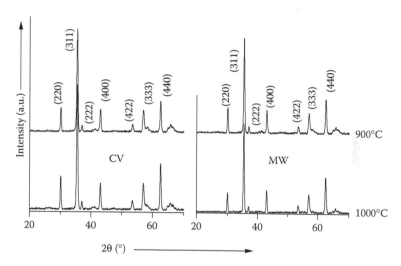

**FIGURE 6.5** XRD patterns of CV- and MW-sintered NZF ceramics. (Data from Shannigrahi S.R. and Tan C.K.I., *Technologies*, 3, 47–57, 2015.)

for NZF ceramics using CV processing and 12%–15% for MW-sintered samples.[27,28] The maximum shrinkage of 15.45% was observed for NZF ceramics sintered at 950°C using MW.

Microstructures of sintered BT and BT + LiF are shown in Figures 6.7 and 6.8. No considerable grain growth is observed up to a sintering temperature of 1000°C for pure BT. The pure BT ceramics could still be in the initial sintering stage; henceforth, the grain sizes are still small. However, the average grain size of the BT + LiF samples sintered using MW at 900°C for 0.5 h is ~3.1 μm, which is approximately twice larger than the BT + LiF samples sintered using CV at 900°C for 2 h, with an average grain

**FIGURE 6.6** Sintered (left) and pre-sintered (right) samples. (From Shannigrahi S.R. and Tan C.K.I., *Technologies*, 3, 47–57, 2015.)

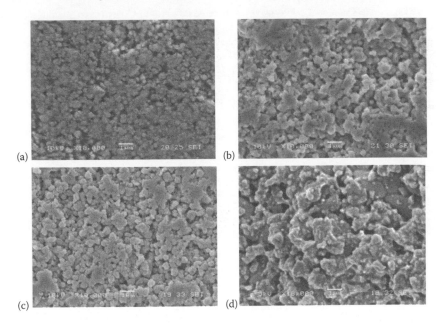

**FIGURE 6.7** SEM morphologies of (a) BT-CV-900°C for 2 h, (b) BT-CV-1000°C for 2 h, (c) BT-MW-900°C for 0.5 h, and (d) BT-MW-1000°C for 0.5 h. (From Shannigrahi S.R. and Tan C.K.I., *Technologies*, 3, 47–57, 2015.)

size of ~1.7 μm. It can be observed that there is a trend of increased grain growth size as the MW sintering temperature increases from 850°C to 1000°C. The average grain size values of MW-sintered samples are also approximately twice larger than those of CV-sintered samples at the same temperature. Grain sizes of MW-sintered NZF at 900°C are nearly twice those of CV-sintered samples (Figure 6.9). As seen from the results, the MW-sintered samples reach the intermediate or final stage of the sintering process much earlier under the impact of the MW field compared to CV-sintered samples. The highest average grain size of ~1.2 μm is measured for MW-sintered samples at 1000°C, and the lowest grain size of 1 μm is measured for CV-sintered samples at 900°C. These results can be backed up from the XRD scans as stated earlier.

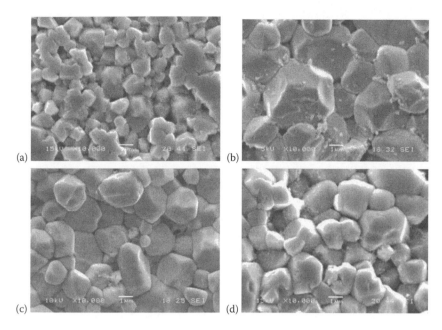

**FIGURE 6.8** Surface morphologies of (a) BT + LiF-CV-900°C for 2 h, (b) BT + LiF-CV-1000°C for 2 h, (c) BT + LiF-MW-900°C for 0.5 h, and (d) BT-LiF-MW-1000°C for 0.5 h. (From Shannigrahi S.R. and Tan C.K.I., *Technologies*, 3, 47–57, 2015.)

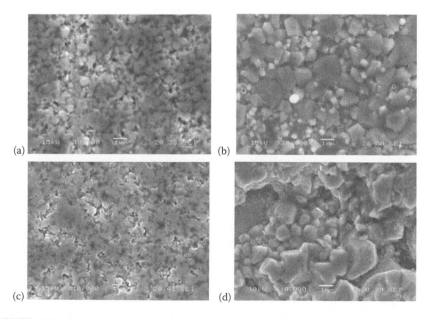

**FIGURE 6.9** Surface morphologies of (a) NZF-CV-900°C for 2 h, (b) NZF-CV-1000°C for 2 h, (c) NZF-MW-900°C for 0.5 h, and (d) NZF-MW-1000°C for 0.5 h. (From Shannigrahi S.R. and Tan C.K.I., *Technologies*, 3, 47–57, 2015.)

Sintering of crystalline materials occurred through several mechanisms such as atomic transport, vapor transport (evaporation/condensation), surface diffusion, lattice (volume) diffusion, grain boundary diffusion, and dislocation motion. Figure 6.10 shows a schematic representation of sintering mechanisms for four particles. During the CV heating process, vapor transport, surface diffusion, and lattice diffusion from the particle surfaces to the neck lead to neck growth and coarsening of the particles without densification. This densification mechanism leads through grain boundary diffusion and lattice diffusion from the grain boundary to the neck.[29] During MW heating, the specimen gets heated from inside to outside just opposite to CV heating. Moreover, the MW absorption mainly depends upon the complex permittivity and permeability of the specimens. As both the specimens already developed a dipole, they responded well to MW and resulted in good densification of the samples at a short sintering period.

The variations of dielectric constant ($\varepsilon$) and dielectric loss (tan $\delta$) for both MW-sintered and CV-sintered BT and NZF ceramics samples are shown in Figures 6.11 and 6.12. For BT, it can be seen that the CV-sintered pellets show the highest dielectric constant at 100 kHz. For the MW-sintered BT + LiF at 1000°C, the value of $\varepsilon$ is 1240, whereas at the same temperature for the CV-sintered sample the value of $\varepsilon$ 800.[30–32] For all MW-sintered samples, it can be observed that the values of $\varepsilon$ are higher compared to the CV-sintered samples; this may be attributed to the higher density in the MW-sintered specimens. The value of $\varepsilon$ for all samples decreases slowly and gradually as the frequency increases from 100 to 10,000 kHz. For NZF, the MW-sintered pellets show the highest value of $\varepsilon$ at 100 kHz. A value of 25 was obtained for NZF MW-sintered sample at 950°C. It is evident that as the sintering temperature for CV methods increases from 900°C to 1000°C, the dielectric constant decreases. As NZF is a magnetic material, the dielectric properties should be poor as shown in the results.

The room-temperature *P–E hysteresis* loops of the BT samples sintered under various conditions are presented in Figure 6.13. It can be seen that the polarization of MW-sintered BT samples at 1000°C is the highest among all the studied specimens as recorded, although all the samples show poor ferroelectric properties. It can be observed that samples sintered at low temperatures end up with a leaky nature. BT + LiF samples also exhibit a leaky nature even when sintered at 900°C, which is attributed to the leakage from lithium itself and/or porosity.

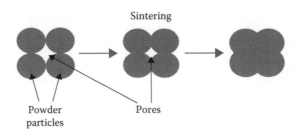

**FIGURE 6.10** Schematic representation of sintering mechanisms for a system of four particles. (From Shannigrahi S.R. and Tan C.K.I., *Technologies*, 3, 47–57, 2015.)

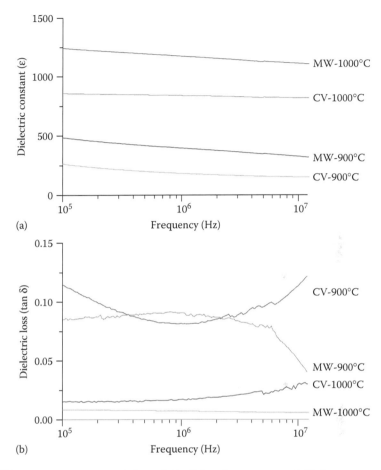

**FIGURE 6.11** Frequency variation of (a) dielectric constant and (b) loss tangent of BT ceramics sintered using CV and MW techniques for 2 h and 0.5 h, respectively. (From Shannigrahi S.R. and Tan C.K.I., *Technologies*, 3, 47–57, 2015.)

Figure 6.14 depicts the room-temperature M–H loops of NZF ceramics sintered using CV and MW techniques. It can be observed that in CV-sintered ceramics, the saturation magnetization ($M_s$) reaches its highest value of 68 emu/g at a sintering temperature of 1000°C, whereas in MW-sintered ceramics, the highest value of $M_s$ observed is 88 emu/g at a sintering temperature of 950°C. Additionally, the $M_s$ values for the MW-sintered samples strongly correlate with the sintering temperature. This higher $M_s$ value can be due to the 2.45 GHz MW field interacting with charged cations, thus causing a change in $Zn^{2+}$ and $Fe^{3+}$ arrangements, which are critical in the alteration of dipole moments.[14,19]

Using AFM topography, we have estimated the statistical value of porosity in the sintered ceramics. Figure 6.15a–c depicts the AFM technique, which includes AFM raw topography, flattened topography (undulation removed) followed by thresholding. The percentage of voids is reflected as the percentage of black pixels. Based on

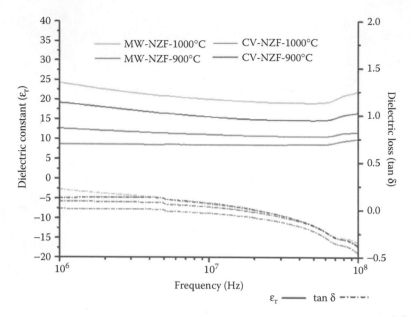

**FIGURE 6.12** Frequency variation of dielectric constant and loss tangent of NZF ceramics sintered using CV and MW techniques for 2 h and 0.5 h, respectively. (From Shannigrahi S.R. and Tan C.K.I., *Technologies*, 3, 47–57, 2015.)

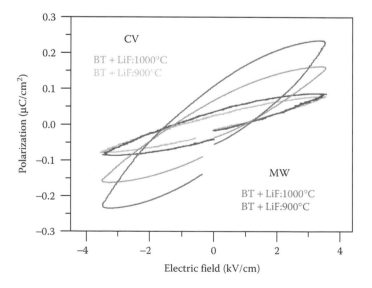

**FIGURE 6.13** Room-temperature P–E hysteresis loops of BT ceramics sintered using CV and MW techniques. (From Shannigrahi S.R. and Tan C.K.I., *Technologies*, 3, 47–57, 2015.)

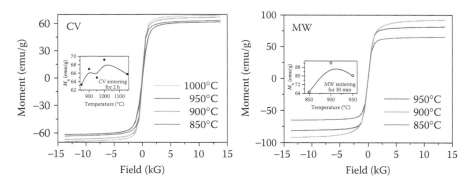

**FIGURE 6.14**   Room-temperature M–H loops of NZF ceramics sintered using CV and MW techniques. (From Shannigrahi S.R. and Tan C.K.I., *Technologies*, 3, 47–57, 2015.)

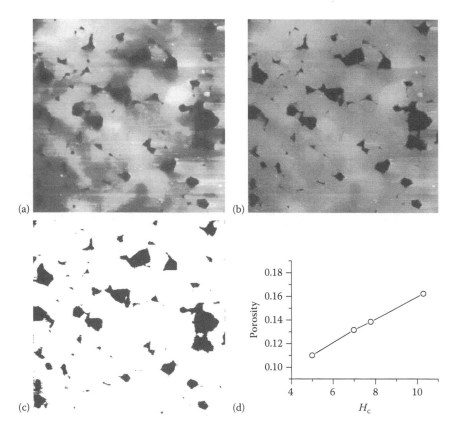

**FIGURE 6.15**   (a) Raw AFM topography; (b) flattened topography (undulations removed); (c) thresholding; and (d) porosity versus magnetic coercive field. (From Shannigrahi S.R. and Tan C.K.I., *Technologies*, 3, 47–57, 2015.)

the estimated statistical data for NZF-sintered ceramics, we have derived the relationship between the porosity and the magnetic coercive field as shown in Figure 6.15d. Coercive force is probably the property most sensitive to porosity and grain size. The increase in coercive force with porosity is linear as expected. Again, this effect may be caused because the high-porosity samples contain smaller particles, which have higher coercive force.

Besides the BT and NZF ceramics, another technologically important materials system namely $BiFeO_3$, popularly known as BFO, is one of the most studied materials for its room temperature multiferrocity. BFO thin films were developed using the sol–gel spin coating technique on $SiO_2$/Pt/Ti substrate. The green gel coated films were heat-treated using MW and CV techniques. Figure 6.16 compares the surface morphologies BFO thin films heat-treated using different recipes.

It is found that the average grain sizes are 105 and 180 nm, respectively, for the BFO samples sintered using CV and MW processes. This clearly defines the advantages of ceramic processing using MW heat treatment techniques. The grain sizes of MW-sintered BT and NZF are evidently larger than those of CV-sintered samples. This improvement in the structural properties influences the electrical and magnetic properties of the samples. The magnetic properties of MW-sintered NZF have significantly improved. However, dielectric results show that BT with LiF samples sintered using CV process portray a leaky nature due to the presence of lithium and/or porosity due to low density, which is similar to that of NZF specimen sintered using CV technique, which exhibits low dielectric properties. Therefore, we can conclude that MW sintering methods do show that at a much shorter processing time for the same temperature, it is able to achieve close to final stages of sintering compared to the sample sintered conventionally at the same temperature. In a close comparison, this information will be very much

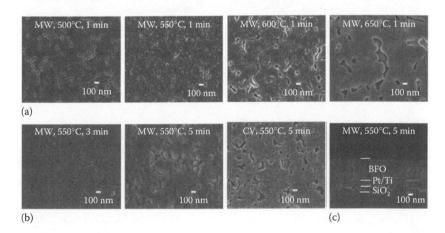

**FIGURE 6.16**  Surface morphologies of the sol–gel processed $BiFeO_3$ thin films developed using microwave (MW) sintering (a) at different temperatures for a fixed time, (b) at different times for fixed temperature in comparison to conventional (CV) sintering, and (c) the cross-sectional morphology of the sol–gel processed BFO films developed on Pt/Ti/$SiO_2$/Si substrate and sintered using microwave.

useful for the BT–NZF composite fabrication for multiferroic research purposes too. Moreover, much research should be conducted on MW sintering to process BT and NZF ceramics using solid-state reaction technique including calcination.

## REFERENCES

1. Chris T. Reading Feynman into nanotechnology: A text for a new science. *Techné*, 2008, 13, 133–168.
2. Eric D. The promise that launched the field of nanotechnology. *Metamodern: The Trajectory of Technology*. Retrieved on May 13, 2011.
3. Greiner A.; Wendorff J.H.; Yarin A.L.; Zussman E. Biohybrid nanosystems with polymer nanofibers and nanotubes. *Appl. Microbiol. Biotechnol.*, 2006, 71, 387–393.
4. Lalwani G.; Henslee A.M.; Farshid B.; Lin L.; Kasper F.K.; Qin Y.-X.; Mikos A.G.; Sitharaman B. Two-dimensional nanostructure-reinforced biodegradable polymeric nanocomposites for bone tissue engineering. *Biomacromolecules*, 2013, 14, 900–909.
5. Roco M.C. Senior Advisor for Nanotechnology, National Science Foundation. Retrieved on November 8, 2009.
6. Bethune D.S.; Klang C.H.; De Vries M.S.; Gorman G.; Savoy R.; Vazquez J.; Beyers R. Cobalt-catalyzed growth of carbon nanotubes with single-atomic-layer walls. *Nature*, 1993, 363, 605–607.
7. Nanotechnology Information Center. *Properties, Applications, Research, and Safety Guidelines*. American Elements. Retrieved on May 13, 2011.
8. Analysis: This is the first publicly available online inventory of nanotechnology-based consumer products. The Project on Emerging Nanotechnologies, 2008. Retrieved on May 13, 2011.
9. Scrinis G. Nanotechnology and the environment: The nano-atomic reconstruction of nature. *Chain Reaction.*, 2007, 97, 23–26.
10. Uskoković V. Nanotechnologies: What we do not know. *Technol. Soc.*, 2007, 29, 43–61.
11. National Research Council. *Research Progress on Environmental, Health, and Safety Aspects of Engineered Nanomaterials*, 2013, Washington, DC: The National Academies Press.
12. National Research Council. *A Research Strategy for Environmental, Health, and Safety Aspects of Engineered Nanomaterials*, 2012, Washington, DC: The National Academies Press.
13. Spillman W.B.; Sirkis J.S.; Gardiner P.T. Smart materials and structures: What are they? *Smart Mater. Struct.*, 1996, 5, 247.
14. Clark D.; Sutton W.H. Microwave processing of materials. *Ann. Rev Mater. Res. Bull.*, 1989, 68, 376–386.
15. Thostenson E.T.; Chou T.-W. Microwave properties: Fundamental and applications. *Composites Part A*, 1999, 30, 1055–1071.
16. Reisel A.D.; Schops S.; Lenk A.; Schmutzler G. Microstructural comparison of conventional and microwave sintered $BaTiO_3$. *Adv. Eng. Mater.*, 2007, 9, 400–405.
17. Sadhana K.; Krishnaveni T.; Praveena K.; Bharadwaji S.; Murthy S.R. Microwave sintering of nanobarium titanate. *Scripta Mater.*, 2008, 59, 495–498.
18. Mahboob S.; Dutta A.B.; Prakash C.; Swaminathan G.; Suryanarayana S.V.; Prasad G.; Kumar G.S. Dielectric behaviour of microwave sintered rare-earth doped $BaTiO_3$ ceramics. *Mater. Sci. Eng. B*, 2006, 134, 36–40.
19. Shannigrahi S.R.; Pramoda K.P.; Nugroho F.A.A. Synthesis and characterization of microwave sintered ferrite powders and their composite films for practical applications. *J. Magn. Magn. Mater.*, 2012, 324, 140–145.

20. Yodoji P.; Peelamedu R.; Agrawal D.; Roy R. Microwave sintering of Ni–Zn ferrites: Comparison with conventional sintering. *Mater. Sci. Eng. B*, 2003, 98, 269–278.

21. Sun W.; Li J.; Liu W.; Li C. Preparation of fine tetragonal barium titanate powder by a microwave–hydrothermal process. *J. Am. Ceram. Soc.*, 2006, 89, 118–123.

22. Peelamedu R.; Grimes C.; Agrawal D.; Roy R. Ultralow dielectric constant nickel–zinc ferrites using microwave sintering. *J. Mater. Res.*, 2003, 18, 2292–2295.

23. Ghasemi A.; Mousavinia M. Structural and magnetic evaluation of substituted $NiZnFe_2O_4$ particles synthesized by conventional sol–gel method. *Cerm. Int.*, 2014, 40, 2825–2834.

24. Pallathadka K.P.; Huang A.; Shannigrahi S.R. On some properties of PZT–NZF composite films manufactured by hybrid synthesis route. *Cerm. Int.*, 2011, 37, 431–435.

25. Walker B.E.; Rice J.R.W.; Pohanka R.C.; Spann J.R. Densification and strength of $BaTiO_3$, with LiF and MgO additives. *Am. Cemm. Soc. Bull.*, 1976, 55, 284–285.

26. Sorescu M.; Diamandescu L.; Peelamedu R.; Roy R.; Yadoji P. Structural and magnetic properties of NiZn ferrites prepared by microwave sintering. *J. Magn. Magn. Mater.*, 2004, 279, 195–201.

27. Shannigrahi S.R.; Tan C.K.I. Comparison of grain structure, electrical and magnetic properties of BaTiO3 and Ni0.5Zn0.5Fe2O4 ceramics sintered using microwave and conventional techniques. *Technologies*, 2015, 3, 47–57.

28. Zadeh H.N.; Glitzky C.; Dorfel I.; Rab T. Low temperature sintering of barium titanate ceramics assisted by addition of lithium fluoride-containing sintering additives. *J. Eur. Ceram. Soc.*, 2009, 30, 81–86.

29. Lange F.F. Liquid phase sintering: Are liquids squeezed out from between compressed particles? *J. Am. Ceram. Soc.*, 65, C23.

30. Hsiang H.I.; His C.S.; Huang C.C.; Fu S.L. Sintering behavior and dielectric properties of $BaTiO_3$ ceramics with glass addition for internal capacitor of LTCC. *J. Alloys Comp.*, 2008, 459, 307–310.

31. Gao L.; Huang Y.; Hu Y.; Du H. Dielectric and ferroelectric properties of (1-x)$BaTiO_3$–x$Bi_{0.5}Na_{0.5}TiO_3$ ceramics. *Ceram. Int.*, 2007, 33, 1041–1046.

32. Gao S.; Wu S.; Zhang Y.; Yang H.; Wang X. Study on the microstructure and dielectric properties of X9R ceramics based on $BaTiO_3$. *Mater. Sci. Eng. B*, 2011, 176, 68.

# 7 Design of Magnetic Semiconductors in Silicon

*Michael Shaughnessy, Liam Damewood,*
*and Ching-Yao Fong*

## CONTENTS

**ABSTRACT**    Controlling magnetic phenomena in semiconductors may allow new technological advances. In this chapter we review theoretical progress in understanding magnetism in silicon by doping transition metal elements. An in-depth study of single dopant energetics and bonding features at different sites is given. Spin-polarized densities of states, spin-polarized charge densities, and structural parameters are calculated and studied to yield a physical picture of the magnetic structure around transition metal dopants in Si. Next, other structures composed of many transition metal atoms in silicon, namely digital ferromagnetic heterostructures (DFHs) with and without defects, and trilayers, are studied to understand properties of extended transition metal structures. In particular, the conditions for half metallicity are described.

## 7.1 INTRODUCTION

The discovery of a magnetic layer structure formed by Fe(100)/Cr(100) exhibiting magnetoresistance (MR) (Baibich et al. 1988) stimulated tremendous interest in developing new devices utilizing spin or spin and charge for information storage.

In 1992, magnetoresistive memory technology was first reviewed (Daughton 1992). Magnetic random access memory (MRAM), composed of layers of transition metal elements (TMEs), such as iron (Fe) and cobalt (Co) separated by a thin insulator, is now a common component in computers (Savtchenko et al. 2003). In 1996, the term *spintronics* was conceived (Wolf et al. 2001) and has since been a popular term in the device materials research community. This term means to integrate the spin of carriers into semiconductors for information storage and transport. In Table 7.1, the advantages of spintronics are compared to those of conventional electronics.

Semiconductors formed by the group IV element silicon (Si) have a simple crystal structure: the diamond structure. Si has been the backbone of various transistors for modern-day electronic applications. Thus, mature technologies for its growth and doping processes have been developed (Voutsas 1997). Si is a natural candidate for doping magnetic transition metal (TM) elements to cultivate spintronic materials for manipulating charge- and spin-carrying information. Among the TM elements, manganese (Mn) is the most effective because of its large atomic magnetic moment, 5 $\mu_B$, where $\mu_B$ is the Bohr magneton. The large moment is a consequence of its electronic configuration. There are five d electrons and two s electrons giving a valence of 7 to the Mn. The s-electron spins antialign and compensate their moments, whereas the five d electrons occupy the five d states (the orbital angular momentum, $l$, is 2) and align their spins under Hund's first rule to reduce the Coulomb repulsions. With the g-factor equal to 2 for the 3d states, the resultant magnetic moment of the atom is 5 $\mu_B$.

The mature technological foundation of Si may facilitate the realization of spintronic devices. Recently, several groups around the world have been exploring Si-based spintronic materials (Zhang et al. 2004; Bolduc et al. 2005) because of predicaments associated with realizations of spintronic devices using other materials, such as Heusler alloys (De Groot et al. 1983) and $CrO_2$ (Ji et al. 2001). In particular, $T_C$ of $CrO_2$, determined by the point-contact Andreev reflection method (Ji et al. 2001), is far below room temperature (4 K). The $T_C$ is 88 K for half-Heusler alloy NiMnSb through photoemission measurements (Hordequin et al. 2000).

In this chapter, we focus on progress in the design of Si-based magnetic semiconductor materials. In Section 7.2, the common influences of the diamond structure on the electronic states of a TM dopant are given. Mn-doped Si will be presented in Section 7.3. Section 7.4 concerns the designs of $Mn_xSi_{1-x}$ in the form of a digital

## TABLE 7.1
## Comparison of Electronics and Spintronics

|                     | Electronics                     | Spintronics                                          |
| ------------------- | ------------------------------- | ---------------------------------------------------- |
| Carrier             | Charge                          | Spin or both charge and spin                         |
| Comparison          | Limited in speed and dissipation | High speed at very low power                          |
| Information storage | 0 and 1 bits                    | More possibilities: Qbits, if angular momentum is involved |
|                     | Volatile                        | MR (0,1)                                              |
|                     |                                 | Non-volatile                                         |

magnetic heterostructure and trilayers. Finally, in Section 7.5, a summary and critical comments on the future prospects of spintronic devices will be made.

## 7.2 FEATURES OF THE DIAMOND STRUCTURE ON THE ELECTRONIC STATES OF A TM DOPANT

When a TME, such as Mn, substitutes a Si in the diamond structure, the surrounding four Si atoms impose the effect of the crystal field with tetrahedral symmetry. Figure 7.1a shows the atoms in a unit-cell with the dopant shown in large shaded circle. The black lines outline a cube to guide the eyes. The cube is the conventional unit-cell of the diamond. The fivefold degeneracy of d states of TM atom are lifted into two groups: The triply degenerate states are composed of the $d_{xy}$, $d_{yz}$, and $d_{zx}$ orbitals, the so-called $t_{2g}$ states, and a doubly degenerate manifold is constituted by $d_{z2}$ and $d_{x2-y2}$, the $e_g$ states (Figure 7.2). The linear combination of $t_{2g}$ states forms orbitals pointing to the four nearest neighbors (nn), whereas the lobes of the $e_g$ orbitals point to the regions between the nearest neighboring atoms. The former orbital interacts with one of the $sp^3$ type of orbitals of a nn Si to form the bonding and antibonding states, separated by a large gap as shown in Figure 7.2. The center section schematically shows the physical situation. Such d and p mixing is called the d–p hybridization. The doubly degenerate states form the nonbonding states located in the gap of the bonding and antibonding states. Without the exchange interaction, the three sets of states have spin degeneracy. To illustrate clearly and convincingly, we use the crystal of MnAs in zinc blende structure to exhibit the charge distribution of the bonding and nonbonding states (Pask et al. 2003). Figure 7.3 gives the charge distributions of $t_{2g}$ and the $e_g$ states of MnAs along the chain of the group III and V atoms. One of the $t_{2g}$ electrons of the Mn and one of the $sp^3$ electrons of As form bond charges, indicated by an arrow. The contours of the $e_g$ states point away from the As atom. They show the nonbonding character. This is the essential picture to

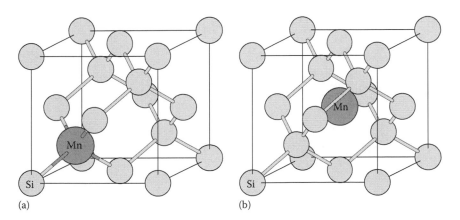

**FIGURE 7.1** (a) Conventional cell of Si with a Mn occupying a substitutional (S) site. (b) Mn occupying an interstitial (I) site. Mn atoms in both cases are shown in large shaded circles and labeled.

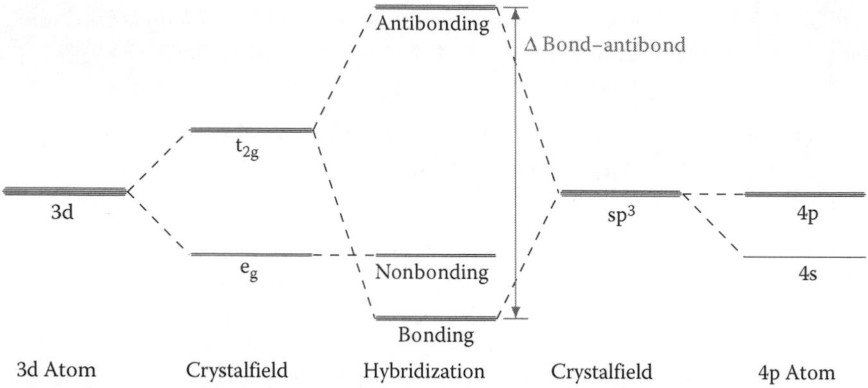
Antibonding
Δ Bond–antibond
$t_{2g}$
3d
sp$^3$
4p
$e_g$
Nonbonding
4s
Bonding
3d Atom
Crystalfield
Hybridization
Crystalfield
4p Atom

**FIGURE 7.2** Schematic diagram of energy levels from atomic (sides) to bonding ($t_{2g}$), antibonding ($t_{2g}^*$) and nonbonding ($e_g$) states (center). The $t_{2g}$ type of orbitals is formed by the linear combination of the degenerate $d_{xy}$, $d_{yz}$, and $d_{zx}$ states of Mn in the tetrahedral environment: the effect of the crystal field. One of these lobes points toward the nn Si. The sp$^3$ is a mixture of the s- and p-like states of Si also under the tetrahedral environment with one lobe pointing toward the Mn. They form d–p hybridization to give the bonding, antibonding, and nonbonding states.

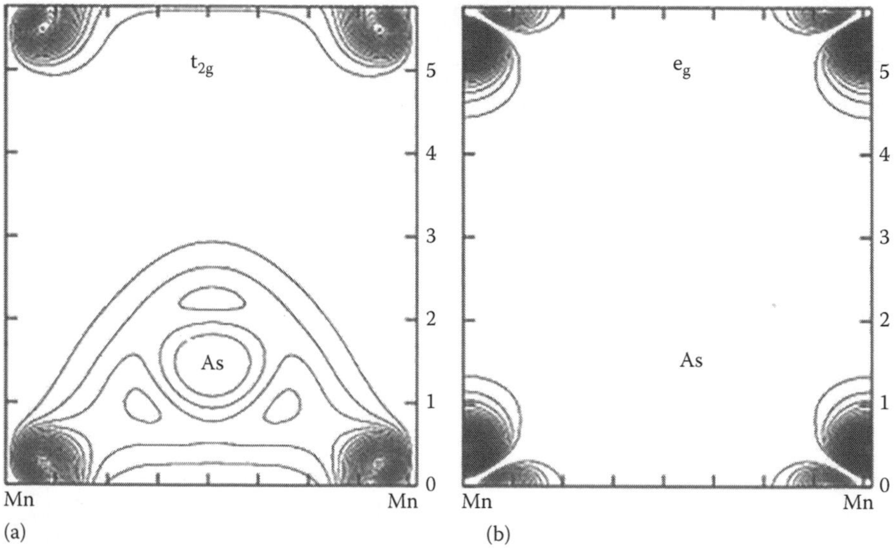
$t_{2g}$
As
Mn
Mn
(a)
$e_g$
As
Mn
Mn
(b)
0
1
2
3
4
5

**FIGURE 7.3** The charge distributions of the $t_{2g}$ (a) and $e_g$ (b) states of MnAs in the section formed by [110] and [001] vectors. The section contains chain of atoms formed by Mn and As.

understand the physical properties of TM-doped Si when the Mn substitutes a Si, a substitutional site (S) (Figure 7.1a). The local magnetic moment at the Mn is formed by the remaining d electrons at the Mn. The action of the exchange interaction causes a difference in occupation of the majority spin and minority spin states, resulting in a net magnetic moment of the doped sample.

The other possible location for the TME impurity in the diamond structure is to occupy an interstitial site (I) (Figure 7.1b). The physical features of this case will be elaborated later in this chapter.

## 7.3 MN-DOPED SI

### 7.3.1 MOTIVATION

Doping Mn into Si can form magnetic Si-based materials. They can be in both clustered and well-mixed alloy forms. In this chapter, we concern ourselves with the bulk alloys with a special quality: the Mn-doped Si alloys exhibit the so-called half-metallicity. In 1984, De Groot et al. (1983) predicted an interesting half-metallic half-Heusler alloy, nickel manganese antimony (NiMnSb), to be a half-metal. This prediction stimulated a worldwide search for half-metals. A half-metal should satisfy two conditions: (1) One of its spin channels exhibits metallic behavior, whereas the oppositely oriented spin channel shows semiconducting properties. A typical density of states (DoSs) of half-metallic CrAs (Pask et al. 2003) is shown in Figure 7.4. The

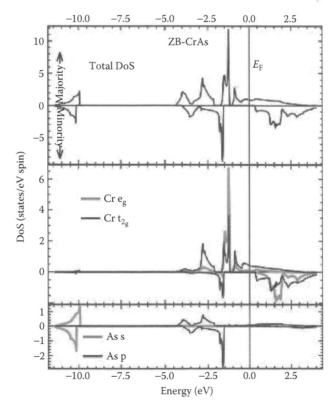

**FIGURE 7.4** DoS of CrAs, a predicted half-metal. The top panel shows the metallic majority spin channel gives the semiconducting properties of the minority spin states. The unique features are as follows: Fermi energy, $E_F$, intersects finite DoS of the spin-up states and is in the gap of the spin-down channel. The bottom panel shows the partial DoS of Cr d states and As s and p states.

unique feature is that the Fermi energy, $E_F$, intersects the finite DoS of the majority (up) spin channel and is located in the gap of the minority (down) spin states. (2) The magnetic moment/unit-cell must be an integer. The reasons are as follows:

1. The total number of electrons/unit-cell, $N_t$, is an integer.
2. The number of electrons/unit-cell filling up to the top of the valence band (VB) in the semiconducting channel, $N_{min}$, is an integer. The moments of these electrons are canceled by an equal number of electrons in the other spin channel as shown in Figure 7.4.
3. For 3d electrons, the g-factor is 2.0.
4. The magnetic moment/unit-cell is $N_t - 2N_{min}$, an integer in units of $\mu_B$.

This integer condition requires that the magnetic moment deduced from the measured saturation magnetization be close to integer value; however, the condition should be strictly satisfied by theoretical calculations.

The usefulness of a half-metal in spintronics is explained below. According to the Julliére formula (Julliere 1975), the MR of a ferromagnetic sample is expressed in terms of $P$, the spin polarization at $E_F$:

$$\text{MR} = \frac{2P^2}{1 - P^2} \tag{7.1}$$

where:

$$P = \frac{\text{DoS}_\uparrow - \text{DoS}_\downarrow}{\text{DoS}_\uparrow + \text{DoS}_\downarrow} \tag{7.2}$$

and the DoSs are evaluated at $E_F$. For MnCrAs, a half-metal, $P = 1$ (as a reference, $P$ of iron is 0.45). In this perfect situation, MR = 100%. Consequently, a half-metal is an ideal candidate for spintronic applications that utilize the MR mode because its conductivity is completely spin polarized.

Various half-metallic, half- and full-Heusler alloys (De Groot et al. 1983; Galanakis and Mavropoulos 2003), oxides, such as $CrO_2$ (Ji et al. 2001) and $Fe_3O_4$ (Dedlov 2004), TM chalcogenides (Galanakis and Mavropoulos 2003), pnictides (Galanakis and Mavropoulos 2003; Pask et al. 2003), and MnC (Pask et al. 2003; Qian et al. 2004) have been investigated theoretically and experimentally. There is, however, no spintronic device made of these half-metallic materials. All half-metallic alloys and compounds mentioned above show serious impediments for making devices operating at room temperature. For example, Heusler alloys are difficult to grow due to their complicated structures (Fong et al. 2013). In addition, half-metallicity vanishes in NiMnSb, a half-Heusler alloy, at 88 K (Hordequin et al. 2000). $CrO_2$ shows half-metallicity at 4 K, but it disappears precipitously as the temperature increases (Ji et al. 2001). For $Fe_3O_4$, it is known to suffer from the Verwey transition (Verwey 1939), a structural phase transition, at 120 K causing the disappearance of its half-metallic properties (Dedkov 2004). Furthermore, there is still no direct experimental evidence showing half-metallicity in the pnictides or MnC, even though thin-film forms of CrAs have been grown and its measured saturation magnetization agrees well with the theoretical prediction (Akinaga et al. 2000; Pask et al. 2003).

Since 2005, groups around the world have begun to focus their efforts to develop magnetoelectronics by leveraging the mature Si technology. The searches began with the explorations of half-metals composed of Si. In Japan, Kubota et al. (2009) reported half-metallicity in film forms of Si-related materials, such as Heusler alloys involving Si, $Co_2Fe_xMn_{1-x}Si$ using magnetic tunneling junction measurement. Sagar et al. (2011) studied the effects of interfaces on magnetic activation volumes in films of $Co_2FeSi$. In China, a group at Fudan University performed molecular beam epitaxial (MBE) growth of TME doped in Si (Su et al. 2008). Mn wires on Si (100) surface have been grown in the United States (Liu and Reinke 2008). However, there is no magnetic measurement on these nanowires. Clusters of $Mn_xSi_y$, where $1 \leq x \leq 34$ and $1.7 \leq y \leq 66$ are magnetic materials, have also attracted much attention (Karhu et al. 2012). On the theory side, Durgun et al. (2008) and Xu et al. (2009) studied half-metallic Si nanotubes. Magnetic properties of Si films doped with Mn and Mn on the Si(100) surface were reviewed by Zhou and Schmidt (2010). We currently have two unpublished cases of Mn on the reconstructed Si(100) $2 \times 1$ surfaces that are magnetic insulators with integer moments.

### 7.3.2 DETERMINATION OF THE DOPING SITE

We started to design half-metallic Si-based alloys in 2005 with a simplest approach: substituting a single Si atom with a single Mn atom, as was done for GaAs. It so happens that there is experimental evidence for this simple scenario (Bolduc et al. 2005). The experiment was carried out by doping Mn in Si using the ion implantation method. The method uses an energetic ionic beam at a base pressure of $5 \times 10^{-6}$ torr and at an energy of 300 keV with doses of $1 \times 10^{15}$ cm$^{-2}$ and $1 \times 10^{16}$ cm$^{-2}$ incident on a bulk Si with either n- or p-type lightly doped commercial samples.

This method is commonly used for doping semiconductors. They reported of growing $Mn_xSi_{1-x}$ with $x = 0.1\%$ and 0.8%, and measured the magnetic moment per Mn, $m$. The value of $m$ is 5 $\mu_B$ for the 0.1% case. It is known that the magnetic moment of a free Mn atom is 5 $\mu_B$ as discussed in Section 7.1. How can a Mn atom in the Si lattice avoid interaction with the nearby Si atoms and exhibit the same moment as a free atom?

Let us apply the so-called ionic model, originally proposed for estimating the magnetic moment/atom in $CrO_2$ (Schwarz 1986), based on the charge transfer from less electronegative element to more electronegative element. The electronegativity is 1.5 for Mn and 1.8 for Si. Thus, there should be charge transfer from Mn to Si in the alloy according to the model. Mn, having seven valence electrons, is surrounded by four nearest neighbor (nn) Si and transfers four of its electrons to each of its nn Si to form the d–p hybridized bonding and antibonding states as shown in Figures 7.2 and 7.3. This leaves three electrons at the Mn. Under Hund's first rule with the g-factor of 2 for a 3d electron, the $m$ value should therefore be 3 $\mu_B$ instead of 5 $\mu_B$.

To resolve this conflict, we started with a series of small Si supercells to model the alloy by stacking up conventional cells of Si along three directions and placing one Mn close to the center. Models with 8, 64, and 216 atoms give $m$ to be less than

5 $\mu_B$ but larger than 3 $\mu_B$. The important results are that the moments are not integer because there are interactions between the center Mn and Mn in the neighboring supercells. Finally, we tried models consisting of 512 Si atoms. In one such model, after substituting the center Si by a Mn, we did not remove the substituted Si. We show, in Figure 7.5, the 513-atom model corresponding to a doping concentration of 0.19%, which simulates the experimental $Mn_x Si_{1-x}$ alloy at $x = 0.1\%$ to within a factor of 2 (Shaughnessy et al. 2009). This model should be sufficient to isolate the Mn atom. Before reducing the forces acting on atoms inside the supercell due to the alloying, we checked that the maximum component of the forces acting on those Si atoms, located farther away from the Mn atom (at the edge of the supercell), is already small: 0.015 eV/Å. The small force indicates that the dopant is isolated and has a small effect on Si atoms that are far from the Mn. In the 216-atom cell, the corresponding force components are 0.08 eV/Å. Due to the doping in the supercell, Si atoms, particularly those near the Mn atom, should experience strains. We carried out ionic relaxation calculations to reduce the forces acting on the atoms to be less than 6.0 meV/Å.

Now, the model we focus on has the substituted Si occupying an interstitial (I) site serving as the second neighbor (sn) of the Mn atom. We call this group of atoms formed by the Mn, one of its nearest neighbors (nn), and the second neighbor Si as the Mn–nn Si–sn Si complex, shown in Figure 7.6.

We began with the experimental Si lattice constant of 5.43 Å and optimized it with the Mn impurity present in its relaxed configuration. The optimized lattice constant is 5.46 Å.

We found the presence of the sn Si weakens the d–p hybridization between the Mn and the nn Si causing the following:

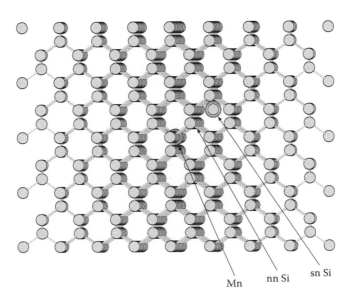

**FIGURE 7.5** The relaxed 513-atom supercell. Mn is shown in large shaded circle, Si in small circle, and the second neighbor (sn) Si at I site in double rings.

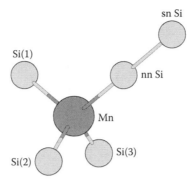

**FIGURE 7.6** The Mn–nn Si–sn Si complex.

1. The d electron, formerly participating in the d–p hybridization between the Mn and its first nn of the complex, retreats back to the Mn and increases the local moment at the Mn from 3 to 4 $\mu_B$ (calculated by integrating the spin density around the atom).
2. The $sp^3$ electron of the nn Si participating in the same d–p hybridization moves closer to the Si. Its spin is now uncorrelated with the d electron that retreated to the Mn. The spin moment at the Mn polarizes this $sp^3$ electron. The total moment of the sample is 5 $\mu_B$ in agreement with the experimental measurements

The spin density with the sn Si in the section containing the complex is given in Figure 7.7.

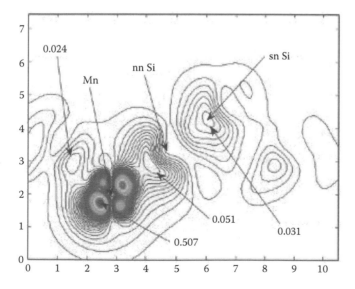

**FIGURE 7.7** The spin density difference of the up and down spin channels. The arrows show the positions of the relevant atoms and selected contour values ($e/\text{Å}3$). All distances are in angstrom. The $x$-axis of the partial cross-section is the [110] direction of the supercell and the $y$-axis.

The agreement between the experimental and theoretical magnetic moment in $Mn_xSi_{1-x}$ with $x = 0.1\%$ provides two pieces of information: (1) The Mn atom doped in Si occupies an S site under the growth method of ion implantation and (2) it is important to build models as close as possible to those in the experiments.

## 7.4 $M_Nx_SI_{1-x}$

### 7.4.1 SI-RELATED DIGITAL FERROMAGNETIC HETEROSTRUCTURE

An intuitively simple doping in Si is in the form of a δ-layer. In fact, Kawakami et al. (2000) attempted to grow δ-layer doping of Mn in GaAs. Even though theoretical (Sanvito and Hill 2001) results show that it is a two-dimensional (2D) half-metal and is now called digital ferromagnetic heterostructure (DFH). Experiment encountered difficulties in controlling the spread of Mn from a single layer for growing a perfect δ-layer doping with MBE method. It is still possible to improve the growth techniques (Kitchen et al. 2006).

We decided to design such a structure in Si by stacking eight conventional cubes of the diamond structure to form a supercell and model the DFH by substituting (S site) one layer of Si by Mn (Qian et al. 2004). Instead of using the usual cubic axes in directions perpendicular to the z-axis for the supercell, we used the vectors along the [110] and [−110] with length of $a/\sqrt{2}$, where $a$ is the lattice constant of the conventional cell so the lattice constants in the plane perpendicular to the z-axis are just the distances from the atom at the corner of the cube to the atoms at one of the face centers of the diamond structure. In this model, the total number of atoms is 32 instead of 64. Only one Mn per supercell is needed to model the δ-layer. The model is shown in Figure 7.8.

The optimized lattice constant is 5.45 Å. The DoS is shown in Figure 7.9. $E_F$ falls in the gap of the semiconducting spin-down channel. The calculated magnetic moment is 3 $\mu_B$. The results satisfy the conditions of a half-metal. Relaxation has several effects, some of which are as follows:

- It decreases the total energy.
- It enhances the DoS at $E_F$.
- It opens the gap in the semiconducting channel.

By examining the band structure, this DFH is a two-dimensional (2D) half-metal as was the DFH formed in GaAs. To see this, we show the band structure (Figure 7.10a) and Fermi surfaces (Figure 7.10b) of the ↑ spin channel inside the projected 2D Brillouin zone. Around Γ, there is no band nearby until $k$ is more than halfway between Γ and $R$, the contour labeled 1 (Figure 7.10a). This band defines the hole surface at Γ. The labels "2" and "3" around $R$ are occupied states: these are the electron surfaces. Along the Γ–$Z$ path, there is no intersection of the bands near $E_F$. One feature of this DFH is the presence of $E_F$ near the top of the VB. Its significance will now be discussed.

In the well-known half-Heusler alloy NiMnSb, experiments show that its half-metallic properties disappear at 88 K (Hordequin et al. 2000). A possible cause is

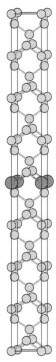

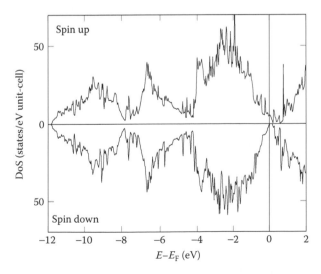

**FIGURE 7.8** The 32-atom supercell. Mn atoms are indicated by the large shaded circles.

**FIGURE 7.9** DoS of the DFH showing the half-metallicity. $E_F$ intersects the DoS in the ↑ spin channel and is in the gap but near the top of the VB of the ↓ spin states.

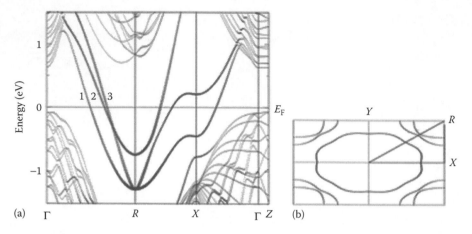

**FIGURE 7.10** The band structure of Si-DFH (a) and the Fermi surfaces of the ↑ spin channel (b) in the 2D Brillouin zone.

thermally excited spin-flip transitions for electrons at $E_F$ in the metallic channel to the bottom of the conduction bands (CBs) in the semiconducting channel because the $E_F$ of NiMnSb lies just below the bottom of the CB. In Figure 7.11a, a schematic energy diagram near $E_F$ for NiMnSb described above and the diagram of the DFH are shown.

The nearness of $E_F$ to the top of the VB in the Si-DFH, however, suggests the spin-flip transitions may not be as effective in destroying the half-metallicity. As shown in Figure 7.11b, the $E_F$ is close to the top of the VB. The gap is 0.25 eV, which is 10 times room-temperature thermal energy (0.025 eV) as shown in Figure 7.11. Thus, the thermal excitation from $E_F$ to the bottom of the CB is ineffective. However,

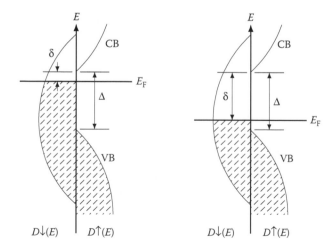

**FIGURE 7.11** Schematic diagrams of the bands near $E_F$ for (a) NiMnSb and (b) the DFH. The energy gap between the $E_F$ and the bottom of the CB is denoted by $\delta$ and $\Delta$ is the gap of the semiconducting channel.

spin-flip transitions from the top of the VB to the metallic states at $E_F$ are greatly reduced by the matrix element effect. The VB states are localized near Si atoms, whereas the metallic states are formed by the d states of the Mn. Therefore, the overlaps between the initial and final states of transitions between the VB and the metallic states are small.

We recognized that it is extremely difficult to realize ideal $\delta$-layer doping experimentally (Kawakami et al. 2000), so we decided to investigate whether the half-metallicity can be sustained under imperfections in the $\delta$-layer. We considered three defects in the $\delta$-layer of the DFH, but only two were reported (Fong et al. 2009). The model of 32 atoms (Figure 7.12b), containing only one Mn in a supercell, cannot be used to examine defects in the $\delta$-layer. We expanded the supercell model to contain a total of four Mn in the $x$–$y$ plane. The expanded ideal model is given in Figure 7.12a. The optimized lattice constant is 5.47 Å. Without relaxation, both the 32- and 64-atom models have metallic properties in the $\uparrow$ channel. The gaps are

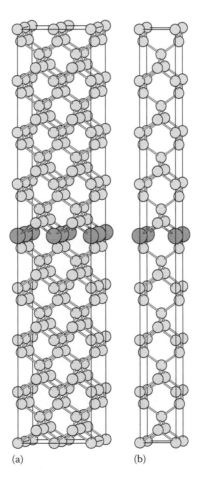

(a)                              (b)

**FIGURE 7.12**   Perfect $\delta$-layers: (a) the 64-atom supercell; and (b) the 32-atom supercell. Mn atoms are shown in large shaded circles.

**TABLE 7.2**

**Summary of the Optimized Lattice Constant, the Gap Energy (eV) in the Semiconducting Channel, and the Magnetic Moment ($\mu_B$/cell) of the 32- and 64-Atom Supercells and the Gap Energy in the Relaxed Models**

| Case | Optimized Lattice Constant (Å) | Energy Gap (eV) | Magnetic Moment ($\mu_B$/Unit-Cell) |
|------|-------------------------------|-----------------|-------------------------------------|
| 32-atom | 5.45 | 0.25 | 3 |
| 64-atom | 5.47 | 0.25 | 12 |

0.21 eV (32-atom) and 0.21 (64-atom) in the $\downarrow$ channel. For the 64-atom model, the total energy after relaxation is lower by 0.141 eV. The $\uparrow$ channel is metallic, whereas the $\downarrow$ states have a gap of 0.25 eV. All of these agree with the relaxed 32-atom case. Because the 64-atom cell has four Mn, its magnetic moment is 12.0 $\mu_B$/unit-cell.

The results of the optimized lattice constants, the gap values in the semiconducting channel, and the magnetic moment are summarized in Table 7.2.

The three defect models are shown in Figure 7.13. One Mn in the $\delta$-layer is replaced by a Si in case (a). In case (b), a vacancy is created in the $\delta$-layer. In case (c), a Mn atom replaces a Si atom above the $\delta$-layer. This latter case simulates the spread of the $\delta$-layer during growth.

The first two cases show similar half-metallic properties as the ideal case. We plot the DoS of case 1 in Figure 7.14. $E_F$ intersects the finite DoS of the majority spin channel and is located in the gap of the semiconducting minority spin channel. The DoS of case 2 shows similar results. The DoS, the gap values, and the magnetic moments of three defective models are given in Table 7.3.

It is clear that half-metallicity is robust in cases 1 and 2. Case 3 (Figure 7.12c) has Mn atoms in the nn configuration of two adjacent layers. In this case, half-metallicity is not predicted because the d–d interaction between the neighboring Mn reduces the bonding and antibonding gaps. We suggest during growth that it is necessary to avoid such Mn–Mn nearest neighbor interactions. In reality, this can be difficult to accomplish. If the MBE growth method (Kawakami et al. 2000; Kitchen et al. 2006) is used, it is necessary to have the beam concentration of Mn less than one monolayer.

A side comment is in order. Another possible method is as follows: in principle, the atomic layer deposition (ALD) method (Leskelä and Ritala 2003) guarantees atomically thin monolayers, which will be worth to try. By using large processing window, the deposition is not sensitive to fluctuations in temperate and rate of flow of the precursors.

The above results for the defective $\delta$-layer are summarized as follows:

- Perfect $\delta$-layer doping of Mn in Si, DFH, is a 2D half-metal. It is unlikely that the Curie temperature, $T_C$, will be lowered by spin flip transitions.
- It is not necessary to have a perfect $\delta$-layer doping for a DFH to have half-metallic properties.
- It is advised to avoid having two Mn atoms forming a nearest neighbor pair.

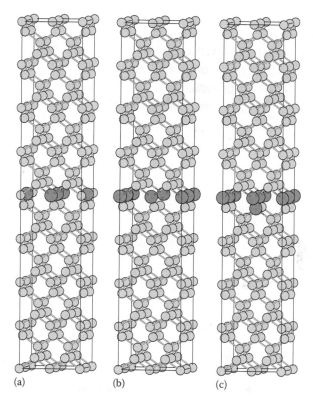

**FIGURE 7.13** Models of three defect cases: (a) Mn replaced by a Si, (b) vacancy where the center Mn is removed, and (c) Mn over the δ-layer, that is, the spread of the δ-layer. Mn atoms are shown in large shaded circles.

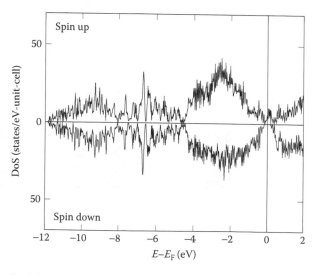

**FIGURE 7.14** DoS for case 1.

**TABLE 7.3**

**Properties Relevant to Half-Metallicity of the Three Defective Cases**

| | DoS at Fermi Energy ($E_F$) (States/eV Unit-Cell) | | | Magnetic Moment |
|---|---|---|---|---|
| Case | ↑ | ↓ | Gap (eV) | ($\mu B$/Unit-Cell) |
| 1 | 5.88 | 0.00 | 0.29 | 9.00 |
| 2 | 2.03 | 0.00 | 0.49 | 9.00 |
| 3 | 3.96 | 3.87 | – | 5.52 |

### 7.4.2 SINGLE DOPING OF FE AND MN IN SI

The S-site doping of Mn has experimental evidence (Bolduc et al. 2005) as supported by our theory (Shaughnessy et al. 2009). Zhu et al. (2008) suggested that codoping with As can lower the S-site doping barrier of Mn and half-metallicity can be preserved. There remains another issue: is this the only possible site for doping? This issue was addressed by Wu et al. (2007) using the Wien2K code (Blaha et al. 2001) and a similar model by Qian et al. (2006). They found that there is an interstitial (I) site with a lower total energy for doping Mn compared to the one for the S site, but they did not report the energy difference. In Figure 7.15, the images of the S and I sites are given. Furthermore, they showed that for one-fourth monolayer coverage, the I-site doped model is a half-metal. At higher concentrations, the half-metallicity disappears due to the interaction between the interstitial Mn atoms.

About the same time, we addressed more basic questions (Shaughnessy et al. 2010):

1. What is the energetic difference between the two doping sites?
2. What is the physical origin causing the difference?
3. What are possible distortions in the immediate neighborhood of the dopant— contraction or expansion? Can they be predicted before growth?

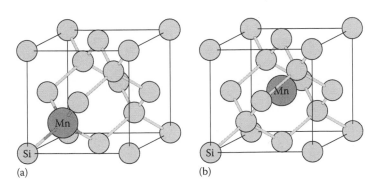

(a)                                (b)

**FIGURE 7.15** Elementary picture of (a) an S site and (b) an I site. Mn atoms are shown in large shaded circles.

These basic questions are answered by examining single dopings of Fe and Mn in Si at the two sites. To answer question 1, we start with the expression of the formation energy:

$$E_{for} = E_{coh} - N\mu_{Si} - \mu_{TME} \qquad (7.3)$$

where:
  $N$ is the number of Si atoms in a supercell
  $\mu_{Si}$ is the chemical potential of a Si atom, −5.433 eV/Si

The chemical potential of the TME is denoted by $\mu_{TME}$ and $E_{coh}$ is the cohesive energy of the system. We then calculated the difference between the formation energies of the TME at the two sites. Because the number of Si atoms in the two cases differs by 1, the difference of the formation energies is

$$\Delta E_{for} = E_{coh}(I) - \left[ E_{coh}(S) + \mu_{Si} \right] \qquad (7.4)$$

The results are given in Table 7.4. As the last column of the table shows, the I site has a lower formation with the TME at the S or I site. Table 7.4 shows that the energy difference is on the order of 0.5 eV. The physical origin can be understood by using Figure 7.16.

In Figure 7.16, the plots provide the answer to question 2. The different character of the bonding at S and I sites leads to the energetic difference. The dark arrow in Figure 7.16a indicates a d–p hybridized bond, whereas in Figure 7.16b the light arrow shows charges of Mn shifted to the open space by the attraction of the nuclear charge of the neighboring Si. There is no bond formed in the case of Figure 7.16b. In order to dope Mn at an S site, there is a bond breaking between the substituted Si and its four neighbors. The energy of new bond formation between the Mn and the same neighboring Si atoms cannot compensate energy required for breaking the bonds. No such bond breaking involved at the I site. Fe doping shows similar features.

---

**TABLE 7.4**

**The Relaxed Formation Energy Difference,**
**$\Delta E_{for} = E_{for}(I) - E_{for}(S)$ of the 8-, 64-, and**
**216-Atom Cases with the TME at the S or I Site**

| Element | Cell | $E_{for}(I) - E_{for}(S)$ (eV) |
|---------|----------|------------|
| Mn | 8-atom | −0.164 |
| | 64-atom | −0.452 |
| | 216-atom | −0.528 |
| Fe | 8-atom | −0.415 |
| | 64-atom | −0.503 |
| | 216-atom | −0.473 |

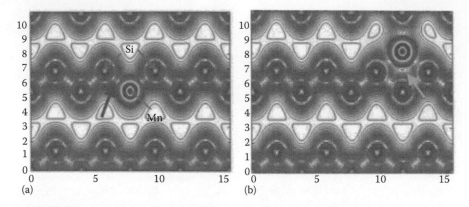

**FIGURE 7.16** Charge density plots. (a) Mn occupies at an S site. A d–p bond, indicated by the dark arrow, is formed between Si and Mn. (b) There is no bond-breaking process between Si and Mn when Mn occupies an I site. Instead, the Mn shifts its charge to the open space toward the second neighbor Si, where the light arrow points.

Question 3 is answered in Table 7.5. For Mn, the lattice expands independent of the doping sites, whereas Fe doping causes lattice contraction at the S site but lattice expansion at the I site. From atomic radii, 1.61Å for Mn, 1.56Å for Fe, and 1.11Å for Si, it is possible to predict the contraction of Fe case at the S site because its radius is smaller than that of the Mn atom.

There are several interesting findings when comparing the magnetic moments of various dopings in Table 7.6. The magnetic moments at both sites in the small 8-atom supercell are not integers. This agrees with the result obtained by Wu et al.

**TABLE 7.5**

**Bond Length (Å) between the Transition Metal Element and Its Nearest Neighbor Si**

| Element | Case | Site | Bond Length (Å) | Bond Relaxation (Å) |
|---------|------|------|-----------------|---------------------|
| Mn | 8-atom | S | 2.38 | 0.02 |
| | | I | 2.41 | 0.05 |
| | 64-atom | S | 2.40 | 0.04 |
| | | I | 2.40 | 0.04 |
| | 216-atom | S | 2.40 | 0.04 |
| | | I | 2.43 | 0.07 |
| Fe | 8-atom | S | 2.32 | −0.04 |
| | | I | 2.36 | 0.00 |
| | 64-atom | S | 2.26 | −0.10 |
| | | I | 2.40 | 0.04 |
| | 216-atom | S | 2.25 | −0.11 |
| | | I | 2.40 | 0.04 |

**TABLE 7.6**

**Comparison of the Calculated Magnetic Moments ($\mu_B$/Mn) for Mn and Fe in 8-, 64-, and 216-Atom Cells and at the S and I Sites**

| Element | Case | Site | Magnetic Moment ($\mu_B$/Mn) |
|---------|------|------|------------------------------|
| Mn | 8-atom | S | 2.97 |
|  |  | I | 3.16 |
|  | 64-atom | S | 3.00 |
|  |  | I | 3.00 |
|  | 216-atom | S | 3.00 |
|  |  | I | 3.00 |
| Fe | 8-atom | S | 1.76 |
|  |  | I | 2.05 |
|  | 64-atom | S | 0.00 |
|  |  | I | 2.00 |
|  | 216-atom | S | 0.00 |
|  |  | I | 2.00 |

(2007), in which no half-metallic properties are found for the I site doping if the layer has more than one-fourth of coverage. In the 8-atom cases, physically there is an interaction between the Mn atoms in the neighboring supercells. For larger supercells, the moment of Mn at the S site is 3 $\mu_B$, which can be characterized by the ionic model. Additionally, Fe is a valence 8 element. In the presence of four nn Si at the S site, Fe shares four of its electrons to form d–p hybridized bonds. Thus, the ionic model predicts the moment to be 4 $\mu_B$. The calculated moments of Fe are smaller than the prediction, in particular at the S site. The contraction around the doping does not provide enough volume for the localized d electrons to obey the Hund's first rule. Some spins flip to have antiparallel configurations. Thus, the net moments are reduced.

The results are summarized as follows:

1. At an I site, both Fe and Mn cause local expansion. This explains qualitatively the results of Fe doping in Si showing an increase in measured lattice constant of the alloy with respect to the pure Si (Su et al. 2008).
2. At an S site, Fe doping shows contraction, whereas Mn doping shows expansion. It is possible to predict this behavior from the atomic radii.
3. At an I site, there is no bond formation. However, due to the open diamond-type structure, some of the d electrons of the TME shift their charges into the open regions as shown in Figure 7.16. Local moment can still be formed by the remaining d electrons at the Mn.
4. The physical origin of the energetic differences between the two sites is due to the requirement that four Si–Si bonds have to be broken at the S site. The energy cost is not fully compensated by the formation of new d–p hybridized bonds. No bond-breaking process happens at an I site. The energy difference between the two sites is about 0.5 eV.

5. The ionic model does not predict the magnetic moment of Fe at the two sites because of the local distortion. The difference is a demonstration of the Pauli exclusion principle.

We emphasize that the Si systems are interesting, because the simple physics, such as the Pauli exclusion principle, the Hund's first rule, and unit-cell sizes, provide the understanding of the calculated results. Based on point 4, it should be energetically possible to dope Mn atoms at I sites in Si.

### 7.4.3 TRILAYER

With the δ-layer results showing half-metallicity and lower doping barrier for the I site, we decided to design a trilayer in which Mn occupy I sites conforming to the metallic device configurations (Baibich et al. 1988). The model without any further doping is shown in Figure 7.17a with the Mn shown in large shaded circles and Si

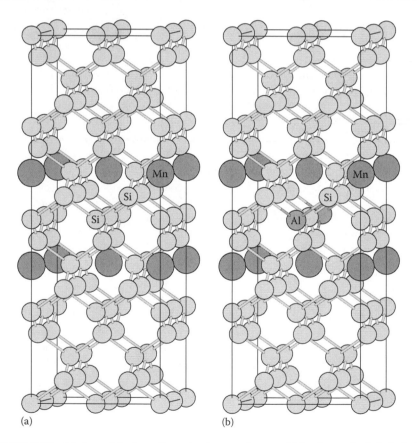

(a)                                                        (b)

**FIGURE 7.17** (a) A model of a trilayer having two layers of Mn (large shaded circles) at I sites. Si atoms are shown as yellow circles. (b) A trilayer with hole doping (small shaded circles; either Al or Ga) in the Si layers between the two Mn layers. The labeled atoms are included in the section of the charge density plots in Figure 7.18.

shown in small circles. As in the 64-atom model, the x, y, and z axes are along [110], [−110], and [001] of the conventional cubic cell, respectively. Any device fabricated from this type of trilayer can operate under the MR mode.

To operate in the MR mode, a possible configuration has the current flowing perpendicular to the layer planes. This is the so-called current perpendicular to plane (CPP) configuration. However, we realize from the model shown in Figure 7.17a that there are Si layers between the Mn (large shaded circles) layers that would inhibit the current flow. To enhance the current, we consider doping the Si layers between the Mn layers. Either n-type or p-type doping in Si can be easily carried out by experiments because of the mature Si technologies. We make a decision so that experimentalists will save their efforts and avoid the time-consuming trial and error process.

For n-type doping, the donor electrons may reduce the gap of the Si. A smaller gap of the host material can reduce the likelihood of getting a half-metal. Furthermore, the donors may overlap with the electronic wave functions of the Mn. This can lead to the sample behaving as a ferromagnetic metal. For a metal, such as Fe, the spin polarization at $E_F$ is in general less than the one formed by local moments. This is not favorable for half-metallicity. Alternatively, p-type doping can possibly increase the gap. In addition, the presence of holes causes some of the electrons of the Mn to shift their charges toward the holes similar to the case of the Mn at I site discussed in Section 7.3.4.

The p-doping may enhance the formation of local moments of the sample. We decided to consider p-doping with either Al or Ga (Yang et al. 2013). The hole-doped model is shown in Figure 7.17b with the Al or Ga in small shaded circles. The relaxed lattice constants, the magnetic moment, and the energy gap in semiconducting channel are summarized in Table 7.7.

Without hole doping, the trilayer is not a half-metal. This result is consistent with the one given by Wu et al. (2007). The gaps for Al and Ga dopings are determined by the generalized gradient approximation (Perdew et al. 1996), which in general underestimates the magnitude of a semiconducting channel but can be improved by using the GW method (Damewood and Fong 2011). At this point, we have not tuned the gap by layer separation.

---

**TABLE 7.7**

**Lattice Constants in the x–y Plane and the z Direction (Specified by the Factor Which Should Be Multiplied against the Lattice Constant in the x–y Plane), Magnetic Moments ($\mu_B$/Unit-Cell), and Gaps for the Al- and Ga-Doped Trilayers**

| Hole Dopant | x–y-Lattice Constant (Å) | Factor | Magnetic Moment ($\mu_B$/Unit-Cell) | Energy Gap (eV) |
|---|---|---|---|---|
| None | 5.473 | 3.970 | 12.3 | NA |
| Al | 5.470 | 3.970 | 14 | 0.030↑ |
| Ga | 5.458 | 3.995 | 14 | 0.044↑ |

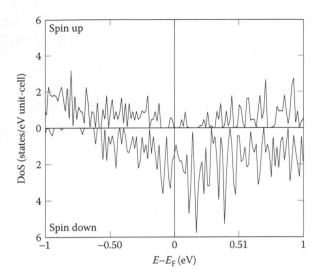

**FIGURE 7.18**    The DoS of the trilayer with additional Al doping. $E_F$ falls within the gap of the spin minority channel, fulfilling the first criteria for a half-metal.

Because the DoSs of both dopants exhibit the same qualitative features, we only plot the DoSs of the Al doping in Figure 7.18. $E_F$ is in the gap of the spin-up channel and intersects the finite DoSs of the spin-down states. Therefore, with integer magnetic moments, both hole-doped trilayers are half-metals.

Based on the results of single doping of Mn in Si, the magnetic moment should be 3 $\mu_B$/Mn. In our model, there are four Mn atoms. The moment/unit-cell should be 12 $\mu_B$/unit-cell; however, the calculated moment is 14 $\mu_B$/unit-cell. The other 2 $\mu_B$/unit-cell comes from the dangling bonds at the site of the substitutional group III element. To show this, we calculated the difference of the spin density without and with hole doping. The results are shown Figure 7.18. The contour labeled (1) has a value of 0.0205 electrons/$\text{Å}^3$ and is close to where Si atoms are located. This contour and the corresponding one on the other side are associated with the dangling bonds of the nn Si to the Al (as shown in Figure 7.19).

The results are summarized as follows:

- With Mn at I sites in the trilayer structure, half-metallicity can be obtained by doping with either Al or Ga in the region between the Mn layers.
- The dangling bonds due to the p-type doping can contribute to the magnetic moment of the sample.

There are several ways to improve the results we have. The first step is recognizing each layer needed to be one-fourth filled. At this moment, we have results on one Mn in each layer (less than 1/4 doping) and three group III elements. The gap is increased to 0.1 eV, even without a GW correction, which is 4 times larger than room temperature. By optimizing the layer spacing, the gap in the semiconducting

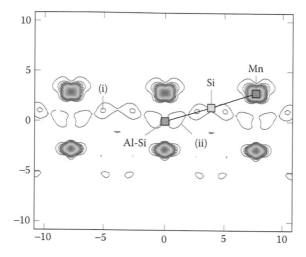

**FIGURE 7.19**  Differences of spin densities with and without hole doping. The section contains the Al, Si, and Mn shown in Figure 7.17b. The label Al-Si indicates the position where Al replaces Si. The Si dangling bonds located near the Al atoms are indicated by (i). The value of the maximum at (i) is 0.0088 electrons/$Å^3$. The regular bond is labeled by (ii) and the value of the contour at (ii) is 0.0027 electrons/$Å^3$.

channel can possibly be improved. Because the type of doping of Mn at the I site does not need to break a bond between pairs of Si, the energy barrier for this doping is at least 0.5 eV per atom less compared to the doping at an S site. Furthermore, the Mn layer needs not be an ideal δ-layer to facilitate the growth and the hole doping in Si should be feasible due to the mature technology. The results of the trilayers are extremely encouraging for realizing spintronic materials using Si. We hope that experimentalists will use layer growth methods, in particular the MBE growth method, to try.

We realized that spin injection is still a challenging research topic (Jansen 2012; Dankert et al. 2013). The general approach is to use ferromagnetic metals to inject and detect spin in Si. This type of setup causes serious impedance mismatch. To address this, $Fe/Al_2O_3$ has been used as an injector to reduce the mismatch. We suggest that besides using trilayers as active devices, they can also be used as spin source for injection because their magnetic moment/Mn is about 4 $μ_B$. The current in a trilayer can be nearly 100% spin polarized. It is also possible to grow Si on top of it. There is no lattice constant mismatch to worry about. If there is a barrier needed in the interface between the trilayer and Si, it can possibly be overcome by growing a thin $SiO_2$ layer, so the injection can be operated under tunneling processes.

Recent work (Naji et al. 2014) has explored phase diagrams and magnetic properties of trilayer structures, using the Ising model and a mean field approximation. Depending on the intra- and interlayer couplings between the spins of the electrons on the atoms in the layers, different collective magnetic states are predicted, including ferromagnetic, ferrimagnetic, and antiferromagnetic. By carefully tuning the layer separation, the doping in the region between the layers, and the type of TM composing the layers, the trilayers (Yang et al. 2013) offer a rich ground to explore the coexistence of phases and other interface magnetic properties.

## 7.5 SUMMARY

Let us first make a few general comments on the lessons learned from first-principles calculations.

- With the dilutely doped Mn in Si, $Mn_xSi_{1-x}$ with $x = 0.1\%$, our predictions of the magnetic moment/Mn do not agree with the measured value of 5 $\mu_B$/Mn until the model of the superlattice has $x$ within a factor of 2 of the experimental value. This implies that to get agreement between experiment and theory, a realistic model is essential. Otherwise, quantitative disagreement is to be expected. For example, if the small supercells alone were tried, the theory would conclude that the measured magnetic moment would be questionable.
- It is advised that one should completely understand the outputs of any first-principles code. We have shown that the results bring new microscopic insights, which are relevant to the physics involved in the Si systems. The Pauli exclusion principle, the Hund's first rule, and the local environments are all manifested in the outputs.

For the Mn doped in Si, we took the following steps:

1. We first explained the measured magnetic moment in dilutely doped Mn in Si to explore the possible doping sites. It suggests that Mn occupies an S site.
2. Next, we showed 2D half-metallicity in δ-layer doping of Mn at S site in Si.
3. The basic properties of doping a TME, in particular the energetic difference between S and I sites, were also studied. The single doping of Fe and Mn in Si shows the physical origin of different properties of the S and I sites.
4. Finally, we designed trilayers with Mn at I sites doped in Si. With an additional hole doping in the Si region between the Mn layers, the trilayers exhibit half-metallic properties.

We realize that there are still challenges ahead to make spintronic devices using the Mn-doped Si. It is hoped that experimentalists will make use of these results, particularly the trilayer, to grow these materials. They are appealing because the spin–orbit effect in these materials is small and the simple crystal field and d–p hybridization can characterize the important physics in them. We have shown that an ideal δ-layer doping of Mn is not necessary. However, it is necessary to avoid Mn atoms as nn. The 2D features of the DFH and trilayers can offer new physics in low-dimensional systems. There remain several difficult issues to be resolved, such as how to avoid Mn atoms occupying nn sites and whether the Curie temperature can be pushed above room temperature.

## ACKNOWLEDGMENT

This work at University of California, Davis, was supported by the US National Science Foundation, Grant No. ECCS-1232275.

## REFERENCES

Akinaga, H., Manago, T., and Shirai, M. (2000). Material design of half-metallic zinc-blende CrAs and the synthesis by molecular-beam epitaxy. *Jpn. J. Appl. Phys. Lett.* **39**, L1118–L1120.

Baibich, M. N., Broto, J. M., Fert, A., Vandau, F. N., Petroff, F., Eitenne, P., Creuzet, G., Freiderich, A., and Chen, I. J. (1988). Giant magnetoresistance of (001)Fe/(001)Cr magnetic superlattices. *Phys. Rev. Lett.* **61**, 2472–2475.

Blaha, P., Schwarz, K., Madsen, G., Kvasnicka, D., and Luitz, J. (2001). *WIEN2k*. Technical University, Vienna, Austria.

Bolduc, M., Awo-Affouda, C., Stollenwerk, A., Huang, M. B., Ramos, F. G., Agnello, G., and LaBella, V. P. (2005). Above room temperature ferromagnetism in Mn-ion implanted Si. *Phys. Rev. B* **71**, 033302-1–033302-4.

Damewood, L., and Fong, C. Y. (2011). Local field effects in half metals—A GW study of zincblende CrAs, MnAs, and MnC. *Phys. Rev. B* **83**, 113102-1–113102-4.

Dankert, A., Dulal R. S., and Dash, S. P. (2013). Efficient spin injection into silicon and the role of the Schottky barrier. *Sci. Rep.* **3**, 3196-1–3196-8.

Daughton, J. M. (1992). Magnetoresistive memory technology. *Thin Solid Films* **216**, 162–168.

Dedkov, Y. 2004. Spin-resolved photoelectroni spectroscopy of oxidic half-metallic ferromagnets and oxide/ferromagnet interfaces. PhD thesis, Technical University of Aachen, Aachen, Germany.

De Groot, R. A., Mueller, F. M., Vanengen, P. G., and Buschow, K. H. J. (1983). New class of materials—Half-metallic ferromagnets. *Phys. Rev. Lett.* **50**, 2024–2027.

Durgun, E., Çakir, D., Akman, N., and Ciraci, S. 2008. Half-metallic silicon nanowires. *Phys. Rev. Lett.* **99**, 256806-1–256806-4.

Fong, C. Y., Pask, J. E., and Yang, L. H. 2013. *Half-Metallic Materials and Their Properties*. Imperial College Press, London.

Fong, C. Y., Shaughnessy, M., Snow, R., and Yang, L. H. (2009). Theoretical investigations of defects in a Si-based digital ferromagnetic heterostructure—A spintronic material. *Phys. Stat. Solid.* **C7**, 747–749.

Galanakis, I., and Mavropoulos, P. (2003). Zinc-blende compounds of transition elements with N, P, As, Sb, S, Se, and Te as half-metallic systems. *Phys. Rev. B* **67**, 104417-1–104417-8.

Hordequin, C., Ristoiu, D., Ranno, L., and Pierre, J. (2000). On the cross-over from half-metal to normal ferromagnet in NiMnSb. *Eur. Phys. J. B* **16**, 287–293.

Jansen, R. 2012. Si spintronics. *Nat. Mater.* **11**, 400–408.

Ji, Y., Strijkers, G. J., Yang, F. Y., Chien, C. L., Byers, J. M., Anguelouch, A., Xiao, G., and Gupta, A. (2001). Determination of the spin polarization of half-metallic CrO₂ by point contact Andreev reflection. *Phys. Rev. Lett.* **86**, 5585–5588.

Julliere, M. (1975). Tunneling between ferromagnetic-films. *Phys Lett A* **54**, 225–226.

Karhu, E. A., Rőβler, U. K., Bogdanov, A. N., Kahwaji, S., Kirby, B. J., Fritzsche, H., Robertson, M. D., Majkrzak, C. F., and Monchesky, T. L. (2012). Chiral modulations and reorientation effects in MnSi thin films. *Phys. Rev. B* **85**, 094429-1–094429-12.

Kawakami, R. K., Johnston-Halperin, E., Chen, L. F., Hanson, M., Guebels, N., Speck, J. S., Gossard, A. C., and Awschalom, D. D. (2000). (Ga, Mn) As as a digital ferromagnetic heterostructure. *Appl. Phys. Lett.* **77**, 2379–2381.

Kitchen, D., Richardella, A., Tang, J. M., Flatté, M, and Yazdani, A. (2006). Atom-by-atom substitution of Mn in GaAs and visualization of their hole-mediated interactions. *Nat. Lett.* **442**, 436–439.

Kubota, T., Tsunegi, S., Oogane, M., Mizukami, S., Miyazaki, T., Naganuma, H., and Ando, Y. (2009). Half-metallicity and Gilbert damping constant in $Co_2Fe_xMn_{1-x}Si$ Heusler alloys depending on the film composition. *Appl. Phys. Lett.* **94**, 122503-1–122503-3.

Leskelä, M., and Ritala, M. (2003). Atomic layer deposition chemistry: Recent developments and future challenges. *Angew. Chem. Int. Edit.* **42**, 5548–5554.

Liu, H., and Reinke, P. (2008). Formation of manganese nanostructures on the Si(100)-(2 × 1) surface. *Surf. Sci.* **602**, 986–992.

Naji, S., Belhaj, A., Labrim, H., Bahmad, L., Benyoussef, A., and El Kenz, A. (2014). Phase diagrams and magnetic properties of tri-layer superlattices: Mean field study. *Physica A*, **399**, 106–112.

Pask, J. E., Yang, L. H., Fong, C. Y., Pickett, W. E., and Dag, S. (2003). Six low-strain zinc-blende half metals: An ab initio investigation. *Phys. Rev. B* **67**, 224420-1–224420-7.

Perdew, J. P., Burke, K., and Ernzerhof, M. (1996). Generalized gradient approximation made simple. *Phys. Rev. Lett.* **77**, 3865–3868.

Qian, M. C., Fong, C. Y., and Yang, L. H. (2004). Coexistence of localized magnetic moment and oppositely-spin itinerant electrons in MnC. *Phys. Rev. B* **70**, 052404-1–052404-4.

Qian, M. C., Fong, C. Y., Liu, K., Pickett, W. E., Pask, J. E., and Yang, L. H. (2006). Half-metallic digital ferromagnetic heterostructure composed of a δ-doped layer of Mn in Si. *Phys. Rev. Lett.* **96**, 027211-1–027211-4.

Sagar, J., Fleet, L. R., Hirohata, A., and O'Grady, K. (2011). Activation volumes in $Co_2FeSi$ thin films. *IEEE Trans. Magn.* **47**, 2440–2443.

Sanvito, S., and Hill, N. (2001). Ab initio transpot theory for digital ferromagnetc heterostructures. *Phys. Rev. B* **87**, 267202.

Savtchenko, L., Engel, B. N., Rizzo, N. D., Deherrera, M. F., and Janesky, J. A. (2003). Method of writing to scalable magnetoresistance random access memory element. Patent 6,545,906 B1.

Schwarz, K. (1986). $CrO_2$ predicted as a half-metallic ferromagnet. *J. Phys. F Met. Phys.* **16**, L211–L215.

Shaughnessy, M., Fong, C. Y., Snow, R., Liu, K., Pask, J. E., and Yang, L. H. (2009). Origin of large moments in $Mn_xSi_{1-x}$ at small x. *Appl. Phys. Lett.* **95**, 022515-1–022515-3.

Shaughnessy, M., Fong, C. Y., Snow, R., Yang, L. H., Chen, X. S., and Jiang, Z. M. (2010). Structural and magnetic properties of single dopants of Mn and Fe for Si-based spintronic materials. *Phys. Rev. B* **82**, 035202-1–035202-6.

Su, W. F., Gong, J. L., Wang, S., Chen, Fan, Y. L., and Jiang, Z. M. (2008). Group-IV-diluted magnetic semiconductor $Fe_xSi_{1-x}$ thin films grown by molecular beam epitaxy. *J. Cryst. Growth* **311**, 2139–2142.

Verwey, E. J. W. (1939). Electronic conduction of magnetite ($Fe_3O_4$) and its transition point at low temperatures. *Nature* **144**, 327–328.

Voutsas, T. (1997). Low temperature polysilicon technology for advanced display systems. http://quattro.co.uk/corporate/info/rd/tj1/pdf/11.pdf.

Wolf, S. A., Awschalom, D. D., Buhrman, R. A., Daughton, J. M., von Molnar, S., Roukes, M. L., Chtchelkanova, A. Y., and Treger, D. M. (2001). Spintronics: A spin-based electronics vision for the future. *Science* **294**, 1488–1495.

Wu, H., Kratzer, P., and Scheffler, M. (2007). Density-functional theory study of half-metallic heterostructures: Interstitial Mn in Si. *Phys. Rev. Lett.* **98**, 117202-1–117202-4.

Xu, Z., Yan, Q. B., Zheng, Q.-R., and Su, G. (2009). Half-metallic silicon nanowires: Multiple surface dangling bonds and nonmagnetic doping. *Phys. Rev. B* **80**, 081306 (R).

Yang, L. H., Shauhgnessy, M., Damewood, L. J., Fong, C. Y., and Liu, K. (2013). Half-metallic hole-doped Mn/Si trilayers. *J. Phys. D: Appl. Phys.* **46**(16), 165502.

Zhang, F. M., Liu, X. C., Gao, J., Wu, X. S., Du, Y. W., Zhu, H., Xiao, J. Q., and Chen, P. (2004). Investigation on the magnetic and electrical properties of crystalline $Mn_{0.05}Si_{0.95}$ films. *Appl. Phys. Lett.* **85**, 786–788.

Zhou, S., and Schmidt, H. (2010). Mn-doped Ge and Si: A review of the experimental status. *Materials* **3**, 5054–5082.

Zhu, W., Zhang, Z., and Kaxiras, E. (2008). Dopant-assisted concentration enhancement of substitutional Mn in Si and Ge. *Phys. Rev. Lett.* **100**, 027205-1–027205-4.

# 8 Solution-Based Fabrication of Carbon Nanotube Thin-Film Logic Gate

*Yan Duan, Jason Juhala, and Wei Xue*

## CONTENTS

**ABSTRACT** Carbon nanotubes (CNTs) have attracted significant attention for numerous research fields due to their exceptional electrical properties. The prospect of creating sophisticated CNT devices offers enormous potential for electronic circuits and systems. A critical step in the development of practical electronics is the creation of complementary logic circuits. Such circuits require both p-channel and n-channel semiconductors. However, the CNT field-effect transistors (FETs) usually exhibit p-channel characteristics under ambient conditions with holes as the majority carriers. Previous approaches such as potassium doping and high-temperature annealing to convert p-channel CNT FETs into n-channel devices rely on expensive systems and are time-consuming. In this chapter, we report the fabrication of low-cost p-channel and n-channel thin-film transistors (TFTs) using single-walled CNT (SWCNT) thin films. The p-type TFTs are fabricated with two solution-based approaches: one device uses aligned SWCNTs prepared by dielectrophoresis, whereas the other contains random-network SWCNTs prepared by layer-by-layer self-assembly.

A comparative analysis of these two TFTs is conducted. Their electrical characteristics are analyzed and compared with a focus on the on/off ratios. After the verification of the p-type characteristics, the devices are converted to n-type with surface coating of an electron-donating polymer polyethylenimine (PEI). The p-type and n-type TFTs are combined to form a logic gate device: a voltage inverter. The results demonstrate low-cost, solution-based methods for fabricating SWCNT-based electronic devices. We believe that the combination of the simple fabrication methods, easy conversion of the transistors, and satisfactory logic gate switching performance can influence fundamental research in nanomaterials and practical applications of nanoelectronics.

## 8.1  INTRODUCTION

Carbon nanotubes (CNTs), as one of the most studied novel materials, have gained significant attention due to their potential as an alternative nanomaterial to be utilized in micro/nanoscale electronics. Their physical and electrical properties make them an excellent candidate for devices of this scale. At room temperature, the intrinsic mobility of an individual semiconducting single-walled CNT (SWCNT) can exceed 1,000,000 cm$^2$/Vs [1], which is greater than any other known semiconductors. Field-effect transistors (FETs) based on SWCNTs have demonstrated high carrier mobility of 3000 cm$^2$/Vs [2]. Therefore, the semiconducting SWCNT-based FETs are considered to be a promising candidate as the building block for future nanoscale electronics. One unique feature of CNTs is that they can behave as a semiconductor or metallic material depending on their physical properties. As a result, they have the potential to be used as the active material for transistors or as interconnects between electronic components. A wide variety of devices, including transistors, integrated circuits (ICs), optoelectronics, and high-frequency electronics, have been successfully developed using CNTs [3,4].

Continuous improvement of the quality and performance of CNT electronics has resulted in a number of novel manipulation methods. In particular, dielectrophoresis, a simple electrical approach using alternating current (AC) signals, has emerged as an effective approach for the precise deposition of aligned CNTs. There has been extensive research on the mechanisms of dielectrophoresis using both experimental and numerical approaches [5–7]. Previous research has indicated that dielectrophoresis has a selection effect on semiconductor CNTS (sCNTs) and metallic CNTs (mCNTs), as they react to an electric field differently [8,9]. Such a phenomenon can be critical for the development of CNT devices, especially of transistors where electrical switching performance is the primary concern. These previous studies have focused on the dielectrophoresis process through numerical analysis and modeling approaches. The experimental investigation of the selection effect on practical electronic devices has been underexplored, which has limited our understanding of the process and the potential of CNTs in practical applications.

With the semiconducting SWCNTs as the active channel material in FETs, the next critical step in assessing the suitability of the resulting devices for nanoelectronics involves the integration of individual SWCNT FETs to form logic gates, which are the basis of today's digital computer technology. Most logic gates require both p-type and n-type FETs. However, the SWCNT FETs usually exhibit p-type characteristics under ambient conditions with holes as the majority carriers. Numerous research efforts have been devoted in the past decade to develop novel and low-cost ICs using CNT FETs and thin-film transistors (TFTs). Complementary logic devices have been fabricated using potassium doping or vacuum annealing methods to create n-type CNT FETs [10–12]. With the use of commercially available high-purity semiconducting nanotube solutions, high-mobility and high-on/off-ratio CNT TFTs were fabricated [13]. Mechanically flexible logic gates such as NOT, NAND, and NOR devices were also demonstrated using CNT TFTs [14]. Combining these basic logic blocks, more sophisticated ICs can be readily constructed by cascading multiple stages of logic gates [15,16]. For example, a 4-to-16 decoder consisting of 88 CNT TFTs was demonstrated [4]. This also represented the first medium-scale ICs using CNTs. However, the p-to-n conversion effects introduced by potassium doping or annealing usually vanish upon air exposure. Consequently, the n-type transistors can only be operated in vacuum (but not in standard temperature and pressure), which is undesirable for practical applications. In addition, the fabrication processes including the special treatment of these CNT-based FETs are usually expensive and time consuming. Reduction in cost and the complexity of production are therefore needed.

In this chapter, we report the experimental evaluation of two solution-based approaches for the development of CNT transistors and logic gates. These two approaches result in two different configurations of CNTs: one transistor uses dielectrophoresis-aligned CNTs, whereas the other contains random-network CNTs using layer-by-layer self-assembly. The CNT random-network assembly has no selection toward mCNTs or sCNTs, resulting in a perfect mixture from the pristine CNT samples. Electrical characteristics of the two transistors are obtained and analyzed comparatively. The comparison of the on/off ratios ($I_{on}/I_{off}$) of the two transistors demonstrates that mCNTs are more responsive to the dielectrophoretic forces during the deposition process. The device fabricated with layer-by-layer self-assembly has a high on/off ratio and is therefore more suitable for electronic switching applications. We further our experiments by combining two transistors to form a logic gate inverter. Both transistors contain random-network SWCNTs with controlled patterns. After the verification of the p-type characteristics, one of the two transistors is converted to n-type with a polyethylenimine (PEI) surface coating. The resulting device, performing as a logic voltage inverter (or NOT gate), is air stable outside a vacuum or an inert environment. Electrical characteristics of the logic gate inverter are obtained and analyzed. Because the functional logic inverter consists of transistors fabricated with low-cost and easy-to-control methods, it has great potential to be used as an alternative to conventional silicon technology.

## 8.2  MATERIALS

The 90% pure semiconducting SWCNT (IsoNanotubes-S) solution used in this project was purchased from NanoIntegris Inc., Skokie, Illinois. It has a CNT concentration of 0.01 mg/ml and a high content (90%) of sCNTs. The sCNTs are 1.2–1.7 nm in diameter and 0.1–4 µm in length. To maintain the original properties of the sCNTs, no chemical or physical modifications were introduced to the solution. The polyelectrolytes used in self-assembly, including poly(dimethyldiallylammonium chloride) (PDDA) (molecular weight 200,000–350,000, polycation) and poly(sodium 4-styrenesulfonate) (PSS, molecular weight 70,000, polyanion) were obtained from Sigma-Aldrich Co, St. Louis, Missouri. The PDDA and PSS were diluted in deionized (DI) water with final concentrations of 15 and 3 mg/ml, respectively. The pH values of PDDA and PSS were measured as 7.5 and 6.5, respectively. In order to increase the ionic strength and enhance the adsorption of the polyions, 0.5 M NaCl was added to both polyelectrolyte solutions [17]. PEI (average molecular weight 25,000) was purchased from Sigma-Aldrich Co., then diluted in methanol to produce a 20 weight percent (wt.%) solution. Low-resistivity silicon wafers (resistivity 0.001–0.005 Ω cm) were purchased from Silicon Inc, Boise, Idaho. The silicon wafers were covered with 100 nm-thick thermal $SiO_2$ layers on their surfaces. They are sputtered with 100 nm of Cr (adhesion metal) and 200 nm of Au (electrode metal) for electrodes.

## 8.3 DEPOSITION OF CNTs

Individual CNT-based devices have been investigated extensively [18,19]. However, the difficulty of manipulating individual nanotubes reduces the device throughput due to the small dimensions of the CNTs. The process of fabricating individual CNT-based devices is time consuming and difficult to control. CNT thin films, in the form of either aligned arrays or random networks, to the contrary can avoid these problems. The process of producing CNT thin films is faster and simpler, and has higher yield. CNT thin films have been readily integrated into many novel devices [20,21].

Alignment of the CNTs is of particular importance in their applications. Well-oriented CNTs can increase the overall quality of the resulting films, such as the thermal and electrical conductivity, and the mechanical strength. Researchers have been investigating various methods to align CNTs in a predefined order. The most common method to align CNTs is to use different catalysts on substrates during the synthesis process through which CNTs can be aligned either horizontally or vertically [22,23]. It is a delicate process where catalysts and accurate control on temperature and gas flow are required. Another method is to implement fluidic microchannels in the aligning step [24–27]. Surface-patterning techniques and fluids make it possible to obtain highly oriented CNT arrays with controlled pitches.

Dielectrophoresis has emerged as an effective approach for the precise deposition of aligned CNTs. It is a simple process conducted at room temperature under low voltages. A number of parameters such as solution concentration, frequency, deposition period, and AC current amplitude can greatly influence the deposition of CNTs. The process of dielectrophoresis commonly described as the motion of polarized particles in an aqueous environment due to an external, nonuniform electric field. The dielectrophoretic force exerted on a CNT is proportional to the real number part of Clausius–Mossotti factor $f_{cm}$, which can be simplified as follows:

$$f_{cm} = \frac{\varepsilon_n^* - \varepsilon_m^*}{\varepsilon_m^*} \qquad (8.1)$$

where:

$\varepsilon^* = \varepsilon - j\sigma/\omega$ is the complex permittivity that contains the information of permittivity $\varepsilon$, conductivity $\sigma$, and angular velocity of the external electric field $\omega$

The subscripts $n$ and $m$ represent the CNT and the medium, respectively

More detailed discussion on the process can be found in our previous publications [28,29].

Previous analytical studies show that at low frequencies, the force is determined by the conductivity $\sigma$, whereas at high frequencies, the force is determined by the permittivity $\varepsilon$ [28]. Conductivity values are estimated as 108 and 105 S/m for mCNTs and sCNTs, respectively, and permittivity values are estimated as $2.5\varepsilon_0$ (sCNTs) and $2000\varepsilon_0$ (mCNTs). In practical operation frequency ranges of dielectrophoresis, the mCNTs are expected to experience stronger dielectrophoretic forces and are more responsive in the deposition process, resulting in a selection effect for separating the mCNTs from the sCNTs. This has been proven by modeling efforts from various groups. For example, Dimaki and Bøggild demonstrated that the force on an mCNT was more than 100 times stronger than that on an sCNT under a frequency between 1 Hz and 1 GHz. Peng et al. further explored the forces in the high-frequency range, showing that mCNTs experienced much stronger forces (>1000 times) than sCNTs in a frequency range of 500 MHz–1 GHz [9]. In addition, there is a significant difference between the mCNTs and the sCNTs in terms of their polarizability, as explained in previous reports [30,31]. When the mCNTs are placed in an alternating electric field, they are aligned parallel to the induced field. This occurs because the mCNTs have a band gap at or near zero, allowing their polarizability to be along the tube axis. Different from the mCNTs, the sCNTs have a nonzero band gap, leading to their semiconducting properties. The band gap creates transverse polarizability, which is comparable to the polarizability along the tube axis. This variation causes the sCNTs to be aligned in a pseudorandom configuration [31].

Various deposition methods have also been used to develop random-network CNT transistors. Such devices can be fabricated using commercially available CNT products, simplifying the fabrication process and lowering the cost for material preparation. The deposited films contain both mCNTs and sCNTs with their predetermined ratios. However, one major disadvantage of this approach is that the nanotube–nanotube cross-junctions in the films introduce additional energy barriers, which can reduce the effective electric current in the devices [32,33]. Research suggests that using shorter CNTs as well as electrical breakdown can improve the performance of random-network CNT transistors [34,35]. In this project, we use the layer-by-layer self-assembly process, which is a solution-based and easy-to-control procedure, to deposit CNTs. This assembly process is based on electrostatic attraction of materials with opposite charges. A blanket of random-network

CNTs can be assembled to cover the entire surface. The assembly process can be combined with lithography and liftoff of sacrificial materials to create complex structures.

## 8.4    CNT THIN-FILM TRANSISTORS

### 8.4.1    TRANSISTOR DESIGN AND FABRICATION

Figure 8.1 illustrates the structures of the CNT TFTs. The call-out box highlights the structural comparison between aligned and random-network CNT films. The device has a back-gate transistor configuration on a low-resistivity silicon substrate, which also acts as the gate terminal to simplify the fabrication and testing procedures. Contact lithography and wet etching techniques are used to fabricate the Cr/Au electrodes on a low-resistivity $Si/SiO_2$ wafer. The patterned Cr/Au electrodes are used as the source and drain terminals for the transistor, whereas the substrate is used as the back-gate terminal. The detailed information of the electrode design and fabrication procedures can be found in our previous publications [36–38].

Two solution-based methods are used to deposit CNT films. The first method is to use dielectrophoresis to align CNTs on the wafer. In this process, the 90% sCNT solution is placed between the source and drain electrodes. The CNTs are aligned across the electrodes by applying an AC voltage source (frequency of 5 MHz and peak-to-peak voltage of 10 V) using a function generator. After 60 s, the AC signal is turned off and a miniaturized vacuum pump is used to remove the CNT solution from the device, leaving only the CNTs in contact with the substrate on the electrodes.

The second method used to fabricate CNT transistors is based on electrostatic self-assembly of CNTs and a liftoff process to remove sacrificial materials [39]. During this process, a random network of CNTs is generated with no selection of mCNTs or sCNTs. After electrode fabrication, a sacrificial layer of photoresist is patterned using spin coating, contact lithography, and development. The wafer is treated

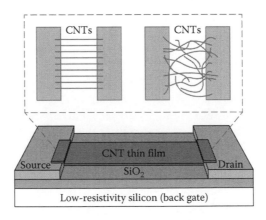

**FIGURE 8.1**    Structure of the CNT TFT. Call-out box shows the top views of aligned CNTs based on dielectrophoresis and a random network of CNTs based on self-assembly.

with polyelectrolytes, PDDA (positive charge) and PSS (negative charge), to obtain desired charges on its surface. The 90% sCNT solution is placed on the device for 30 min to form a random-network CNT film. A liftoff step in ultrasonic is used to remove the photoresist sacrificial layer, along with the other films on top. The surface charging properties of CNTs (negative charge) and PDDA (positive charge) have been determined with a zeta potential instrument. The positive PDDA molecules are used as electrostatic *glues* for uniform deposition of the negatively charged SWCNTs on the substrate.

A sample die containing an array of eight CNT transistors is shown in Figure 8.2 (top). Structural characterization of the devices is carried out using scanning electron microscopy (SEM). SEM images of aligned CNTs and random-network CNTs between the source and drain electrodes are shown in Figure 8.2 (bottom). As depicted, the CNTs are successfully positioned between the electrode teeth patterns, in a 3 μm gap. Based on the measurement of five arbitrarily selected devices which have the same channel length and width, we estimate the numbers of CNTs between the two electrodes under the SEM. The results show that the densities of CNTs from dielectrophoresis and self-assembly are on the same order of magnitude (on the order of 100), suggesting that these two films are comparable for this study.

**FIGURE 8.2** An optical image of a fabricated device on a one-dime coin (top) and SEM images of aligned SWCNTs (bottom left) and random-network SWCNTs on top of a PDDA/PSS bilayer (bottom right) between the source and drain electrodes.

## 8.4.2  COMPARATIVE ANALYSIS

The electrical characteristics of both transistors are obtained using a semiconductor device analyzer. The testing conditions for both transistors are as follows: the drain voltage is swept from 0 to $-5$ V in $-100$ mV increments, with the gate voltage sweeping from 0 to $-5$ V in $-1$ V increments. The current–voltage ($I$–$V$) characteristics of a dielectrophoresis-aligned CNT transistor are shown in Figure 8.3a. The transistor demonstrates p-type characteristics with an explicit field effect. However, the on/off ratio, defined as the high current ($I_{on}$) over the low current ($I_{off}$) at a drain voltage of $-5$ V, is low (~1.20). Figure 8.3b illustrates the $I$–$V$ characteristics of a transistor using the random-network CNTs. The curves show a greater field effect (larger separation between adjacent curves) compared with the results of the aligned CNT transistor. The on/off ratio obtained at a drain voltage of $-5$ V is found to be ~16.62.

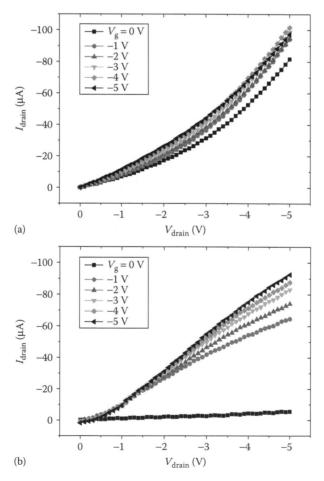

(a)

(b)

**FIGURE 8.3**  Current–voltage characteristic of transistors using (a) aligned CNTs and (b) random-network CNTs.

This difference in on/off ratios is not a unique case as it has been frequently observed in our characterization of CNT transistors. The statistical comparison between the two groups of transistors is illustrated in Figure 8.4a. Group 1 represents transistors using aligned CNTs, whereas group 2 contains devices using random-network CNTs. Each group contains five arbitrarily selected transistors. For the aligned CNT transistors, the average on/off ratio and the standard deviation are obtained as 1.60 and 0.49, respectively. For the random-network CNT transistors, the average on/off ratio and the standard deviation are 19.56 and 2.90, respectively. The apparent difference in on/off ratios indicates that the dielectrophoresis process attracts more mCNTs than the electrostatic self-assembly method.

The transfer characteristics of the CNT transistors are also investigated to provide an in-depth understanding of their electrical properties. Figure 8.4b shows the $I_{drain}$ versus $V_{gate}$ (drain current vs. gate voltage) characteristics of a CNT transistor

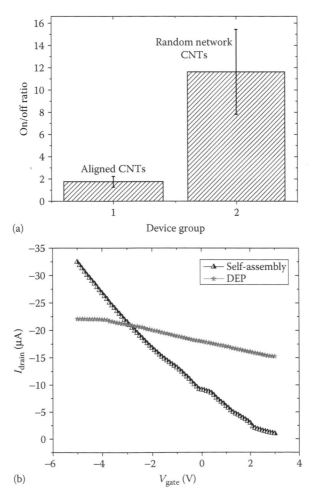

**FIGURE 8.4** Comparison of (a) on/off ratios of the two groups of transistors and (b) the $I_{drain} - V_{gate}$ characteristics of two transistors.

fabricated using the dielectrophoresis process as well as a CNT transistor fabricated using the self-assembly process. The drain voltages are fixed at −4.1 V for both devices. The number of majority charge carriers, holes, is directly proportional to the current flowing between the source and drain electrodes. The slopes of the two transistors from the dielectrophoresis and the self-assembly processes are estimated as −0.9375 and −4.0625, respectively. This indicates that random-network CNT transistors demonstrate a better switching performance than aligned CNT transistors. These results are consistent with the estimation of the Clausius–Mossotti factor $f_{cm}$ expressed in Equation 8.1 and the polarizability difference between the mCNTs and the sCNTs.

Therefore, when fabricating transistors with the dielectrophoresis method, a higher concentration of mCNTs is aligned between the source and drain electrodes. The electrical properties, especially the on/off ratio, are shifted toward the metallic region. This effect limits the applications of these transistors, especially when they are used as electrical switches. Further research will entail reducing the concentration of mCNTs. One method to accomplish this is to use a higher concentration sCNT solution, which contains a lesser percentage of mCNTs. There are commercially available 99% sCNT solutions that could potentially increase the on/off ratio of transistors. Burning off mCNTs through electrical breakdown can also be used to reduce the concentration of mCNTs [40]. In order to selectively break the mCNTs, a positive $V_{drain}$ is applied to maintain the sCNTs in the off state. Although this method is effective, it can lead to a large reduction in current drive.

## 8.5    CNT-BASED VOLTAGE INVERTER

### 8.5.1    INVERTER DESIGN AND FABRICATION

Functional CNT transistors can be used for logic circuits. Figure 8.5 shows the overall fabrication process of a single voltage inverter. Contact lithography and wet etching techniques are used to fabricate the device structure on a low-resistivity 4 inch Si/SiO$_2$ wafer, as shown in Figure 8.5a. The Si/SiO$_2$ wafer is sputtered with 100 nm of Cr (adhesion metal) and 200 nm of Au (electrode metal). The wafer is cleaned with acetone and isopropyl alcohol followed by thorough rinsing with DI water. An adhesion promoter (MCC primer 80/20) and a positive photoresist (Shipley S1805) are applied on the surface of the wafer using a spin coater (300 rpm for 10 s followed by 4000 rpm for 30 s). UV exposure is carried out in a hard contact aligner with an exposure time of 20 s. The exposed photoresist, Cr, and Au are removed using a diluted 351 developer (volume ratio of 20% in water) for 45 s, a Cr etchant for 90 s, and a Au etchant for 9 s. The unexposed photoresist is removed using a 1:1 mixture of H$_2$SO$_4$ and H$_2$O$_2$ at 110°C for 10 min. The patterned Cr/Au electrodes are used as the source and drain terminals for the transistor. Another layer of photoresist is spin coated on the wafer surface. A second photomask is used in the UV exposure step to open up two areas where the SWCNTs can be deposited onto the wafer. The layer-by-layer self-assembly process method is used to deposit the semiconducting SWCNTs solution as the channel material for transistors at room temperature [41].

Next, the photoresist sacrificial structures are removed with acetone in an ultrasonic bath using a liftoff process. As a result, two individual transistors are fabricated, as

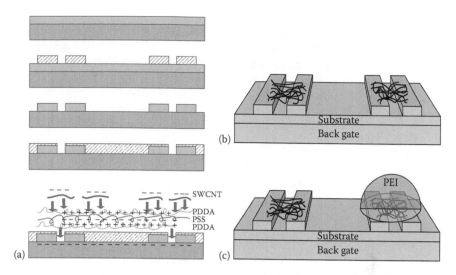

**FIGURE 8.5** Fabrication process of a logic gate inverter. (a) Fabrication flow using lithography and self-assembly. (b) Device structure after fabrication. The device contains two p-type FETs. (c) Device structure after PEI coating.

shown in Figure 8.5b. Both transistors are expected to exhibit p-type characteristics because they are exposed in air after fabrication [42]. The 4 inch Si/SiO$_2$ wafer is cut into 1 cm × 1 cm dies with a dicing saw. Each die contains four logic gate inverters with two transistors for each inverter. The p-type characteristics of each transistor are tested to confirm that they are functional and can be used to form logic gate devices. A precision syringe is used to place a small drop of 20 wt.% PEI solution on the transistor, which is connected to voltage source supply ($V_{ss}$) during the testing process, as illustrated in Figure 8.5c. The PEI coating is left on the device overnight, followed by thorough rinsing with methanol. This rinsing step removes PEI molecules that are loosely adsorbed on the sample surface, leaving a thin layer of PEI on the logic gate inverter.

An optical image of a fabricated device after the first lithography step is shown in Figure 8.6a. Four logic gate inverters are contained on a single die. Their dimensions are relatively large for reliable PEI coating without cross-contamination between the transistors. Future miniaturization can be easily achieved using sacrificial structures in a liftoff process. A SEM image of a logic gate inverter after the second lithography step is shown in Figure 8.6b. The channel areas between the source and drain electrodes are open for the SWCNT deposition. Figure 8.6c shows an optical image of the PEI coating on one of the two transistors of an inverter. A SEM image of the random-network SWCNTs between the source and drain electrodes is shown in Figure 8.6d.

The fabricated logic gate inverter is tested on a probe station. Two metal probes are used to supply the positive ($V_{dd}$) and negative ($V_{ss}$) reference voltages; a small piece of copper tape is attached to the backside of the substrate as the input terminal ($V_{in}$). The circuit diagram of the logic gate inverter is shown in Figure 8.7a. This logic gate contains a p-type FET and an n-type FET; it can be turned on and off based on the applied conditions. Figure 8.7b shows the device design using SWCNT FETs. When a negative input voltage (digital "0") is applied to $V_{in}$, the p-FET is turned

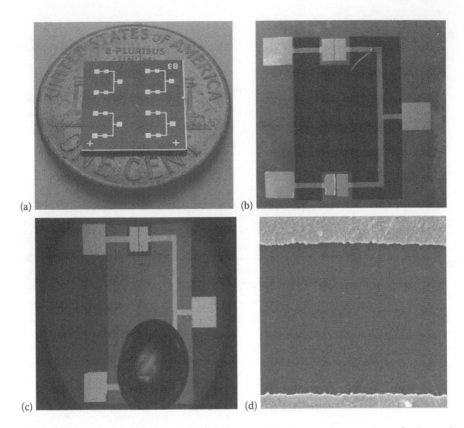

**FIGURE 8.6**  (a) Optical image of a fabricated device on a one-cent coin (penny). (b) SEM image of a single inverter after the second UV exposure and liftoff steps. (c) Optical image of the device with PEI coating on one transistor. (d) SEM image of random-network SWCNTs on top of a PDDA + PSS bilayer between the source and drain electrodes.

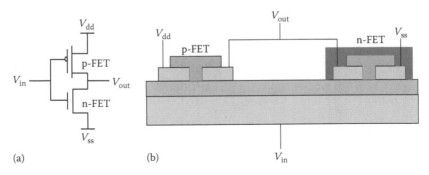

**FIGURE 8.7**  (a) Circuit diagram and (b) the device schematic of a voltage inverter (NOT gate) using two SWCNT FETs.

on and the n-FET is turned off. As a result, the output terminal $V_{out}$ is connected to the positive reference voltage $V_{dd}$, providing a positive output voltage (digital "1"). Similarly, a positive input (digital "1") at $V_{in}$ turns the p-FET off and the n-FET on, resulting in a negative output (digital "0") at $V_{out}$.

## 8.5.2 CHARACTERIZATION

The SWCNT FETs usually exhibit p-type characteristics under ambient conditions. The p-type behavior is due to the adsorbed oxygen ($O_2$) from the ambient as each $O_2$ molecule withdraws 1/10 of an electron from the SWCNTs [42], leaving holes as the majority carriers.

The two transistors in a logic gate inverter are tested after the deposition of the SWCNT films to ensure that both of them can function as p-type FETs. Their electrical characteristics are obtained using a semiconductor device analyzer using the following measurement conditions: The drain voltage is swept from 0 to −5 V in −100 mV increments, the gate voltage is swept from 0 to −5 V in −1 V increments, the positive reference voltage $V_{dd}$ is fixed at 10 V, and the negative reference voltage is $V_{ss}$ is fixed at −10 V. The output characteristics (drain current $I_{drain}$ vs. drain voltage $V_{drain}$) of a self-assembled SWCNT transistor are shown in Figure 8.8a. The transistor

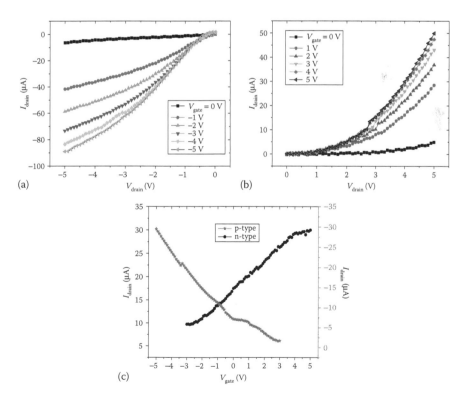

**FIGURE 8.8** Output characteristics of (a) a p-type FET and (b) an n-type FET based on self-assembled random-network SWCNT thin films. (c) Transfer characteristics of an SWCNT transistor before and after the p-to-n conversion using the PEI coating.

demonstrates p-type characteristics with an explicit field effect. The on/off current ratio at a drain voltage of −5 V is found to be ~13.89.

After the fabrication of the devices and the verification of p-type characteristics of both transistors on a single inverter device, PEI surface coating is introduced to convert the p-type transistor (connected to $V_{ss}$) to an n-type transistor. PEI contains amine groups which have high electron-donating ability, causing hole depletion in p-type SWCNTs [43,44]. In addition, a thin layer of PEI on the device prevents the SWCNTs from being exposed to air to overcome the adverse effect of p-doping by $O_2$ adsorption. With the PEI coating step, a p-type FET can be easily converted to an air-stable n-type device. Due to the high electron-donating ability of the PEI molecules, the SWCNTs now have electrons as the majority charge carriers. The output characteristics of an n-type FET using self-assembled SWCNTs and PEI coating are shown in Figure 8.8b. Both the drain voltage and the gate voltage are in the range of 0–5 V. The on/off ratio is obtained as ~10.14. More information about the electrical characteristics of both the p-type and n-type SWCNT FETs can be found in our previous publications [37,38].

The two plots in Figure 8.8c show the transfer characteristics (drain current $I_{drain}$ vs. gate voltage $V_{gate}$) of the same transistor before and after the p-to-n conversion with the PEI coating. The drain voltages are fixed at −4.1 V for the initial p-type device and 4.1 V for the resulting n-type device after the PEI coating. From the figure, we can roughly estimate the threshold voltages as $V_{tp} = 1.5$ V for the p-type FET and $V_{tn} = −4.5$ V for the n-type. For both cases, when the gate voltage is increased, the number of the majority charge carriers, holes for the p-type and electrons for the n-type, in the channel increases. This, in turn, increases the current flowing between the source and drain electrodes. The polarity difference between the two plots clearly demonstrates the conversion of the transistor from p-type to n-type with PEI coating.

After verifying the functions of the p-type FET (connected to $V_{dd}$) and the converted n-type FET (connected to $V_{ss}$), we perform the electrical characterization to obtain the behavior of the logic gate inverter. In this logic device, the input voltage is applied simultaneously to the back gates of the two complementary SWCNT FETs. The device is tested by connecting the p-type FET with $V_{dd}$ for a positive bias and connecting the n-type FET with $V_{ss}$ for a negative bias. This configuration allows the output bias of the inverter to be controlled by changing the input voltage of the two transistors. The static transfer characteristics of the logic gate inverter are shown in Figure 8.9. By sweeping the input voltage from −25 to 25 V while maintaining the reference voltages constant at 10 and −10 V, respectively, the switching of the output signal between two logic states is observed. A positive input voltage turns the n-type FET on and the p-type FET off, resulting in the transmission of the negative polarization voltage to the output. Alternatively, a negative input turns the p-type FET on and the n-type FET off, inducing a positive output.

Based on our experimental conditions, the logic gate inverter shows a clear and satisfactory switching performance outside the low-power range. However, in the lower voltage range between −4 and 4 V, the inverter device is insensitive to the input voltage. We believe that such a phenomenon is caused by un-optimized

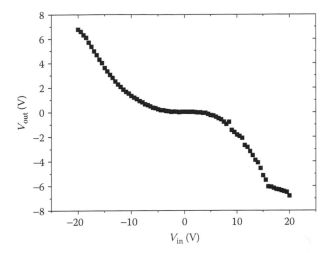

**FIGURE 8.9**    Characteristics of the resulting logic gate inverter.

threshold voltages of the two transistors. The threshold voltages play critical roles in device operation and the optimized values can significantly improve the device performance. A number of approaches that have been demonstrated for other types of TFTs can also be used in our SWCNT FETs. The variation of threshold voltage as a function of the gate oxide thickness and the effect on TFTs were previously reported [45]. Motivated by this phenomenon, the thickness of $SiO_2$ in our devices can be modified for improved device performance. The influence of the channel layer thickness on the electrical performance of TFTs was investigated by several groups [46,47]. This approach can be easily adopted into our design by controlling the thickness of the SWCNT thin films to fine-tune the device operation. Furthermore, by introducing a reactive interfacial layer between the $SiO_2$ layer and the SWCNT film in TFTs, the threshold voltage can also be tuned for optimizing the device performance [48]. To take advantage of the entire voltage range, especially in the low-voltage range, these approaches will be explored and used in our future experiments to advance our understanding in SWCNT-based logic gate devices and circuits.

### 8.5.3    LONG-TERM STABILITY

After the successful fabrication and characterization of the logic gate inverter, another important characteristic we have investigated is the long-time stability of functional devices. The fabricated devices are placed in a glass container in a drawer under normal laboratory conditions. The electrical characteristics of the functional logic gate inverter are tested 7, 14, 30, and 60 days after they were initially fabricated. As demonstrated in Figure 8.10, an arbitrarily selected inverter shows relatively stable on/off ratios with insignificant changes for its p-type FET and n-type

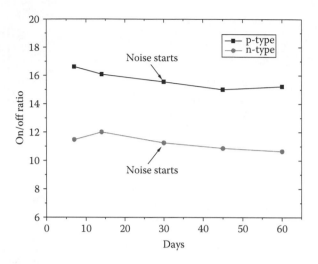

**FIGURE 8.10**   On/off ratio changes for the p-type and n-type FETs in a logic gate inverter over a 60-day period.

FET during a 60-day period. Although noise starts to appear on the resulting *I–V* curves 30 days after the device is fabricated, the inverter and its two transistors continue to function and their performance remains reasonably consistent. This demonstrates that the fabrication methods used in this research provides a reliable approach to produce air-stable logic gate devices that can remain stable for at least 2 months. Further experimentation is needed to explore the long-time stability of the fabricated devices and potential methods to maintain their long-time performance.

## 8.6   CONCLUSION

We have investigated the thin-film deposition of SWCNTs using two solution-based deposition methods: dielectrophoresis and layer-by-layer self-assembly. The electrical characteristics of SWCNT TFTs from both methods are analyzed and compared with a focus on the on/off ratios. The characterization results show that the aligned SWCNT FET has a lower on/off ratio, indicating a selection effect to separate mCNTs from sCNTs in dielectrophoresis under our experimental condition. The as-fabricated SWCNT TFTs exhibit p-type characteristics. The n-doping of SWCNTs using a polymer with high electron-donating ability results in air-stable n-type SWCNT TFTs. By integrating p-type and n-type SWCNT TFTs into one design, we are able to demonstrate a functional logic gate inverter with satisfactory output characteristics. The long-term stability testing shows that the device remains functional for at least 2 months. The results obtained from our experiments provide crucial information on how SWCNTs can be used in future research and practical devices. We believe that the results can lead to the development of more sophisticated logic circuits in the near future.

## REFERENCES

1. Xue, W. and Cui, T., Carbon nanotube micropatterns and cantilever arrays fabricated with layer-by-layer nano self-assembly. *Sensors and Actuators A: Physical*, 2007. **136**(2): pp. 510–517.
2. Javey, A., Kim, H., Brink, M., Wang, Q., Ural, A., Guo, J., McIntyre, P., McEuen, P., Lundstrom, M., and Dai, H., High-κ dielectrics for advanced carbon-nanotube transistors and logic gates. *Nature Materials*, 2002. **1**(4): pp. 241–246.
3. Bandaru, P.R., Electrical properties and applications of carbon nanotube structures. *Journal of Nanoscience and Nanotechnology*, 2007. **7**(4): pp. 1239–1267.
4. Cao, Q. and Rogers, J.A., Ultrathin films of single-walled carbon nanotubes for electronics and sensors: A review of fundamental and applied aspects. *Advanced Materials*, 2009. **21**(1): pp. 29–53.
5. Monica, A.H., Papadakis, S.J., Osiander, R., and Paranjape, M., Wafer-level assembly of carbon nanotube networks using dielectrophoresis. *Nanotechnology*, 2008. **19**(8): p. 085303.
6. Lu, Y., Chen, C., Yang, L., and Zhang, Y., Theoretical simulation on the assembly of carbon nanotubes between electrodes by AC dielectrophoresis. *Nanoscale Research Letters*, 2008. **4**(2): pp. 157–164.
7. Sarker, B.K., Shekhar, S., and Khondaker, S.I., Semiconducting enriched carbon nanotube aligned arrays of tunable density and their electrical transport properties. *ACS Nano*, 2011. **5**(8): pp. 6297–6305.
8. Dimaki, M. and Bøggild, P., Dielectrophoresis of carbon nanotubes using microelectrodes: A numerical study. *Nanotechnology*, 2004. **15**(8): pp. 1095–1102.
9. Peng, N., Zhang, Q., Li, J., and Liu, N., Influences of ac electric field on the spatial distribution of carbon nanotubes formed between electrodes. *Journal of Applied Physics*, 2006. **100**(2): p. 024309.
10. Kong, J., Zhou, C., Yenilmez, E., and Dai, H., Alkaline metal-doped n-type semiconducting nanotubes as quantum dots. *Applied Physics Letters*, 2000. **77**(24): pp. 3977–3979.
11. Derycke, V., Martel, R., Appenzeller, J., and Avouris, P., Carbon nanotube inter- and intramolecular logic gates. *Nano Letters*, 2001. **1**(9): pp. 453–456.
12. Javey, A., Tu, R., Farmer, D.B., Guo, J., Gordon, R.G., and Dai, H., High performance n-type carbon nanotube field-effect transistors with chemically doped contacts. *Nano Letters*, 2005. **5**(2): pp. 345–348.
13. Ha, M., Xia, Y., Green, A.A., Zhang, W., Renn, M.J., Kim, C.H., Hersam, M.C., and Frisbie, C.D., Printed, sub-3V digital circuits on plastic from aqueous carbon nanotube inks. *ACS Nano*, 2010. **4**(8): pp. 4388–4395.
14. Wang, C., Chien, J.-C., Takei, K., Takahashi, T., Nah, J., Niknejad, A.M., and Javey, A., Extremely bendable, high-performance integrated circuits using semiconducting carbon nanotube networks for digital, analog, and radio-frequency applications. *Nano Letters*, 2012. **12**(3): pp. 1527–1533.
15. Ko, H.C., Stoykovich, M.P., Song, J., Malyarchuk, V., Choi, W.M., Yu, C.-J., Geddes III, J.B. et al., A hemispherical electronic eye camera based on compressible silicon optoelectronics. *Nature*, 2008. **454**(7205): pp. 748–753.
16. Sun, D.M., Timmermans, M.Y., Tian, Y., Nasibulin, A.G., Kauppinen, E.I., Kishimoto, S., Mizutani, T., and Ohno, Y., Flexible high-performance carbon nanotube integrated circuits. *Nature Nanotechnology*, 2011. **6**(3): pp. 156–161.
17. Xue, W. and Cui, T., Characterization of layer-by-layer self-assembled carbon nanotube multilayer thin films. *Nanotechnology*, 2007. **18**(14): p. 145709.
18. Li, S., He, P., Dong, J., Guo, Z., and Dai, L., DNA-directed self-assembling of carbon nanotubes. *Journal of the American Chemical Society*, 2004. **127**(1): pp. 14–15.

19. Besteman, K., Lee, J.-O., Wiertz, F.G.M., Heering, H.A., and Dekker, C., Enzyme-coated carbon nanotubes as single-molecule biosensors. *Nano Letters*, 2003. **3**(6): pp. 727–730.

20. Baughman, R.H., Zakhidov, A.A., and de Heer, W.A., Carbon nanotubes—The route toward applications. *Science*, 2002. **297**(5582): pp. 787–792.

21. Liu, T., Phang, I.Y., Shen, L., Chow, S.Y., and Zhang, W.-D., Morphology and mechanical properties of multiwalled carbon nanotubes reinforced nylon-6 composites. *Macromolecules*, 2004. **37**(19): pp. 7214–7222.

22. Suh, J.S. and Lee, J.S., Highly ordered two-dimensional carbon nanotube arrays. *Applied Physics Letters*, 1999. **75**(14): pp. 2047–2049.

23. Liu, H., Takagi, D., Chiashi, S., and Homma, Y., The controlled growth of horizontally aligned single-walled carbon nanotube arrays by a gas flow process. *Nanotechnology*, 2009. **20**(34): p. 345604.

24. Li, S., Liu, N., Chan-Park, M.B., Yan, Y., and Zhang, Q., Aligned single-walled carbon nanotube patterns with nanoscale width, micron-scale length and controllable pitch. *Nanotechnology*, 2007. **18**(45): p. 455302.

25. Huang, Y., Duan, X., Wei, Q., and Lieber, C.M., Directed assembly of one-dimensional nanostructures into functional networks. *Science*, 2001. **291**(5504): pp. 630–633.

26. Salalha, W. and Zussman, E., Investigation of fluidic assembly of nanowires using a droplet inside microchannels. *Physics of Fluids* (1994-present), 2005. **17**(6): p. 063301.

27. Smith, B.W., Benes, Z., Luzzi, D.E., Fischer, J.E., Walters, D.A., Casavant, M.J., Schmidt, J., and Smalley, R.E., Structural anisotropy of magnetically aligned single wall carbon nanotube films. *Applied Physics Letters*, 2000. **77**(5): pp. 663–665.

28. Li, P. and Xue, W., Selective deposition and alignment of single-walled carbon nanotubes assisted by dielectrophoresis: From thin films to individual nanotubes. *Nanoscale Research Letters*, 2010. **5**(6): pp. 1072–1078.

29. Li, P., Lei, N., Xu, J., and Xue, W., High-yield fabrication of graphene chemiresistors with dielectrophoresis. *IEEE Transactions on Nanotechnology*, 2012. **11**(4): pp. 751–759.

30. Blatt, S., Hennrich, F., Löhneysen, H.v., Kappes, M.M., Vijayaraghavan, A., and Krupke, R., Influence of structural and dielectric anisotropy on the dielectrophoresis of single-walled carbon nanotubes. *Nano Letters*, 2007. **7**(7): pp. 1960–1966.

31. Padmaraj, D., Zagozdzon-Wosik, W., Xie, L.M., Hadjiev, V.G., Cherukuri, P., and Wosik, J., Parallel and orthogonal E-field alignment of single-walled carbon nanotubes by ac dielectrophoresis. *Nanotechnology*, 2009. **20**(3): p. 035201.

32. Fuhrer, M.S., Nygård, J., Shih, L., Forero, M., Yoon, Y.-G., Mazzoni, M.S.C., Choi, H.J. et al., Crossed nanotube junctions. *Science*, 2000. **288**(5465): pp. 494–497.

33. Yao, Z., Postma, H.W.C., Balents, L., and Dekker, C., Carbon nanotube intramolecular junctions. *Nature*, 1999. **402**(6759): pp. 273–276.

34. Ishida, M. and Nihey, F., Estimating the yield and characteristics of random network carbon nanotube transistors. *Applied Physics Letters*, 2008. **92**(16): p. 163507.

35. Kumar, S., Pimparkar, N., Murthy, J.Y., and Alam, M.A., Theory of transfer characteristics of nanotube network transistors. *Applied Physics Letters*, 2006. **88**(12): p. 123505.

36. Duan, Y., Juhala, J.L., Griffith, B.W., and Xue, W., A comparative analysis of thin-film transistors using aligned and random-network carbon nanotubes. *Journal of Nanoparticle Research*, 2013. **15**(3): p. 1478.

37. Duan, Y., Juhala, J.L., Griffith, B.W., and Xue, W., Solution-based fabrication of p-channel and n-channel field-effect transistors using random and aligned carbon nanotube networks. *Microelectronic Engineering*, 2013. **103**: pp. 18–21.

38. Duan, Y., Holmes, N.E., Ellard, A.L., Gao, J., and Xue, W., Solution-based fabrication and characterization of a logic gate inverter using random carbon nanotube networks. *IEEE Transactions on Nanotechnology*, 2013. **12**(6): pp. 1111–1117.

39. Xue, W., Liu, Y., and Cui, T., High-mobility transistors based on nanoassembled carbon nanotube semiconducting layer and $SiO_2$ nanoparticle dielectric layer. *Applied Physics Letters*, 2006. **89**(16): p. 163512.

40. Collins, P.G., Arnold, M.S., and Avouris, P., Engineering carbon nanotubes and nanotube circuits using electrical breakdown. *Science*, 2001. **292**(5517): pp. 706–709.

41. Xue, W. and Cui, T., Thin-film transistors with controllable mobilities based on layer-by-layer self-assembled carbon nanotube composites. *Solid-State Electronics*, 2009. **53**(9): pp. 1050–1055.

42. Shim, M., Javey, A., Shi Kam, N.W., and Dai, H., Polymer functionalization for air-stable n-type carbon nanotube field-effect transistors. *Journal of the American Chemical Society*, 2001. **123**(46): pp. 11512–11513.

43. Collins, P.G., Bradley, K., Ishigami, M., and Zettl, A., Extreme oxygen sensitivity of electronic properties of carbon nanotubes. *Science*, 2000. **287**(5459): pp. 1801–1804.

44. Kong, J. and Dai, H., Full and modulated chemical gating of individual carbon nanotubes by organic amine compounds. *The Journal of Physical Chemistry B*, 2001. **105**(15): pp. 2890–2893.

45. Navneet, G., Threshold voltage modelling and gate oxide thickness effect on polycrystalline silicon thin-film transistors. *Physica Scripta*, 2007. **76**(6): p. 628.

46. Ye, W., Xiao Wei, S., Goh, G.K.L., Demir, H.V., and Hong Yu, Y., Influence of channel layer thickness on the electrical performances of inkjet-printed in-Ga-Zn oxide thin-film transistors. *IEEE Transactions on Electron Devices*, 2011. **58**(2): pp. 480–485.

47. Boudinet, D., Benwadih, M., Altazin, S., Gwoziecki, R., Verilhac, J.M., Coppard, R., Le Blevennec, G., Chartier, I., and Horowitz, G., Influence of the semi-conductor layer thickness on electrical performance of staggered n- and p-channel organic thin-film transistors. *Organic Electronics*, 2010. **11**(2): pp. 291–298.

48. Etschmaier, H., Pacher, P., Lex, A., Trimmel, G., Slugovc, C., and Zojer, E., Continuous tuning of the threshold voltage of organic thin-film transistors by a chemically reactive interfacial layer. *Applied Physics A*, 2009. **95**(1): pp. 43–48.

# 9 On the Possibility of Observing Tunable Laser-Induced Band Gaps in Graphene*

*Hernán L. Calvo, Horacio M. Pastawski,
Stephan Roche, and Luis E.F. Foa Torres*

## CONTENTS

**ABSTRACT** Graphene research is a fast growing field of opportunities covering a transversal research landscape from basic science to technology. Here, we focus on the possibility of tuning the electronic and transport properties of graphene through illumination with a laser field. Our driving questions are, Is it possible to induce band gaps in graphene through laser illumination? Can those gaps be observed? In this chapter, we extend previous works on this subject while giving a broader and updated introduction to this fascinating field.

## 9.1 INTRODUCTION

Graphene-based devices offer a wealth of fascinating opportunities for the study of truly two-dimensional physics [1]. The last years have seen an unprecedented pace of development regarding the study of its electronic, mechanical, and optical properties, among many others. More recently, the interplay between these properties has been at the forefront of graphene research [2–5] opening up many promising paths for technology. In this chapter, we give a brief overview to the interplay between

---

* Reprinted and updated; originally published as Chapter 3 of *Graphene, Carbon Nanotubes, and Nanostructures: Techniques and Applications*, CRC Press, 2013.

optical and electronic properties in graphene and we address the issue of tunability of the electronic structure using a laser field.

In graphene-based $p–n$ junctions, Syzranov and collaborators [6,7] predicted in 2008 the emergence of depletion regions in the density of states (DoS) as a result of the interaction with a linearly polarized laser of frequency $\Omega$. These regions, located at energies $\pm\hbar\Omega/2$ above/below the charge neutrality point, are formed by the opening of dynamical gaps in each transversal mode. However, after summing up all band contributions, instead of a gap, one gets a single point of vanishing DoS. Strikingly, this resembles the usual Dirac point at zero energy. Throughout the chapter, we will refer to these regions as *dynamical gaps*.

Later on, Oka and Aoki [8] predicted that the depletion regions at $\pm\hbar\Omega/2$ turn into a gap when changing the polarization from linear to circular; this was confirmed in Reference [9]. Furthermore, they showed that an additional gap opens up the Dirac point due to a higher order process. Similar results were also presented for bilayer graphene [10,11]. In relatively small graphene ribbons, additional laser-induced gaps at other energies not commensurate with half the photon energy emerged as a consequence of the finite size of the sample [12]. These gaps showed to vanish in the bulk limit [13].

In what follows, we will show that these strong renormalizations of the electronic structure of graphene can be rationalized in a transparent way using Floquet theory, which allows to explore electronic and transport properties of materials in the presence of oscillating electromagnetic fields. The origin of the laser-induced band gaps will be shown to be related to an inelastic Bragg's scattering mechanism occurring in a higher dimensional Hilbert space, the Floquet space. In Reference [14], we followed this line of analysis trying to answer questions such as the following: What is the role of the laser polarization on these dynamical gaps, and which experimental setup would be needed to observe them? Our results showed that the polarization could be used to strongly modulate the associated electrical response. Furthermore, the first atomistic simulations of the electrical response (dc conductance) of a large graphene ribbon (of about 1 μm lateral size) were presented. Our results hinted that a transport experiment carried out while illuminating with a laser in the mid-infrared could unveil these unconventional phenomena. There are also additional ingredients which add even more interest to this proposal: The current interest in finding novel ways to open band gaps in graphene and the promising prospects for optoelectronics applications.

Other recent studies have also contributed to different aspects of this field including the following: The analogy between the electronic spectra of graphene in laser fields and those of static graphene superlattices [15]; the possible engineering of the Dirac points [16], similarly to those obtained through a superimposed lattice potential [17]; a study of the effects of radiation on graphene ribbons showing that for long ribbons there is a ballistic regime where edge transport dominates [18]; a proposal to unveil the dynamical gaps using photon echo experiments [19]; and the effects of radiation on tunneling [20], or tunneling times [21]. The influence of mentioned laser-induced band gaps on the polarizability has also been recently discussed in Reference [22]. Laser-induced pumping was studied in Reference [23]. On the experimental side, there is a surge of activity on graphene optoelectronics [2,24,25] and we expect further experiments along this line.

Another captivating possibility is the generation, through a circularly polarized laser [8,26], of Floquet topological insulators [27–29], whose states are much alike those of a topological insulator. The advantage of such a way of generating a topologically protected phase is clear: although topological insulators are usually exotic materials, a laser could bring similar states to widely available materials such as graphene and, more importantly, it could allow for the tuning of their properties (bulk gap, velocity of the propagating edge states, etc.) [11,16,18,28,30]. Reference [30] presents a proposal for achieving this in graphene and offers an analytical solution for the resulting Floquet chiral edge states. Other studies focus on the classification of the associated topological invariants for this nonequilibrium phase [31,32] and the bulk-boundary correspondence [33]. The experimental realization of these phenomena is also a very active area of research in photonic crystals [34], where scattering-free edge transport was demonstrated in a graphene-like optical lattice. Finally, the ac generation of Majorana modes is also a topic of current interest [35–37]. We expect that these will become fruitful research lines in the field of ac transport. Recently, laser-induced band gaps have been experimentally revealed at the surface of a topological insulator using angle-resolved photoemission spectroscopy [38]. The laser-assisted control of a topological insulator is a hot topic [39,40]. The physics is quite similar to the one that is reviewed here and therefore allows for cross-links between these areas.

In the following sections, we discuss the interplay of a laser field in both the electronic structure and transport properties of graphene. In Section 9.2, we introduce the so-called **k.p** model to describe in detail the underlying mechanisms for the opening of energy gaps both in the vicinity of the low-energy Dirac point and at the energies $\varepsilon = \pm\hbar\Omega/2$, where new symmetries are superimposed by the laser field. We derive approximate formulas for the new dynamical gaps and contrast them with numerical data. In Section 9.3, we turn to a $\pi$-orbitals tight-binding (TB) model where the explicit structure of the lattice is taken into account. Under this scenario, we first incorporate the time-dependent field by using Floquet theory and then we calculate the local DoS for a linearly polarized laser. When studying the electronic properties of a graphene layer weakly interacting with a boron nitride substrate, new Dirac points are found to occur at higher energies related with the wavelength of the Moiré pattern induced by the graphene/boron nitride mismatch as reported in Reference [17]. In our case, laser illumination induces a superimposed spatial pattern [15], which may lead to a similar phenomenon when the laser polarization is linear. Departure from linear polarization leads to a suppression of these new Dirac points and dynamical gaps develop. Furthermore, we analyze the transport properties of the model by calculating the dc conductance for laser wavelengths in the mid-infrared domain and explore the laser-induced DoS dips for different choices of the parameters. Finally, in Section 9.4, we summarize the obtained results and discuss about their impact on possible applications.

## 9.2 DIRAC FERMIONS: k.p APPROACH

Because of the honeycomb lattice symmetry of graphene, the low-energy states that are relevant to electronic transport are located close to the two independent Dirac points (valleys) $K$ and $K'$ alternately displayed at the edges of the first Brillouin zone.

In clean samples (i.e., without impurities or distortions) and for external ac fields that do not couple the valleys, we can treat these two degeneracy points separately.

For each valley, the electronic states can be accurately described within the low-energy approximation (also known as the **k.p** method) where the spectrum is assumed to depend linearly on the electronic momentum **k**. The electronic states are then computed by an envelope wave function $\Psi = (\Psi_A, \Psi_B)^T$, where each component refers to the two interpenetrating sublattices $A$ and $B$ [41–43], respectively. This is usually called the sublattice pseudospin degree of freedom, owing to its analogy with the actual spin of the charge carriers. The energies $\varepsilon$ and wave functions are thus solutions of the Dirac equation:

$$H\Psi = \varepsilon\Psi \tag{9.1}$$

where $H$ is the Hamiltonian operator of the system. The time-periodic electromagnetic field, with period $T = 2\pi/\Omega$, is created by a monochromatic plane wave traveling along the $z$-axis, perpendicular to the plane defined by the graphene sheet. The vector potential associated with the laser field is thus written as

$$A(t) = \mathrm{Re}[A_0 e^{-i\Omega t}] \tag{9.2}$$

where $\mathbf{A}_0 = A_0 (1, e^{i\varphi})$ refers to the intensity $A_0 = E/\Omega$ and $\varphi$ to the polarization of the field.

In this choice of the parameters, $\varphi = 0$ yields a linearly polarized field $\mathbf{A}(t) = A_0 \cos \Omega t(\mathbf{x} + \mathbf{y})$, whereas $\varphi = \pi/2$ results in a circularly polarized field $\mathbf{A}(t) = A_0 (\cos \Omega t \, \mathbf{x} + \sin \Omega t \, \mathbf{y})$. Notice here that the **k.p** approach preserves circular symmetry around the degeneracy point and, in consequence, there is no preferential choice for the in-plane directions **x** and **y**. As we shall see in Section 9.3, this is not the case in the TB model for which an explicit orientation of the lattice has to be fixed. In the presence of the ac field, the graphene electronic states are encoded through the Hamiltonian:

$$H(t) = v_F \boldsymbol{\sigma} \cdot \left[ \boldsymbol{p} - eA(t) \right] \tag{9.3}$$

where:

$v_F \sim 10^6$ m/s is the Fermi velocity in graphene

$\sigma = \left( \hat{\sigma}_x, \hat{\sigma}_y \right)$ is the vector of Pauli matrices describing the pseudospin degree of freedom

We operate in the nonadiabatic regime in which the electronic *dwell* time, namely, the traverse time along the laser's spot, is larger than the period $T$ of the laser, such that the electron experiences several oscillations of the field before leaving the illuminated region. In this situation, the Floquet theory represents a suitable approach to describe such electron-photon scattering processes. The Hamiltonian presented in Equation 9.3 is thus expanded into a composite basis between the usual Hilbert space and the space of time-periodic functions. This new basis defines the so-called Floquet pseudostates $|k, n\rangle_{\pm}$, where **k** is the electronic wave vector, characterized by its momentum, and $n$ is the Floquet index associated with the $n$-mode of a Fourier decomposition of the modulation. The subindex refers to the alignment of the pseudospin with respect to the momentum. Considering the $K$-valley, a $+ (-)$ labeled state has its pseudospin oriented parallel (antiparallel) to the electronic momentum. In this representation of the Hamiltonian, the time dependence introduced by the external potential is thus replaced by a series of system's replicas arising from the Fourier decomposition. In this Floquet space, one has to solve a time-independent Schrödinger equation with Floquet Hamiltonian $H_F = H - i\hbar\partial_t$. Recursive Green's function techniques [43] can be exploited to obtain both the dc component of the conductance and the DoS from the Floquet Green's functions [44].

The matrix elements of the resulting Floquet Hamiltonian are then computed as [45]

$$H_{i,j}^{(m,n)} = \frac{1}{T}\int_0^T dt\, H_{i,j}(t)e^{i(m-n)\Omega t} + m\hbar\Omega\delta_{mn}\delta_{ij} \tag{9.4}$$

where:
δ stands for the Kronecker symbol
$i$ and $j$ indicate the pseudospin orientation with respect to the momentum
$m$ and $n$ are Fourier indices related to the field's modulation

Therefore, for a given vector $\mathbf{k} = k\,(\cos\alpha, \sin\alpha)$, the $m$-diagonal block matrix contains both the electronic kinetic terms $\pm\hbar v_F\, k$ and the $m$-component of the modulation of the field $m\hbar\Omega$, that is,

$$H^{m,m} = \hbar v_F \begin{pmatrix} k + m\Omega/v_F & 0 \\ 0 & -k + m\Omega/v_F \end{pmatrix} \tag{9.5}$$

where we have transformed the standard basis (the one related to the nonequivalent sublattices) into a diagonal basis for the diagonal block. Due to the particular sinusoidal time dependence of the field, only inelastic transitions involving the absorption or emission of a single photon are allowed, that is, $\Delta m = \pm 1$. Therefore, the off-diagonal block matrix connecting Floquet states with a different number of photons reads

$$H^{(m,m+1)} = \begin{pmatrix} \gamma_1 & \gamma_2 \\ -\gamma_2 & -\gamma_1 \end{pmatrix} \tag{9.6}$$

This term is responsible for the photon-assisted tunneling processes. Here, the direct hopping term $\gamma_1 = eA_0 v_F/2\,(\cos\alpha + e^{-i\varphi}\sin\alpha)$, with $\alpha = \tan^{-1}(k_y/k_x)$, sets the transition amplitude between Floquet states with same pseudospin character. On the contrary, the off-diagonal term $\gamma_2 = ieA_0 v_F/2\,(e^{-i\varphi}\cos\alpha - \sin\alpha)$ introduces an inelastic backscattering process that enables pseudospin transitions through photon excitation. This leads to the analog of an inelastic Bragg reflection as found, for example, in other context for inelastic scattering in carbon nanotubes [46,47].

In Figure 9.1a, we provide a sketch of the Floquet spectrum for a given orientation of the momentum. The complete spectrum involves an average over all possible orientations and it looks like a series of superimposed Dirac cones touching at energies $m\hbar\Omega$. The effect that the ac field induces on the electronic structure manifests

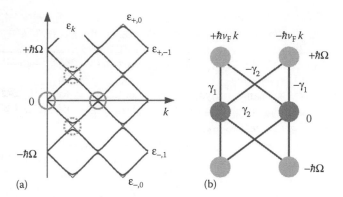

**FIGURE 9.1**   (a) Scheme of the quasi-energies as a function of the electronic momentum $k$. The relevant crossing regions, encircled by dotted and solid lines, yield the dynamical and central gaps, respectively. (b) Representation of the Floquet Hamiltonian for the **k.p** approach. Circles correspond to Floquet states and lines denote inelastic hopping elements.

whenever states with different pseudospin cross. Due to the hopping elements $\gamma_1$ and $\gamma_2$, a new family of gaps will open in the vicinity of the encircled crossings of Figure 9.1a. This defines a region where no states are available and the transport becomes drastically suppressed. As we will discuss in the next lines, the width of these gaps strongly depends on the intensity, frequency, and polarization of the field. For $eA_0v_F < \hbar\Omega$, these effects are clear for energies close to a half-integer of $\hbar\Omega$, where a *dynamical gap* [6] opens as $\varphi$ increases. However, as pointed out by Oka and Aoki [8], around the Dirac point another gap emerges and becomes pronounced in the circularly polarized case. As we will make clear later, the crucial difference between these two is the involved number of (photon-assisted) tunneling processes it takes to backscatter the conduction electrons. For the considered frequencies and intensities of the field, processes beyond the second order can be ignored.

In the mentioned studies [6,8], the authors have considered lasers in two ranges: the far infrared with $\hbar\Omega \sim 29$ meV [6] and the visible range [8]. In the former, the gaps were predicted to be of about 6 meV for a photocurrent generated in a *p–n* junction, whereas in the latter, the photon energy of about 2 eV is much larger than the typical optical phonon energies 170 meV, and severe corrections to the transport properties due to dissipation of the excess energy via electron–(optical)phonon interactions can be expected. Moreover, as pointed out in [14], appreciable effects in this last case would require a power above 1 W/µm², which could compromise the material stability. To overcome both limitations, we quantitatively explore the interaction with a laser in the mid-infrared range $\lambda = 2$–$9$ µm, where photon energies can be made smaller than the typical optical phonon energy while keeping a much lower laser power.

## 9.2.1   DYNAMICAL GAPS UNVEILED THROUGH THE **k.p** MODEL

The leading order process in the interaction with the electromagnetic field occurs at energies $\varepsilon \sim \pm\hbar\Omega/2$ above/below the charge neutrality point. Focusing on the region $\varepsilon \sim \hbar\Omega/2$, we can consider an effective two-level system in which the relevant states are $|k,0\rangle_+$ and $|k,1\rangle_-$. The difference between the eigenenergies is estimated as twice the hopping between the mentioned states and the resulting gap is

$$\Delta_{k=\Omega/2v_F} \approx eA_0v_F\sqrt{1-\cos\varphi\sin\alpha} \tag{9.7}$$

The same analysis can be done around $\varepsilon \sim -\hbar\Omega/2$, where the two-level system is now given by the states $|k,0\rangle_-$ and $|k,1\rangle_+$ (see Figure 9.1a). These yield the same value for the energy gap.

The width of the dynamical gap clearly shows a linear dependence with the field strength and, for a fixed value of $A_0$, is independent of the photon frequency. Furthermore, it is interesting to note that when averaging over all orientations of the wave vector, no net gap opens in the linearly polarized case ($\varphi = 0$) because for the particular value $\alpha = \pi/4$, the hopping vanishes and the backscattering mechanism is suppressed. For the calculation of the DoS, we use the Floquet Green's function technique, in which we define the following Green's function in the product space as

$$G(\varepsilon,k) = \left[\varepsilon I - H_F(k)\right]^{-1} \tag{9.8}$$

where the **k**-vector dependence of the Floquet Hamiltonian enters in both the diagonal blocks (amplitudes) and the off-diagonal ones (directions). The DoS related with the $m = 0$ level includes the sum over all possible values of **k** and writes

$$\nu_0(\varepsilon) = -\frac{1}{\pi}\int_0^\infty k\,dk \int_0^{2\pi} \frac{d\alpha}{2\pi} Im[G_{+,+}^{0,0}(\varepsilon,k)+G_{-,-}^{0,0}(\varepsilon,k)] \tag{9.9}$$

In Figure 9.2a, we show an example of the calculated DoS around the dynamical gap region for different values of the polarization. As can be seen, even in the linearly polarized case there is a strong modification in the DoS that would resemble the usual Dirac point for a Fermi energy around $\varepsilon \sim \pm\hbar\Omega/2$.

The decreasing number of allowed states for different orientations of the electronic momentum becomes evident through the depletion region. By increasing the polarization, one immediately finds that for finite values of $\varphi$ a gap is opened and reaches its maximum for the circularly polarized case. Under this situation, it is clear from Equation 9.7 that there is no dependence with the orientation of the wave vector, which can be interpreted in terms of the recovered axial symmetry, characterized by the standard one-dimensional van Hove singularities emerging at the edges of the gap.

## 9.2.2 ENERGY GAP TUNING AROUND ENGINEERED LOW-ENERGY DIRAC CONES

To describe this mechanism, we refer to the schematic representation of the Floquet Hamiltonian depicted in Figure 9.1b. Here, the leading contribution around the Dirac point ($\varepsilon \sim 0$ and $k \sim 0$) comes from the four paths connecting the crossing states $|k,0\rangle_+$ and $|k,0\rangle_-$ through the first neighboring states. For the calculation, we observe that the mixing between these states is produced by the presence of the

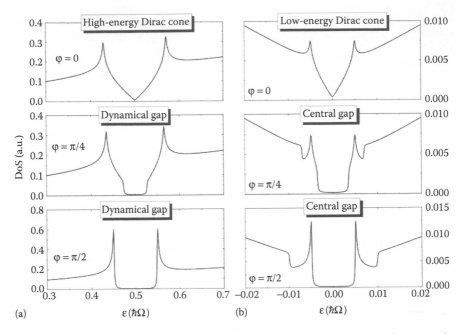

**FIGURE 9.2**  DoS in graphene for several values of the laser polarization. (a) Lowest-order effect around $\varepsilon \sim \hbar\Omega/2$, $k \sim \Omega/2v_F$. (b) Second-order effects around $\varepsilon \sim 0$, $k \sim 0$. The chosen parameters are given by $eA_0v_F/\hbar\Omega = 0.1$.

neighboring states $|k,\pm1\rangle_\pm$. The effective Hamiltonian can be reduced by means of a decimation procedure [43] in which the Floquet space dimension is reduced to the relevant states $|k,0\rangle_\pm$ at $m = 0$ and the other states only enter as a correction term accounting for the interaction with the photonic field. This correction to the diagonal terms cancels out exactly, that is, no energy shift is introduced by the laser. However, this correction does yet contribute through their off-diagonal elements, such that an effective mixing between the two pseudospins is now present and produces an energy difference:

$$\Delta_{k=0} \approx 2\frac{(eA_0v_F)^2}{\hbar\Omega}\sin\varphi \tag{9.10}$$

The quadratic dependence with the field strength is due to the fact that the mixing between these two states involves a twofold process where both the absorption and the emission of a single photon are needed. Additionally, the inverse dependence with the frequency quantifies the amount of energy absorbed and emitted during the tunneling event. For the constrain $eA_0v_F < \hbar\Omega$ considered here, this gap is much smaller than the dynamical gap discussed before. Note also that there is no dependence of the gap with the orientation of the $k$-vector, because we are assuming $k = 0$. By inspecting Equation 9.10, it is easy to observe that the maximum value of the gap is reached in the circularly polarized case, whereas no net gap opens for the linearly polarized case. According to the present approximation, we notice that in this last case no ac effects should be observed in the DoS. However, as can be seen in Figure 9.2b, there is a strong modification around the Dirac point region reflected by an increased slope in comparison with the usual DoS without any laser field. To explain this, it is convenient to include the effect induced by the crossing between states $|k,-1\rangle_+$ and $|k,1\rangle_-$ around $\varepsilon = 0$ and $k = \Omega/v_F$ (right circle with solid line in Figure 9.1a). Although in this case the total energy difference going from $|k,0\rangle_+$ to $|k,0\rangle_-$ is $2\hbar\Omega$, we are in a resonant condition where transitions via virtual states $|k,\pm 1\rangle_\pm$ involve the same energy difference $\hbar\Omega$. The presence of states $|k,2\rangle_\pm$ also contributes to the self-energy correction and the effective hopping term connecting the $m = 0$ states originates a gap:

$$\Delta_{k=\Omega/v_F} \approx \frac{(eA_0 v_F)^2}{2\hbar\Omega}[5 - 2\cos\varphi\sin 2\alpha - 3\cos^2\varphi\sin^2 2\alpha]^{1/2} \tag{9.11}$$

This equation drops to zero for $\varphi = 0$ and $\alpha = \pi/4$ (analogous to the dynamical gap), and reaches its maximum when $\varphi = \pi/2$. This contribution, on the same order than $\Delta_{k=0}$, is responsible for the modification of the DoS around the Dirac point even for $\varphi = 0$. It is interesting to observe that, in addition to the gap, this transition yields two surrounding peaks that persist even in the linear case.

In addition, we observe that this effect reduces by approximately a half the predicted value of Oka and Aoki [8]. Here, we neglect higher order contributions because we consider that the strength of the laser is small compared to the driving frequency. However, as recently discussed in Reference [48], in the opposite limit $eA_0 v_F > \hbar\Omega$, the number of photon transitions becomes large, allowing additional scattering processes that enable the existence of states around the studied regions: In this limit, then, more and more states within the gaps are available and the gap is effectively closed.

## 9.3 TIGHT-BINDING MODEL

In Section 9.2, we have shown how the laser field modifies the electronic structure of 2D graphene in any experiment carried out over a time much larger than the period $T$. A natural question is if these effects would be observable in a transport experiment and how. To answer this, we turn now to the calculation of the transport response using Floquet theory applied to a tight-binding $\pi$-orbitals Hamiltonian. As we shall see, the correspondence between this model and the **k.p** approach becomes evident in the *bulk limit* where the width of the ribbon is on the order of the laser's spot.

In the real space defined by the sites of the two sublattices $A$ and $B$, the electromagnetic field can be accounted for through an additional phase in the hopping $\gamma_{ij}$ that connects two adjacent sites $\mathbf{r}_i$ and $\mathbf{r}_j$, that is, the Peierls substitution:

$$\gamma_{ij} = \gamma_0 \exp\left(i\frac{2\pi}{\Phi_0}\int_{r_i}^{r_j} \mathbf{A}(t)\cdot d\mathbf{r}\right) \tag{9.12}$$

where:

    $\gamma_0 \sim 2.7$ eV is the hopping amplitude at zero field

    $\Phi_0$ is the quantum of magnetic flux

The intensity of the magnetic vector potential is assumed to be constant along the whole sample where the irradiation takes place. Therefore, the computation of such a phase is simply the scalar product between the vector potential and the vector connecting the two sites.

For numerical convenience, we consider an armchair edge structure for the lattice and the vector potential is defined as $\mathbf{A}(t) = A_x \cos \Omega t \, \mathbf{x} + A_y \sin \Omega t \, \mathbf{y}$, where $x$ and $y$ coordinates are on the plane of the nanoribbon. It is important to observe that the particular choice of the edges does not restrict the outcoming results once the bulk limit is reached. For the armchair structure, we can distinguish three principal orientations according to the angle between $\mathbf{r}_j - \mathbf{r}_i$ and $\mathbf{A}(t)$. By considering the scheme in Figure 9.3b, we define the hopping elements $\gamma_{+,\pm}(t)$ and $\gamma_{2+,0}(t)$ where the subindices denote the $x$ and $y$ components of two adjacent sites. In this representation, we have

$$\gamma_{+,\pm}(t) = \gamma_0 \exp\left[ i \frac{a\pi}{\Phi_0} \left( A_x \cos \Omega t \pm \sqrt{3} A_y \sin \Omega t \right) \right] \qquad (9.13)$$

for those terms with simultaneous $x$ and $y$ components and

$$\gamma_{2+,0}(t) = \gamma_0 \exp\left[ i \frac{2a\pi}{\Phi_0} A_x \cos \Omega t \right] \qquad (9.14)$$

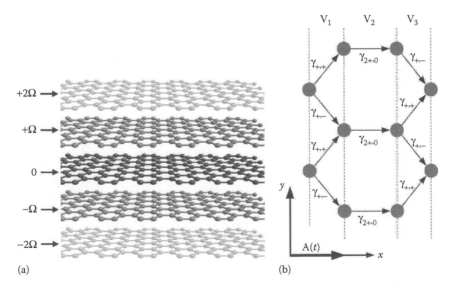

(a)                                    (b)

**FIGURE 9.3** (a) Schematic view of the Floquet space for the TB model in real space. (b) Opening, longitudinal, and closing hopping matrices according to the relative direction between the adjacent sites and the vector potential in an armchair edge structure.

for the hopping element with only the $x$ component, respectively (see Figure 9.3b). Here, $a \sim 0.142$ nm is the nearest carbon–carbon distance. The Fourier decomposition of these expressions is based on the Anger–Jacobi expansion that yields an $m$-transition amplitude in terms of Bessel functions:

$$\gamma_{+,\pm}^{m} = \gamma_0 \sum_{k=-\infty}^{\infty} i^k J_k(z_x) J_{m-k}(\pm z_y) \tag{9.15}$$

$$\gamma_{2+,0}^{m} = \gamma_0 i^m J_m(2z_x) \tag{9.16}$$

where:

$$z_x = \frac{\pi a A_x}{\Phi_0}$$

and

$$z_y = \frac{\sqrt{3}\pi a A_y}{\Phi_0}$$

In contrast to the **k.p** approximation, in this situation we may have processes involving the absorption or emission of several photons at once. However, they decay rapidly with the number of photons and the main contribution still comes from the renormalization at $m = 0$ and the leading inelastic order $m = \pm1$.

We compute the hopping matrices as those connecting two adjacent transverse layers such that the difference in Floquet indices is $m$. If $N$ denotes the number of transverse states, that is, the number of electronic sites in each vertical layer, the hopping matrices are $N \times N$ blocks expressed as

$$\mathbf{V}_1^m = \begin{pmatrix} \gamma_{+,+}^m & \gamma_{+,-}^m \\ 0 & \gamma_{+,+}^m \\ & & \ddots \end{pmatrix}, \mathbf{V}_2^m = \begin{pmatrix} \gamma_{2+,0}^m & 0 \\ 0 & \gamma_{2+,0}^m \\ & & \ddots \end{pmatrix}, \mathbf{V}_3^m = \begin{pmatrix} \gamma_{+,-}^m & 0 \\ \gamma_{+,+}^m & \gamma_{+,-}^m \\ & & \ddots \end{pmatrix} \tag{9.17}$$

where we coin them *opening*, *longitudinal*, and *closing* matrices, respectively, according to the connectivity they induce (see Figure 9.3b). Therefore, in the natural basis defined in real space, the total Floquet Hamiltonian is composed by a periodic block tridiagonal structure where the off-diagonal block matrices contain the $\mathbf{V}_i^m$ terms and the diagonal block only accounts for the site energies $E_m = m\hbar\Omega$ because no gate voltages are applied to the sample. For the proposed values of the field and a typical energy uncertainty of $\eta \sim 30$ $\mu$eV [43], the bulk limit is reached for $N > 10^4$. Hence, it is necessary to decompose the Floquet Hamiltonian in terms of transverse modes because otherwise one should deal with $O(N^3)$ operations, increasing enormously the computation time. For this reason, we limit ourselves to the linearly polarized case in which $A_y = 0$, which is already enough to give a hint on the transport effects. Other choices of the laser orientation and/or polarization would be subjected to the use of better adapted lattice bases or the employment of parallel computing techniques which are beyond the scope of this work.

The total Hamiltonian is transformed according to the normal modes of the sublattices $A$ and $B$. The rotation matrix results in a trivial expansion of that in the appendix of Reference [49] on the composite Floquet basis. The spatial part of the Hilbert space is thus reduced and the hopping matrices have now the dimension $2N + 1$ of the (truncated) Fourier space. The resulting single-mode layers preserve the same Hamiltonian structure with $\mathbf{V}_i^m(q) = 2\mathbf{V}_i^m \cos[\pi q/(2N+1)]$ for $i = 1, 3$ and $\mathbf{V}_2^m(q) = \mathbf{V}_2^m$, where $q$ is the mode number.

The DoS is therefore obtained through the Floquet Green's functions defined as

$$G_F^q(\varepsilon) = \left[ \varepsilon I - H_F^q - \sum_F^q(\varepsilon) \right]^{-1} \tag{9.18}$$

where $H_F^q$ is the $q$-mode Hamiltonian containing the diagonal block of the Fourier components and is written as the following matrix:

$$H_F^q(\varepsilon) = \begin{pmatrix} \ddots & & & \\ & -\hbar\Omega & 0 & 0 \\ & 0 & 0 & 0 \\ & 0 & 0 & +\hbar\Omega \\ & & & & \ddots \end{pmatrix} \tag{9.19}$$

The self-energy $\Sigma_F^q$ is a correction term arising from the presence of the neighbor sites on both sides of the chain. The effective Hamiltonian $H_{\text{eff}}^q(\varepsilon) = H_F^q + \Sigma_F^q(\varepsilon)$ for a transverse layer can be calculated recursively by using a decimation procedure [43] and the DoS is thus obtained from the resulting Green's function as

$$v_0(\varepsilon) = -\frac{1}{\pi} \sum_q \text{Im}\left[ G_F^q(\varepsilon) \right]_{0,0} \tag{9.20}$$

where the subindex 0,0 corresponds to the zero photon state

Note the similarity of this expression with Equation 9.9, where the amplitude and direction of the electronic momentum are discretized in the $q$ modes.

By employing this numerical strategy, we compute the DoS for an armchair edge structure and compare it with the one obtained from the **k.p** approach. As we can observe in Figure 9.4, there is in general a good agreement between both results. In particular, the depletion region around the resonances $\varepsilon \sim \pm\hbar\Omega/2$ is also reproduced by the TB model. However, a closer look of the peaks reveals small differences between both curves. For the TB method, we see two peaks that surround the peak obtained in the **k.p** model (see inset of Figure 9.4). The origin of the splitting of the peaks in the DoS can be attributed to a trigonal warping effect that is inherently included in the TB Hamiltonian, where the explicit structure of the lattice is taken into account.

The peak splitting is also present in the region close to the Dirac point. Here, both the **k.p** and TB results reproduce an increased slope compared with the bare result

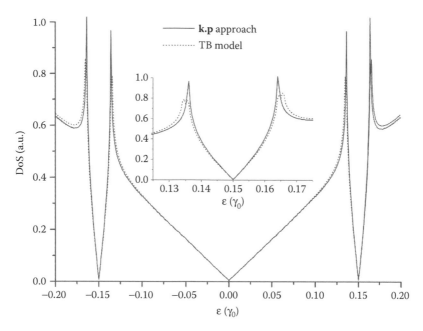

**FIGURE 9.4** DoS for linear polarization of the laser field calculated by both the **k.p** approach (solid line) and the TB model (dotted line). The chosen parameters of the laser are $\hbar\Omega = 0.3\gamma_0$ and $A_0 = 0.01\Phi_0/\pi a$. *Inset*: DoS around the depletion region.

without a laser. This, however, is not visible in Figure 9.4 because the intensity of the laser is too small compared with the frequency. To estimate the relevance of this effect, we explore the DoS around the Dirac point for several values of the intensity (Figure 9.5a) and frequency (Figure 9.5b) of the laser field. In the left panel, we fixed the frequency at $\hbar\Omega = 0.3\gamma_0$ and increased the intensity ($A_0 = E/\Omega$) from 0 to $0.05\Phi_0/\pi a$. As we can observe, the effect the intensity induces on the distance between the peaks is approximately linear, as well as the width of the depletion region. In Figure 9.5b, we fixed the intensity at this final value and decreased the frequency from $0.5\gamma_0$ to $0.2\gamma_0$. In this situation, the distance between the two peaks remains approximately the same while changing the frequency, thus revealing a small (even negligible) dependence.

The numerical analysis of the DoS can be complemented by considering the mode decomposition shown in Figure 9.6. This representation allows an alternative inspection of the different effects we have been discussing throughout this section. Here, we plot the DoS in grayscale as a function of the energy for each transverse mode.

The sum of all these contributions gives the standard DoS. In the figure, each panel represents a fixed value of the laser intensity, starting from $A_0 = 0$ to $0.05\Phi_0/\pi a$. In the absence of a laser field, we can observe (see the upper left panel) the linear behavior of the single-mode band gaps as we move away from the central mode where no gaps occur. This central mode is defined via the identity $V_1^m(N_C) = V_2^m$ in which the system would be homogeneous; therefore, no gap should be opened

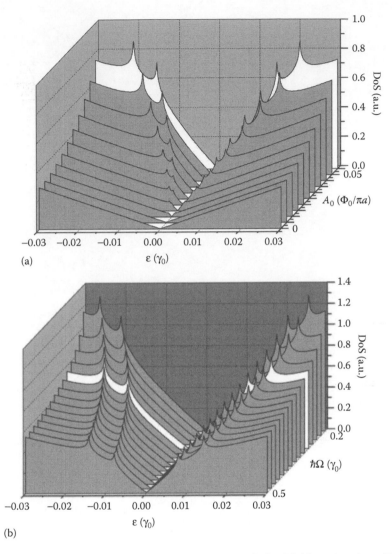

**FIGURE 9.5** DoS around the Dirac point for a linearly polarized field as a function of (a) the laser intensity ($\hbar\Omega = 0.3\gamma_0$, left panel) and (b) laser frequency ($A_0 = 0.05\Phi_0/\pi a$, right panel). The white curve in both figures corresponds to the same choice of parameters ($\hbar\Omega = 0.3\gamma_0$ and $A_0 = 0.01\Phi_0/\pi a$).

for this mode. In the considered example, this results to be $N_C \sim 6600$. This point constitutes the symmetry axis where both contributions on the left and right sides are the same. As we turn on the laser, a gap is suddenly opened at $\hbar\Omega/2 = 0.15\gamma_0$ in each mode (except the central one) and its width increases linearly with the intensity of the laser.

In addition, we observe that the emergence of the splitting of the peaks is manifested by a slight asymmetry in their position with respect to the central mode.

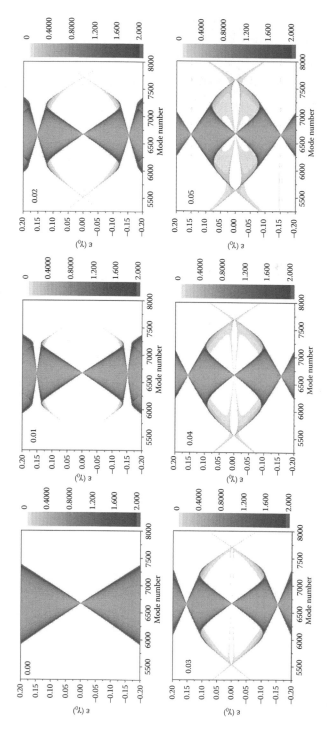

**FIGURE 9.6** Transverse-mode decomposition of the DoS (grayscale, arbitrary units) for a linearly polarized field. The frequency of the laser is fixed at $\hbar\Omega = 0.3\,\gamma_0$, whereas the intensity varies from 0 to 0.05 in units of $\Phi_0/\pi a$. The central mode in this example is $N_C \sim 6600$.

For $A_0 > 0.03 \, \Phi_0/\pi a$, the contribution from side modes originates the increased slope around the Dirac point region. The DoS in this region also shows an asymmetry in the size of the lobes, which originates the splitting of the peaks.

### 9.3.1 Electronic Transport through Irradiated Graphene

In order to estimate the influence of the laser field in a transport experiment, we calculate the two-terminal dc conductance through a graphene stripe of 1 μm × 1 μm in the presence of a linearly polarized laser. The sample is connected to two leads on both sides, considered a prolongation of the stripe to regions where no laser is applied.

By following the same strategy we used before in the calculation of the DoS, we obtain the dc conductance from the recursive calculation of the Floquet Green's functions. In this sense, the mode decomposition is again a key tool in the numerical implementation of the recursive formula for the self-energy correction. In particular, the semi-infinite leads are incorporated through an initial self-energy where the opening, longitudinal, and closing hopping matrices result to be diagonal. Because in this region no ac fields are applied, the elements of the above hopping matrices are simply $\gamma^m_{+,\pm} = \gamma^m_{2+,0} = \gamma_0 \delta_{m,0}$. Therefore, as shown in Figure 9.7, the resulting lattice structure for a single mode in the leads is the one of a dimer with alternating hoppings $\gamma_0$ and $2\gamma_0 \cos[\pi q/(2N+1)]$. In the sample region, however, the effect of the time-dependent field is described by the hopping elements connecting the nearest transverse layers at different Fourier levels. According to Equations 9.15 and 9.16, these hoppings attenuate fast with the difference between the involved Fourier replicas. For the considered examples in which $eA_0v_F/\hbar\Omega \sim 0.1$, it usually converges for a small number of photons (~3). The dc conductance is thus calculated through the transmittance via the Floquet Green's functions (see appendix in Reference [44]):

$$T_{RL}(\varepsilon) = \sum_{q,n} 2\Gamma^q_{(R,n)}(\varepsilon) \left| G^q_{(R,n)\leftarrow(L,0)}(\varepsilon) \right|^2 2\Gamma^q_{(L,0)}(\varepsilon) \tag{9.21}$$

where

$$\Gamma^q_{(\alpha,n)}(\varepsilon) = -\mathrm{Im} \sum_\alpha^q (\varepsilon + n\hbar\Omega)$$

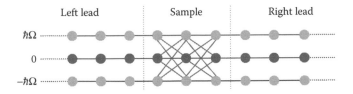

**FIGURE 9.7** Scheme of the transport setup of the transverse-mode decomposition of the Floquet Hamiltonian. The central region (sample) contains additional hopping elements connecting different Fourier sheets due to the presence of the time-dependent field.

The dc conductance is then obtained from Landauer's formula [50] and we consider that the system preserves space inversion symmetry, that is, $T_{RL}(\varepsilon) = T_{LR}(\varepsilon) = T(\varepsilon)$. Therefore, the linear conductance reads

$$G(\varepsilon_F) = -\frac{2e^2}{h} \int_{-\infty}^{\varepsilon_F} d\varepsilon\, T(\varepsilon) \frac{d}{d\varepsilon} f(\varepsilon) \tag{9.22}$$

with $f(\varepsilon) = 1/\left[1 + \varepsilon^{(\varepsilon - \varepsilon_F)/kT}\right]$ the Fermi's distribution function. In the zero-temperature limit, the derivative of $f(e)$ results to be a Dirac delta and we recover a linear relation between these two functions:

$$G(\varepsilon_F) = \frac{2e^2}{h} T(\varepsilon_F) \tag{9.23}$$

For the numerical calculation of the conductance, we consider a laser field whose wavelength lies within the mid-infrared region. We show in Figure 9.8 the resulting conductance at zero temperature for different values of the laser power. We consider $\lambda = 10\ \mu m$ ($\hbar\Omega \sim 140$ meV) in Figure 9.8a and $\lambda = 2\ \mu m$ ($\hbar\Omega \sim 620$ meV) in Figure 9.8b, respectively. For the case where no external laser is applied (gray dashed lines in Figure 9.8a), the conductance shows a linear dependence with the Fermi energy. In this situation, there is a perfect transmission and each channel contributes a unit of the conductance quantum $G_0 = 2e^2/h$. This effect becomes evident in the small plateaus along the whole curve, whose widths depend on the total number of transverse modes, that is, the width of the sample. While increasing the energy of the carriers, more channels participate in the transport and the resulting conductance increases. However, when we turn on the laser, the interaction with the field induces a significant depletion around the region $\varepsilon_F \sim \hbar\Omega/2$. In the depletion region, we can observe that the minimum value at $\varepsilon_F = \hbar\Omega/2$ decreases with the laser power. Additionally, this minimum value approaches to zero by increasing the size of the sample (i.e., the irradiated area) because the larger is the time spent

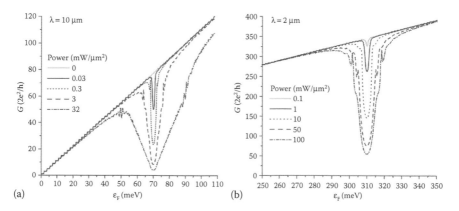

**FIGURE 9.8**  dc conductance through a graphene stripe of 1 μm × 1 μm in the presence of a linearly polarized laser as a function of the Fermi energy and laser power for a laser wavelength of (a) λ = 10 μm and (b) λ = 2 μm.

by the carriers the higher the backscattering probability. Notice that there are no visible effects close to the Dirac point $\varepsilon_F = 0$. This is due to the fact that the laser polarization is linear, and the position of the channel bands remains unaffected. By comparing the two plots, one can observe that the effect of the laser is relatively more significant in the left panel. For instance, we observe that for $P \sim 0.03$ mW/$\mu$m$^2$, the depletion is still visible, whereas in the right panel, for $P \sim 0.1$ mW/$\mu$m$^2$, it is hard to distinguish. This is related to the slope of the gapped modes in the DoS of Figure 9.6 (see, for example, upper central panel), which approximately goes like

$$\left( \frac{d\Delta}{dN} \right)_{\varepsilon = \hbar\Omega} \propto \frac{\sqrt{P}}{\Omega} \qquad (9.24)$$

and therefore for smaller frequencies the number of modes affected by the field is increased, and thus the stronger is the conductance suppression. We emphasize that in the above equation the dependence of the slope (and hence the width and depth of the depletion) with the laser power is given by a square root, and in consequence there are still visible effects when decreasing this value even by 4 orders of magnitude.

Regarding the effect of static disorder, it is usually found that the presence of local impurities or topological defects in the lattice tends to broaden energy gaps produced by confinement (as in a graphene nanoribbon) [51]. *A priori*, a similar effect is to be expected here and further studies are needed. In order to estimate the robustness of the depletion regions, we calculate the conductance for $\lambda = 10$ $\mu$m and explore its behavior for different values of the temperature. This is shown in Figure 9.9, where each panel corresponds to a fixed value of the laser power. As we take a small finite temperature (~5 K), the plateaus structure is no longer visible and the curve is smoothed. The depletion regions are almost the same for the upper panels (large power) but slightly reduced in the lower ones. Nevertheless, this is still visible for the case $P = 0.03$ m$\Omega$/$\mu$m$^2$. By increasing the temperature, the width of the depletion region also increases, whereas its height decreases to the bare value of zero field.

We emphasize that the dip in the conductance persists for small values of the laser power. In particular, even for temperatures around the 20 K, the depletion region in the case of a laser power $P = 0.03$ mW/$\mu$m$^2$ is on the order of several units of $G_0$; therefore, this should be observable in a clean sample. Such features in the conductance could also be resolved in its derivative, where a peak emerges in the region of the depletion.

## 9.4 CONCLUSIONS

In sum, in this chapter we have shown that a circularly polarized laser field strongly modifies the electronic structure of graphene by generating a series of energy band-gaps and, importantly, it also impacts on the transport response (dc conductance) of the irradiated material. We conclude that, for these features to be observable, experiments with lasers in the mid-infrared would be particularly timely. The modifications are predicted to arise both around the Dirac point and at $\hbar\Omega/2$ leading to results that are strongly dependent on the laser polarization, thereby enabling its use as a control parameter.

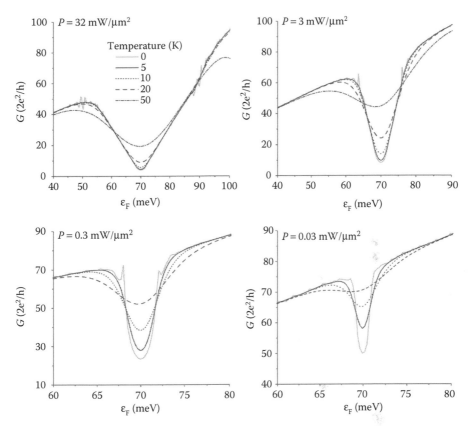

**FIGURE 9.9**  dc conductance as a function of the Fermi energy and temperature for different values of the laser power. The chosen wavelength of the laser is $\lambda = 10$ μm.

As pointed out already in the introduction, the potential technological impact and the high level of recent activity in this field, makes of this an outstanding area for further research and much needed experiments.

## ACKNOWLEDGMENTS

The authors acknowledge discussions and correspondence from Junichiro Kono, Gonzalo Usaj, Pablo Perez-Piskunow, and Frank Koppens, as well as support from CONICET, ANPCyT, and SeCyT-UNC. SR acknowledges the Spanish Ministry of Economy and Competitiveness for national project funding (MAT2012-33911). LEFFT acknowledges support from ICTP-Trieste and the Alexander von Humboldt Foundation.

## REFERENCES

1. A. H. C. Neto, F. Guinea, N. M. R. Peres, K. S. Novoselov, and A. K. Geim, The electronic properties of graphene, *Rev Mod Phys*, vol. 81, p. 109, 2009.
2. F. Bonaccorso, Z. Sun, T. Hasan, and A. C. Ferrari, Graphene photonics and optoelectronics, *Nat Photonics*, vol. 4, p. 611, 2010.

3. F. Xia, T. Mueller, Y.-M. Lin, A. Valdes-Garcia, and P. Avouris, Ultrafast graphene photodetector, *Nat Nanotechnol*, vol. 4, p. 839, 2009.

4. N. M. Gabor, J. C. W. Song, Q. Ma, N. L. Nair, T. Taychatanapat, K. Watanabe, T. Taniguchi, L. S. Levitov, and P. Jarillo-Herrero, Hot carrier-assisted intrinsic photoresponse in graphene, *Science*, vol. 334, no. 6056, p. 648, 2011.

5. D. Sun, G. Aivazian, A. M. Jones, J. S. Ross, W. Yao, D. Cobden, and X. Xu, Ultrafast hot-carrier-dominated photocurrent in graphene, *Nat Nanotechnol*, vol. 7, p. 114, 2012.

6. S. V. Syzranov, M. V. Fistul, and K. B. Efetov, Effect of radiation on transport in graphene, *Phys Rev B*, vol. 78, p. 045407, 2008.

7. S. Mai, S. V. Syzranov, and K. B. Efetov, Photocurrent in a visible-light graphene photodiode, *Phys Rev B*, vol. 83, p. 033402, 2011.

8. T. Oka and H. Aoki, Photovoltaic hall effect in graphene, *Phys Rev B*, vol. 79, p. 081406, 2009.

9. O. V. Kibis, Metal-insulator transition in graphene induced by circularly polarized photons, *Phys Rev B*, vol. 81, p. 165433, 2010.

10. D. S. L. Abergel and T. Chakraborty, Generation of valley polarized current in bilayer graphene, *Appl Phys Lett*, vol. 95, p. 062107, 2009.

11. E. Suárez Morell and L. E. F. Foa Torres, Radiation effects on the electronic properties of bilayer graphene, *Phys Rev B*, vol. 86, no. 12, p. 125449, 2012.

12. H. L. Calvo, P. M. Perez-Piskunow, S. Roche, and L. E. F. Foa Torres, Laser-induced effects on the electronic features of graphene nanoribbons, *Appl Phys Lett*, vol. 101, no. 25, p. 253506, 2012.

13. H. L. Calvo, P. M. Perez-Piskunow, H. M. Pastawski, S. Roche, and L. E. F. F. Torres, Non-perturbative effects of laser illumination on the electrical properties of graphene nanoribbons, *J Phys Condens Matter*, vol. 25, no. 14, p. 144202, 2013.

14. H. L. Calvo, H. M. Pastawski, S. Roche, and L. E. F. Foa Torres, Tuning laser-induced band gaps in graphene, *Appl Phys Lett*, vol. 98, p. 232103, 2011.

15. S. E. Savelev and A. S. Alexandrov, Massless Dirac fermions in a laser field as a counterpart of graphene superlattices, *Phys Rev B*, vol. 84, p. 035428, 2011.

16. P. Delplace, Á. Gómez-León, and G. Platero, Merging of Dirac points and Floquet topological transitions in ac-driven graphene, *Phys Rev B*, vol. 88, no. 24, p. 245422, 2013.

17. M. Yankowitz, J. Xue, D. Cormode, J. D. Sanchez-Yamagishi, K. Watanabe, T. Taniguchi, P. Jarillo-Herrero, P. Jacquod, and B. J. LeRoy, Emergence of superlattice Dirac points in graphene on hexagonal boron nitride, *Nat Phys*, vol. 8, p. 382, 2012.

18. Z. Gu, H. A. Fertig, D. P. Arovas, and A. Auerbach, Floquet spectrum and transport through an irradiated graphene ribbon, *Phys Rev Lett*, vol. 107, p. 216601, 2011.

19. O. Roslyak, G. Gumbs, and S. Mukamel, Trapping photon-dressed Dirac electrons in a quantum dot studied by coherent two dimensional photon echo spectroscopy, *J Chem Phys*, vol. 136, p. 194106, 2012.

20. A. Iurov, G. Gumbs, O. Roslyak, and D. Huang, Anomalous photon-assisted tunneling in graphene, *J Phys Condens Matter*, vol. 24, p. 015303, 2012.

21. J.-T. Liu, F.-H. Su, H. Wang, and X.-H. Deng, Optical field modulation on the group delay of chiral tunneling in graphene, *New J Phys*, vol. 14, p. 013012, 2012.

22. M. Busl, G. Platero, and A.-P. Jauho, Dynamical polarizability of graphene irradiated by circularly polarized ac electric fields, *Phys Rev B*, vol. 85, p. 155449, 2012.

23. P. San-Jose, E. Prada, H. Schomerus, and S. Kohler, Laser-induced quantum pumping in graphene, *Appl Phys Lett*, vol. 101, no. 15, p. 153506, 2012.

24. M. M. Glazov and S. D. Ganichev, High frequency electric field induced nonlinear effects in graphene, *Phys Rep*, vol. 535, no. 3, pp. 101–138, 2014.

25. R. R. Hartmann, J. Kono, and M. E. Portnoi, Terahertz science and technology of carbon nanomaterials, *Nanotechnology*, vol. 25, p. 322001, 2014.

26. J. Inoue and A. Tanaka, Photoinduced transition between conventional and topological insulators in two-dimensional electronic systems, *Phys Rev Lett*, vol. 105, p. 017401, 2010.
27. N. H. Lindner, G. Refael, and V. Galitski, Floquet topological insulator in semiconductor quantum wells, *Nat Phys*, vol. 7, p. 490, 2011.
28. T. Kitagawa, T. Oka, A. Brataas, L. Fu, and E. Demler, Transport properties of non-equilibrium systems under the application of light: Photoinduced quantum hall insulators without Landau levels, *Phys Rev B*, vol. 84, p. 235108, 2011.
29. J. Cayssol, B. Dóra, F. Simon, and R. Moessner, Floquet topological insulators, *Phys Status Solidi RRL*, vol. 7, p. 101, 2013.
30. P. M. Perez-Piskunow, G. Usaj, C. A. Balseiro, and L. E. F. F. Torres, Floquet chiral edge states in graphene, *Phys Rev B,* vol. 89, p. 121401(R), 2014.
31. T. Kitagawa, E. Berg, M. Rudner, and E. Demler, Topological characterization of periodically driven quantum systems, *Phys Rev B*, vol. 82, p. 235114, 2010.
32. A. Gómez-León and G. Platero, Floquet-Bloch theory and topology in periodically driven lattices, *Phys Rev Lett*, vol. 110, no. 20, p. 200403, 2013.
33. M. S. Rudner, N. H. Lindner, E. Berg, and M. Levin, Anomalous edge states and the bulk-edge correspondence for periodically driven two-dimensional systems, *Phys Rev X*, vol. 3, no. 3, p. 031005, 2013.
34. M. C. Rechtsman, J. M. Zeuner, Y. Plotnik, Y. Lumer, D. Podolsky, F. Dreisow, S. Nolte, M. Segev, and A. Szameit, Photonic Floquet topological insulators, *Nature*, vol. 496, no. 7444, pp. 196–200, 2013.
35. A. A. Reynoso and D. Frustaglia, Unpaired Floquet Majorana fermions without magnetic fields, *Phys Rev B*, vol. 87, no. 11, p. 115420, 2013.
36. A. Kundu and B. Seradjeh, Transport signatures of Floquet Majorana fermions in driven topological superconductors, *Phys Rev Lett*, vol. 111, no. 13, p. 136402, 2013.
37. M. Thakurathi, K. Sengupta, and D. Sen, Majorana edge modes in the Kitaev model, *Phys Rev B*, vol. 89, p. 235434, 2014.
38. Y. H. Wang, H. Steinberg, P. Jarillo-Herrero, and N. Gedik, Observation of Floquet-Bloch states on the surface of a topological insulator, *Science*, vol. 342, no. 6157, pp. 453–457, 2013.
39. B. Dóra, J. Cayssol, F. Simon, and R. Moessner, Optically engineering the topological properties of a spin hall insulator, *Phys Rev Lett*, vol. 108, p. 056602, 2012.
40. P. Rodriguez-Lopez, J. J. Betouras, and S. E. Savelev, Dirac fermion time-Floquet crystal: Manipulating Dirac points, *Phys Rev B*, vol. 89, p. 155132, 2014.
41. P. R. Wallace, The band theory of graphite, *Phys Rev*, vol. 71, p. 622, 1947.
42. T. Ando, Theory of transport in carbon nanotubes, *Semicond Sci Technol*, vol. 15, p. 13, 2000.
43. H. M. Pastawski and E. Medina, Tight binding methods in quantum transport through molecules and small devices: From the coherent to the decoherent description, *Rev Mex Phys*, vol. 47 S1, p. 1, 2001.
44. L. E. F. Foa Torres, Mono-parametric quantum charge pumping: Interplay between spatial interference and photon-assisted tunneling, *Phys Rev B*, vol. 72, p. 245339, 2005.
45. J. H. Shirley, Solution of the Schrödinger equation with a Hamiltonian periodic in time, *Phys Rev*, vol. 138, p. 979, 1965.
46. S. Roche and L. E. F. Foa Torres, Inelastic quantum transport and Peierls-like mechanism in carbon nanotubes, *Phys Rev Lett*, vol. 97, p. 076804, 2006.
47. L. E. F. Foa Torres, R. Avriller, and S. Roche, Nonequilibrium energy gaps in carbon nanotubes: Role of phonon symmetries, *Phys Rev B*, vol. 78, p. 035412, 2008.
48. Y. Zhou and M. W. Wu, Optical response of graphene under intense terahertz fields, *Phys Rev B*, vol. 83, p. 245436, 2011.

49. C. G. Rocha, L. E. F. F. Torres, and G. Cuniberti, ac transport in graphene-based Fabry-Pérot devices, *Phys Rev B*, vol. 81, p. 115435, 2010.

50. S. Kohler, J. Lehmann, and P. Hänggi, Driven quantum transport on the nanoscale, *Phys Rep*, vol. 406, p. 379, 2005.

51. A. Cresti, N. Nemec, B. Biel, G. Niebler, F. Triozon, G. Cuniberti, and S. Roche, Charge transport in disordered graphene-based low dimensional materials, *Nano Res.*, vol. 1, p. 361, 2008.

# 10 Applications of Nanocarbons for High-Efficiency Optical Absorbers and High-Performance Nanoelectromechanical Systems

*Anupama B. Kaul, Jaesung Lee,*
*and Philip X.-L. Feng*

## CONTENTS

**ABSTRACT**   In this chapter, some key characteristics of carbon-based nanomaterials and the role they play in enabling high-efficiency optical absorbers and mechanical resonators for nanoelectromechanical systems (NEMSs) are highlighted. In the first case, optical absorbers based on vertically aligned multiwalled carbon nanotubes (MWCNTs) are described that show an ultralow reflectance, which is 2 orders of magnitude lower compared to a reference material, Au-black, from a wavelength range of $\lambda \sim 350$–2500 nm. Reflectance measurements on the MWCNT absorbers after heating them in air to 400°C showed negligible changes in reflectance. The high optical absorption efficiency of the MWCNT absorbers over a broad spectral range, coupled with their thermal ruggedness, suggests that they have promise in solar energy harnessing applications, as well as thermal detectors for radiometry. In the second case, it has been shown that the mechanical and electrical properties of *as-grown*, vertically oriented carbon nanofibers (CNFs) are ideally suited for NEMS applications that derive benefit from the rugged and resilient mechanical properties of these nanostructures. The CNFs are synthesized using a plasma-enhanced chemical vapor deposition (PECVD) process and high-sensitivity optical interferometry is used to conduct mechanical resonance measurements. These resonator measurements show that the flexural resonances in the CNFs are in the very-high-frequency (VHF) regime up to 15 MHz, based on the typical geometries of the CNFs considered in these experiments. Our proof-of-concept measurements on the mechanical resonance characteristics of the CNFs suggest that they have exciting prospects for the resonant sensing and detection of radiation, adsorption, and other physical and/or biochemical processes.

## 10.1   INTRODUCTION

Research on nanocarbons has been at the forefront over the past several decades and has unarguably contributed tremendously to the birth of the field of nanotechnology. These efforts include providing insights into the novel properties of carbon nanomaterials; development of large-area, wafer-scale synthesis techniques; design of new characterization tools for understanding the role of imperfections in these materials; and development of processing methods to integrate these materials into novel device architectures. In particular, in the context of applications of interest to the electronics industry, carbon nanomaterials have been proposed as viable alternatives to traditional materials used in Si integrated circuits (ICs), which include their exploration in nanoscale transistors [1–6], interconnects [7,8], field emission displays [9], biosensors [10], heat transporting assemblies [11], and thermoelectric [12], photovoltaic [13], and optical materials [14] platforms, as well as nanoelectromechanical systems (NEMSs) [15–17] given their remarkable mechanical properties [18].

The origin of the extraordinary material properties [19] prevalent in carbon-based nanomaterials arises from the structural arrangement of carbon atoms in the crystalline lattice. A diverse spectrum of allotropes is possible with these materials, which display a rich variety of physical properties. Even for the commonly known three-dimensional (3D) form of carbon, namely, diamond and graphite, properties differ from the $sp^3$-bonded diamond to the $sp^2$-bonded graphene from which graphite is derived. A single layer of these carbon atoms in graphite results in the two-dimensional (2D) form of carbon, namely, graphene, which is a honeycomb arrangement of carbon atoms. The lowest dimensional forms of carbon, namely, zero-dimensional (0D) buckyballs or buckminsterfullerene C60 spheres, were discovered in 1985 by Smalley and coworkers [20], for which they received a Nobel Prize in Chemistry in 1997.

When a 2D graphene sheet is rolled into a cylinder, a one-dimensional (1D) or quasi-1D form of carbon results, namely, carbon nanotubes (CNTs). This 1D form of carbon has captured the interest of scientists and engineers for nearly two decades since its discovery in 1991 [21]. This first form of CNTs synthesized using an arc discharge process was for MWCNTs. Two years later, another form of CNTs, namely, single-walled CNTs (SWCNTs) [22], was discovered which is a single rolled-up sheet of graphene having a typical diameter of 1–2 nm. MWCNTs consist of concentric cylinders that have an interlayer spacing of 0.3–0.4 nm and diameters that are at least an order of magnitude larger than SWCNTs between 10 and 30 nm. When a 2D graphene sheet is patterned into strips, another 1D or quasi-1D form of carbon results, namely, graphene nanoribbons (GNRs) [23]. An additional quasi-1D structure, carbon nanofibers (CNFs), comprises graphene layers inclined to the central axis, commonly referred to as the fishbone or herringbone crystal structure.

Graphitic carbon nanomaterials such as 0D fullerenes, 1D SWCNTs, quasi-1D MWCNTs, and CNFs, and 3D graphite are thus all derivatives of graphene's 2D honeycomb lattice. The $sp^2$ bonding of carbon atoms in graphitic carbon nanomaterials generally affords them exceptional material properties, such as a high-charge carrier mobility, high-thermal conductivity, and excellent mechanical flexibility, which make attractive for a range of applications [24–27]. It is interesting to note that fullerenes—the lowest dimensional form of carbon nanomaterials—were the first to be discovered among the graphitic carbon nanomaterial family by Kroto et al. [28] in 1985, followed by CNTs in 1991 by Iijima [21], whereas graphene has remained elusive until recently despite being the basis for the other forms of carbon nanomaterials.

Because the mechanical exfoliation of graphene in 2004 [29], there has been a surge in research for unveiling the remarkable properties of this single sheet of carbon atoms and applying these properties to a wide range of devices and technological applications. As early as 1962, Boehm et al. [30] reported on the synthesis of single- and multilayer graphene through the reduction of graphene oxide, and this technique has gained renewed interest recently. In addition, graphene was also suggested to form on transition metals such as Pt (100) and Ni (111) by hydrocarbon dissociation [31–33], and other techniques such as metal intercalation [34]. Although these surface science investigations of graphene on metals [35] occurred several decades ago, the exploration of fundamental properties, such as the quantum hall effect and other interesting phenomena [36–38], as well as the ensuing applications of graphene, did not occur until after 2004, following the seminal experiments of Novoselov and Geim on mechanically exfoliated graphene that earned them the 2010 Nobel Prize in Physics.

Up until recently, 2D atomic monolayers have been presumed to generally exist only as parts of larger 3D structures, such as epitaxially grown crystals on lattice-matched substrates [39]. The existence of free-standing graphene was dismissed for some time as not being thermodynamically stable by Landau [40] and then several decades later by Mermin [41]. At the same time, the electronic structure and properties of graphene were analyzed theoretically by Wallace as early as 1947 [42] who showed that a single sheet of $sp^2$-hybridized carbon would have a linear energy dispersion relation at the *K-point* of the Brillouin zone.

Due to its flexibility, strength, high conductivity, transparency, and low cost, graphene has been proposed for a number of applications reviewed recently [43], such as a replacement for indium tin oxide (ITO) [44]. It also appears attractive for organic light-emitting diodes, as well as displays, touch screens [45], and solar cells [46]. The large surface area-to-volume ratio of graphene suggests that it also has promise in ultracapacitor applications [47] or chemical sensors [48,49], and coupled with its remarkable electronic properties, it is a strong contender for next-generation electronic devices in nanoelectronics [50,51]. Graphene and graphene-like materials have also been applied to composite materials applications [24,52,53].

In this chapter, we present an overview of nanocarbon materials, focusing specifically on 1D CNTs, CNFs, and 2D graphene. In Section 10.2, we discuss the application of 1D CNTs for optical absorbers. In Section 10.3, we highlight the use of 1D CNFs and 2D graphene for NEMS, specifically for resonator applications for mass and strain sensing.

## 10.2 NANOCARBONS FOR OPTICAL ABSORBERS

### 10.2.1 NANOMATERIALS FOR OPTICAL ABSORBERS

The ability of nanomaterials to trap light effectively has important implications for their use in energy harnessing, optical blacks for radiometry, as well as detectors. A survey of a host of nanomaterials, such as CdSe nanocrystals [54], graphene [55], and surface plasmon modes in metallic nanoparticles [56], reveals the promise such materials have in a wide range of optical applications. In this section, we describe another type of nanomaterial, which is exceptional at trapping incoming light as a result of its unique physical structure. Such a structure is composed of porous arrays of thin (10–15 nm diameter), vertically oriented MWCNTs. Interestingly, porosity in ordered nanostructures has been theoretically [57] shown to enhance optical absorption properties, which has also been verified experimentally in Si-based nanomaterials [58,59], and similar conclusions have also been drawn from theoretical calculations made on low-density arrays of CNTs [60].

The CNTs synthesized here for the optical absorber applications suggests that the mechanism of transducing optical energy into thermal energy may be an attractive platform for using such absorbers in energy harnessing, high-sensitivity thermal detectors, radiative cooling, thermography, antireflection coatings, and optical baffles to reduce scattering. In Section 10.2.2, the synthesis and optical characterization of the absorbers is discussed, followed by the ruggedness of these absorbers when they are exposed to high temperatures in Section 10.2.3.

### 10.2.2 SYNTHESIS AND OPTICAL CHARACTERIZATION

Previously synthesized MWCNTs [61] and SWCNTs [62] for optical absorber applications used water-assisted thermal chemical vapor deposition (CVD) [63], which yields exceptionally high growth rates with CNT lengths >100 μm; here alignment occurs primarily via the crowding effect [64,65]. Although thermal CVD is generally considered ineffective in aligning short CNTs (<10 μm), here we show that a glow discharge leads to vertically aligned CNTs with lengths more than an order of magnitude lower, which nonetheless still yield broadband, high-efficiency optical absorption characteristics in the ultraviolet (UV)-to-infrared (IR) range. This vastly extends the previously reported measurements on MWCNTs that were conducted in the visible [61], to well into the IR regime where it is increasingly difficult to find suitable black and opaque coatings. A thin and yet highly absorbing coating with absorptance A is valuable for thermal detector applications in the IR, for radiometry in order to enhance sensitivity, because the detectivity $D^* \propto A$ [66], and there is also a concomitant reduction in the response time.

Besides the importance of vertical alignment in yielding highly absorbing materials, the other structural trait for enhancing optical absorption efficiency is a high site density. Unlike earlier reports where the CNTs were synthesized directly on Si or $SiO_2$ [61,62], we demonstrate the growth of high-efficiency MWCNT absorbers directly on metallic substrates with site densities as high as ~$4 \times 10^{11}$/cm$^2$. In many other applications, it is desirable to grow CNTs directly on metals for lowering contact resistance, but the challenges in stabilizing catalyst particles on metallic surfaces at high temperatures have generally reduced site densities of CNTs manyfold (up to 100 times). In addition, prior attempts at growing MWCNTs for optical absorber applications on substrates other than Si, such as $LiNbO_3$, yielded an absorption efficiency of ~85% from $\lambda \sim 600$ to 1800 nm [67], whereas the CNT absorbers synthesized here on metallic substrates are shown to have an absorption efficiency >99.98% from $\lambda \sim 350$ to 2500 nm [68]. Even cermet-based materials, currently used for solar selective coatings on metallic substrates such as Cu and Al [69], have absorption efficiencies that are several orders of magnitude (up to $10^4$ times) lower than that reported in this chapter. A plasma-based process increases the potential of forming these absorbers at lower synthesis temperatures compared to thermal CVD, increasing future prospects of integrating such absorbers with a wider range of materials such as low-cost, flexible substrates for solar cells or with thermoelectrics, as well as fragile, temperature-sensitive micro-machined structures for IR sensing.

The initial substrate for the synthesis of the MWCNTs was a <100>-oriented Si wafer on which a layer of 100–200 nm-thick NbTiN—a refractory, high-temperature conducting nitride—was deposited reactively in a $N_2$ and Ar ambient using dc magnetron sputtering at a power of ~220 W and 5 m torr. Bimetallic layers of Co (thickness range 0.6–6 nm) and 2.5 nm-thick Ti were e-beam evaporated and served as the catalyst. Besides the Co/Ti/NbTiN/Si templates, control samples of Co/Ti/Si, Co/NbTiN/Si, and Co/Si were also prepared. Multiple samples (area ~ 4 cm$^2$) were placed on a wafer during plasma-enhanced CVD (PECVD) growth so that comparative analysis could be performed for different combinations of templates under similar synthesis conditions. At temperatures in the range of 550°C–750°C, $H_2$ was flowed into the chamber for several minutes, and the growth gases acetylene ($C_2H_2$) and ammonia ($NH_3$) were then introduced to a typical pressure of ~5 torr and the discharge was then ignited.

The choice of the template for PECVD synthesis of our MWCNTs was vital in synthesizing a high-density array of CNTs, and directly impacts the optical absorption characteristics. For example, the scanning electron microscope (SEM) image in Figure 10.1a shows amorphous carbon deposits when Co/Ti was placed directly on Si at 750°C, exhibiting a largely reflective surface (inset of Figure 10.1a). However,

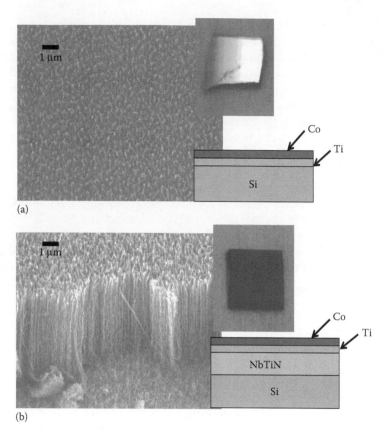

**FIGURE 10.1** (a) SEM micrograph of a Co/Ti/Si sample after dc PECVD growth. Top right inset shows an optical image of the sample depicting a reflective surface. (b) SEM micrograph of a Co/Ti/NbTiN sample after growth, depicting a high-density carpet of MWCNTs. Bottom right inset shows an optical image of the sample depicting a visually black sample to the naked eye. The spatial uniformity of the MWCNT ensembles is high over large length scales.

using a Co/Ti/NbTiN template yielded a visually black sample to the naked eye (inset of Figure 10.1b), and the SEM image (Figure 10.1b) depicts a high-density array of MWCNTs, which traps incoming light and suppresses reflection. The lack of growth of MWCNTs on Co/Ti/Si templates (Figure 10.1a) suggests that the presence of a refractory metallic nitride, such as NbTiN, is important in stabilizing the catalyst nanoparticles to prevent diffusion and alloying of the catalyst with the underlying Si at high temperatures. In addition, the density of MWCNTs in the absence of the Ti layer on the Co/NbTiN templates was low. It is speculated that the Ti may enable the Co to fragment into nanoparticles, similar to the role of Mo in the Co–Mo bimetallic catalyst system [70]. The Ti–Co system also appears to incorporate a larger fraction of C compared to Co alone, enhancing CNT growth [71]. Besides being of interest as absorbers in solar photothermal applications, the high areal density of MWCNTs on reflective, low-resistivity (~110 $\mu\Omega$ cm) metallic substrates may substantially reduce the CNT-to-substrate contact resistance.

The high magnification image in Figure 10.2a shows the surface of the MWCNT arrays which is rough, a factor that also contributes to scattering the incoming light diffusively. Figure 10.2b shows the SEM image of our benchmark, a Au-black absorber, which was synthesized using approaches similar to prior reports [72], where the percolated, random network of such a diffuse metal black should be apparent.

The optical measurements on the samples were conducted from $\lambda \sim 350$ to 2500 nm using a high-resolution, fiber-coupled, spectroradiometer (ASD Inc., Fieldspec Pro, Boulder Co) where a standard white light beam was focused at normal incidence to the sample, as shown by the schematic in the inset of Figure 10.2c. The bare fiber connector of the spectroradiometer was oriented at ~40° from the normal. Relative

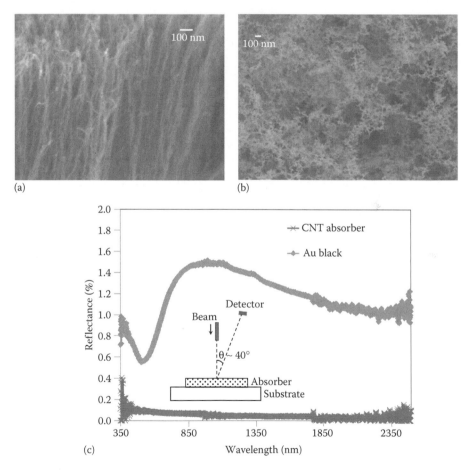

(a)

(b)

(c)

**FIGURE 10.2** (a) High-magnification SEM image shows the porous, vertically aligned morphology of the CNT absorbers, in contrast to the reference benchmark Au-black absorber sample and (b) which depicts a percolated, randomly aligned network of fibers. SEMs taken at 30° viewing angle. (c) Reflectance measurement from $\lambda \sim 350$–2500 nm for the MWCNT absorber and a Au-black absorber sample. The measurement setup is illustrated in the bottom inset. The Au-black reference sample has $R$ ~100 times larger where $R \sim 0.02\%$ at $\lambda \sim 2000$ nm for the CNT sample compared to 1.1% for the Au-black.

reflectance spectra were obtained by first white referencing the spectroradiometer to a 99.99% reflective spectralon panel. The reflected light intensity from the sample was then measured and the spectra were compared for samples synthesized under different growth conditions.

The optical reflectance response of the CNT absorber is shown in Figure 10.2c, where the spectra are compared to the reference Au-black absorber. The reflectance $R$ of the CNT absorber is nearly 2 orders of magnitude lower than that of the Au-black absorber, for example, ~0.02% at $\lambda \sim 2000$ nm compared to 1.1% for Au-black absorber. Other commonly used absorbers, such as NiP, have higher $R \sim 0.5\%$–1% for $\lambda \sim 320$–2140 nm [73], whereas ultra-black NiP alloy has $R \sim 0.16\%$–0.18% from $\lambda \sim 488$ to 1500 nm [74] and black paint has $R > 2.5\%$ from $\lambda \sim 600$ to 1600 nm. Top-down synthesized Si nanotips exhibit $R \sim 0.09\%$ at $\lambda \sim 1000$ nm [75], whereas bottom-up synthesized nanocone arrays have been reported to have an absorption efficiency of ~93% between $\lambda \sim 400$ and 650 nm [76].

The catalyst thickness appears to be an important synthesis parameter that impacted the optical absorption efficiency in these carbon-based nanoabsorbers. Figure 10.3a shows the reflectance spectra taken for two samples synthesized at Co catalyst thicknesses $c \sim 5$ and 0.9 nm (Ti thickness fixed at 2.5 nm). The sample with $c \sim 0.9$ nm has a wavelength-independent response from $\lambda \sim 350$ to 2500 nm with $R$ in the 0.02%–0.03% range. The sample with $c \sim 5$ nm, synthesized at identical conditions, has a wavelength-dependent $R$ which decreased from 0.94% at $\lambda \sim 400$ nm to ~0.33% at $\lambda \sim 2000$ nm.

Tentatively, the decreased reflectance/increased absorption may be expressed through an exponential decrease of the transmitted intensity $I(x)$ following a simple Lambert–Beer law formulation, that is, $I(x) = I_0 \exp(-\alpha x)$, where $I_0$ is the initial intensity of the incoming light and $\alpha$ is the absorption coefficient. The schematic in Figure 10.3b shows the geometry of the optical interrogation, as the incoming light traverses through the sparse aerogel-like structure of CNTs with intensity $I(x)$ at

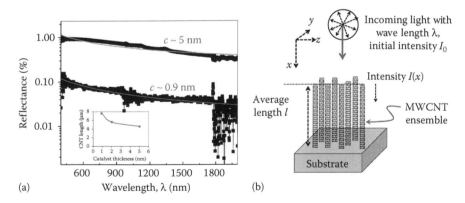

(a)                                                                              (b)

**FIGURE 10.3** (a) Optical reflectance spectra taken for two samples with $c \sim 0.9$ and 5 nm. The sample with $c \sim 5$ nm shows a wavelength-dependent $R$. Inset shows the variation of $l$ with $c$; superimposed in the reflectance vs. wavelength data are theoretical fits from which the ratio of $\kappa$ at $c \sim 0.9$ and 0.5 nm was determined. (b) The geometry used for the modeling analysis.

any vertical location inside the CNT array. The typical absorption coefficient with $\alpha \sim 10^4$ cm$^{-1}$ for $I(x = 8$ $\mu$m$)$ is ~30 times the $I(x = 4.5$ $\mu$m$)$. Now, the reflectance seems to decrease to the same order, that is, on the average $R$ drops from ~0.94 to ~0.03, ~30 times as well. The SEM images of the samples with $c \sim 0.9$, 2, and 5 nm are shown in Figure 10.4a–c, respectively. This yielded an MWCNT site density of ~4 × 10$^{11}$/cm$^2$ with MWCNT diameters $d \sim 10$–15 nm for $c \sim 0.9$ and a site density of ~6 × 10$^9$/cm$^2$ with $d \sim 80$–100 nm for $c \sim 5$ nm. Although the length $l$ of the MWCNTs decreased as $c$ increased (inset of Figure 10.3a), with $c \sim 5$ nm, $l$ was still >5 $\mu$m, well above $\lambda$ in these measurements, suggesting that the reduced absorption from the thicker catalyst is likely a result of changes in the fill fraction. The ability to engineer optical absorption efficiency by controlling the catalyst thickness is an attractive feature in tuning the optical absorption properties of the MWCNT ensembles.

A mechanism by which porous objects suppress reflection is through a reduction in the effective refractive index $n$. However, porosity alone may not necessarily be the primary factor involved because the Au-black absorber samples, a largely porous structure (SEM in Figure 10.2b) had higher reflectance compared to the MWCNT samples. This enhanced absorption may arise from the weak coupling of electrons in the vertically oriented CNTs to the incoming, normally incident radiation, with

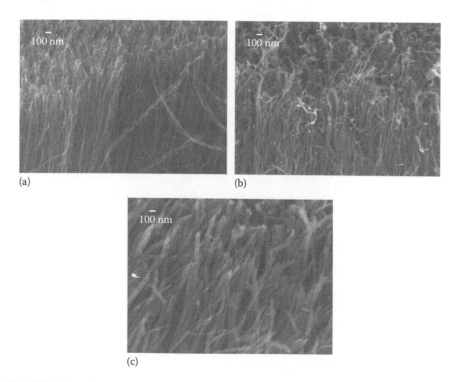

(a)                          (b)

(c)

**FIGURE 10.4**   The morphology of the MWCNTs is depicted in (a), (b), and (c) for $c \sim 0.9$, 2, and 5 nm, respectively. In (a), thin, vertically aligned tubes are depicted that have a high fill fraction, with a site density of ~4 × 10$^{11}$/cm$^2$ and MWCNT diameters of ~10 − 15 nm, whereas in (c) the site density decreases to ~6 × 10$^9$/cm$^2$ with MWCNT diameters of ~80–100 nm.

minimal backscattering and enable light to propagate into the long pores within the arrays until it is finally absorbed. A phenomenological model for absorption was developed using a formulation where the ensembles were treated as a composite medium consisting of nanostructures and air. The intensity at any given point $x$ in Figure 10.2b is given by $I(x) = I \exp(-\alpha x)$, where $\alpha = 4\pi\kappa/\lambda$ and $\kappa$ is the extinction coefficient. Assuming that there is no effective transmission through the substrate, $R(x) \sim \left[ I_0 - I(x) \right]$. The corresponding variation of $R$ with $\lambda$ was then fit to $\sim a_1 e^{a_2/\lambda} + a_3$, where $a_1$ is related to the incident intensity $I_0, a_2$ is a measure of the optical absorption length $(=\kappa \cdot l)$ and $a_3$ is a constant. The fit to the data is shown in Figure 10.3a for $c \sim 0.9$ and 5 nm. From the fits, the values of $a_2$ were determined to be ~0.025 and ~0.026 for $c \sim 0.9$ and 5, respectively, and given that the ratio, $\left[ (a_2)_{0.9}/(a_2)_5 \right] = \left[ (\kappa \cdot l)_{0.9}/(\kappa \cdot l)_5 \right]$ and that $l$ is 8 and 5 μm, respectively, we obtain a ratio of the extinction coefficients, $\kappa_{0.9}/\kappa_5$ of ~0.6.

It is interesting that the extinction ratio is smaller for the MWCNTs grown with $c \sim 0.9$ nm compared to the CNTs grown with $c \sim 5$ nm. This observation can be rationalized on the basis of a smaller area fraction in the former case, that is, 0.31 versus 0.47. Although such a rationalization does not explicitly consider the volume absorption due to a larger $l$ in the former case (i.e., 8 vs. 5 μm), it is justified because it has previously been shown that for the case of absorption in Si nanowires [77], the absorption in a thin film over a wide energy range comparable to the one used in the present work is on the average equivalent to the absorption in the nanowires. The larger absolute magnitude of $R$ for the sample with $c \sim 5$ nm compared to the sample with $c \sim 0.9$ nm may indicate an influence of the substrate in the latter, the effect of which is more pronounced due to a shorter $l$ for $c \sim 5$ nm (Figure 10.4c).

A more detailed analysis of the impact of catalyst thickness on the optical reflectance properties of the MWCNT absorbers was conducted for a wide range of catalyst thicknesses (Figure 10.5). This data (at $\lambda \sim 1500$ nm) shows a minimum of $R$ at $c \sim 1$ nm. However, $R$ increases when $c \sim 0.6$ nm due to the inability to nucleate a

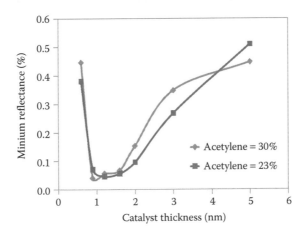

**FIGURE 10.5**    The reflectance measurement as a function of $c$ (taken at $\lambda \sim 1.5$ μm) for two acetylene gas ratios (30% and 23%). Growth conditions were 750°C, 172 W of plasma power, 30% $C_2H_2$, and 5 torr.

high enough areal density of MWCNTs; such behavior was consistent for two different acetylene gas concentrations, as indicated in Figure 10.5.

### 10.2.3 High-Temperature Optical Performance

We now present data that demonstrate the exceptionally low $R$ of the MWCNT absorbers even after they were exposed to temperatures as high as 400°C in air under an oxidizing environment, as might be expected with incident solar radiation. By comparison, the structural characteristics of the Au-black absorber reference gradually deteriorate with increasing temperature. However, the structural characteristics of the MWCNT absorber samples are largely unchanged when heated from 25°C to 400°C, as can be inferred from the data in Figure 10.6.

From the optical spectra (Figure 10.6), it is apparent that $R$ of the Au-black absorber sample increases as it is heated from 25°C to 200°C (2%) and it is ~23% at 400°C (at $\lambda \sim 2000$ nm). However, the $R$ of the CNT absorbers is still very low, ~0.022% after heating to 200°C (inset of Figure 10.6), and remains unchanged after exposure to temperatures as high as 400°C, which can be correlated to the structural integrity of the CNT absorbers to temperatures as high as 400°C.

In conclusion, we have successfully shown that, through catalyst engineering, PECVD-synthesized MWCNTs yield a high site density on metallic substrates which exhibit ultralow reflectance (~0.02%) over a wide spectral range from UV to IR for relatively thin (<10 µm) absorber ensembles. Their highly efficient optical absorption properties and exceptional ruggedness at high temperatures suggest their promise in solar photothermal applications and IR thermal detectors for radiometry

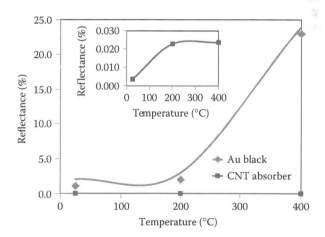

**FIGURE 10.6** Optical reflectance measurements were made in the $\lambda \sim 350$–2500 nm range for the Au-black and CNT absorber samples at 25°C and after subjecting them to temperatures up to 400°C. The figure shows $R$ of the Au-black and CNT absorber samples as a function of temperature. The Au-black absorber shows $R$ increases up to 23% after heating to 400°C. The inset shows $R$ of the CNT sample that increases slightly after exposure to 200°C, but it is still very low, ~0.022% at $\lambda \sim 2000$ nm, and remains unchanged after exposure to temperatures as high as ~400°C.

applications. In addition, the use of a plasma-based process increases the potential for synthesizing the absorbers at lower temperatures in the future, increasing the likelihood of integrating the absorbers with low-cost flexible substrates, potentially for solar cell applications, as well as thermoelectrics and micromachined structures for enabling new classes of IR sensors, particularly for rugged environments.

Besides the use of nanocarbons for optical absorbers, we now turn to their use as mechanical resonators for NEMS applications, as described in Section 10.3.

## 10.3 NANOCARBONS FOR NEMS

NEMSs are gaining increasing attention as viable alternatives for beyond-complementary metal–oxide semiconductor (CMOS) architectures, for enabling zero-leakage, ultralow-power, abrupt switching devices due to their inherently mechanical construction [78]. Such nanomechanical structures also offer advantages as electronic components operational in harsh environments, such as high radiation and high temperatures, in contrast to conventional solid-state devices which are more vulnerable to failure in hostile environments. In addition to their utility as potential beyond-CMOS switches, dynamical and resonant properties of NEMS has allowed these characteristics to be exploited for various sensing applications, such as mass sensing [79], force detection [80], and measurements in the quantum-mechanical regime [81].

Although many materials have been explored for NEMS devices, including conventional materials such as Si [82] and SiC [83], carbon-based materials have also been noted as another promising candidate for NEMS. Nanocarbons, such as graphene [84], CNTs [85–87], and CNFs [88,89], exhibit ultralow density and an exceptionally high Young's modulus. Coupled with their high elasticity and mechanical resilience, nanocarbon materials appear to be ideally suited for enabling high-performance, high-longevity NEMS resonators. In Section 10.3.1, we discuss the application of quasi-1D CNFs for NEMS resonator applications, followed by a discussion on the use in 2D graphene for NEMS discussed in Section 10.3.2.

### 10.3.1 CNF-BASED NEMS RESONATORS

Our CNFs were synthesized using electric field assisted growth. Electric fields applied during CVD growth of nanotubes serve to preferentially orient them in the direction of the field [90], which has also been observed in the PECVD synthesis of MWCNTs and fibers [91,92]. We have also applied the inherent E-fields present in plasma growth to develop processes for forming vertically aligned MWCNTs using dc PECVD, where a gas mixture of $NH_3$ and $C_2H_2$ was used at ~700°C. A stable and uniform dc glow discharge results over an area as large as ~75 mm. The plasma is more intensely concentrated toward the base electrode as pressure is increased, and the self-charging voltage also saturates after the first few minutes of growth under constant current control.

The vertically oriented CNFs were synthesized with pre-patterned Ni catalyst islands. Catalyst islands in the 300 nm range were formed in-between the electrodes using top-down wafer-scale approaches. After the liftoff process of the e-beam evaporated catalyst islands, high-purity acetylene ($C_2H_2$) and ammonia ($NH_3$) were introduced at 700°C, which served as the carbon feedstock and diluent gas, respectively, for the bottom-up synthesis of the CNFs using dc PECVD. When the desired growth pressure had been attained (~5 torr), the dc discharge was ignited at a power of ~200 W, and the growth was carried out for a fixed duration to the desired length of the CNFs. The CNFs ranged in length from 1 to 10 μm.

Given that the CNFs synthesized here using PECVD are targeted for NEMS applications, it is important to gain some insights into their mechanical properties. In this regard, bending tests were conducted on individual CNFs using a nanoprobe inside a SEM. The nanoprobe was mechanically manipulated so that it physically deflected the CNF to the right, as shown in Figure 10.7a. The CNF sustained bending angles φ as large as φ ~ 70° (Figure 10.7b), and it then returned elastically to its initial position (Figure 10.7c). The CNFs were able to tolerate such severe strains over tens of cycles without detachment from the substrate or fracture within the tube body, suggesting that their mechanical resilience and elasticity are extremely attractive for NEMS applications.

The resonance measurements are conducted using an optical interferometry setup shown in Figure 10.8, where a 633 nm He–Ne laser with a spot size of ~1 μm is

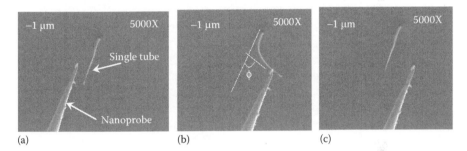

**FIGURE 10.7** (a) Bending test on a single CNF with a nanoprobe reveals that the CNF is extremely resilient and able to tolerate bending angles φ as large as 70° as shown in (b). The CNF returns elastically to its initial position as shown in (c).

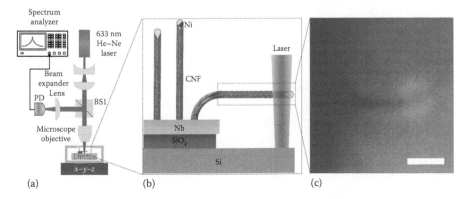

**FIGURE 10.8** Implementing laser optical interferometry on individual CNFs. (a) An optical interferometry displacement detection system. (b) A schematic illustration of a focused laser beam interrogating a single CNF. (c) An optical micrograph of a CNF under a laser spot focused near the tip of the CNF. Scale bar: 1 μm.

focused on a single CNF inside a vacuum chamber equipped with an $x$–$y$–$z$ stage. The $x$–$y$–$z$ stage is manipulated so that the laser spot is placed precisely on the tip of the CNF, as shown by the schematic in Figure 10.8b.

Figure 10.8c depicts the optical microscope image of a single CNF captured within the focused laser spot. In prior measurements, high laser power was shown to melt and oxidize the Ni catalyst, leading to structural deformations in the CNFs [93]. It appears that at 500 µW, the laser power is low enough to prevent damage to the Ni catalyst, and yet high enough to enable resonance measurements with high precision. As the reflected light from the CNF and the substrate interferes, the Brownian motion arising from fundamental thermal fluctuations in the CNF is detected by a photodetector, and the readout is captured by a spectrum analyzer that displays the fundamental thermomechanical noise spectrum.

The undriven Brownian motion of a number of CNFs is recorded, with representative examples shown in Figure 10.9a and b, where the top insets depict the corresponding scanning SEM images of the CNFs under consideration. The $Q$ is extracted by fitting the measured thermomechanical noise spectrum to a damped harmonic resonator model. Figure 10.9a exhibits a measured resonance at 3.490 MHz with $Q$ of 250, whereas Figure 10.9b shows a resonance frequency of 3.789 MHz with $Q$ of 300. Most of the devices tested had CNF lengths and diameters in the range of $L \approx 2$–$6$ µm and $d \approx 40$–$100$ nm, respectively, and the corresponding fundamental mode of the resonance frequency was within ~3–7 MHz.

The resonance characteristics of the CNFs is examined before and after exposing the CNFs to electron radiation inside a SEM operated at relatively low acceleration voltage (3 kV), high current (1.9 nA), and long irradiation time. The electron beam

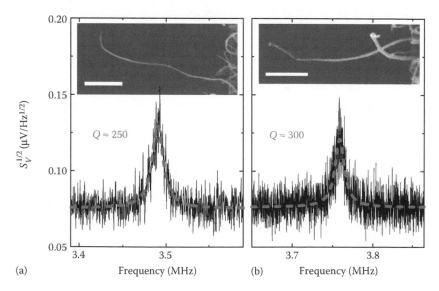

**FIGURE 10.9**  Resonances measured from two individual CNFs (a and b). Dashed line shows fitting results for determining $Q$ factors of the resonances. Insets are SEM images of the CNFs. Scale bars: 1 µm.

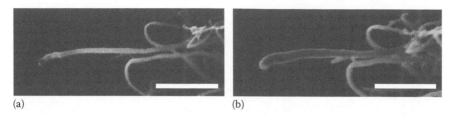

(a)                                                                              (b)

**FIGURE 10.10** SEM images of a CNF cantilever resonator showing the effects of electron beam-induced carbon deposition on surface. (a) Top-view SEM image before electron beam shower. (b) After ~15 min of 3 kV, 1.9 nA electron beam irradiation. Scale bar: 1 μm.

inside a SEM can easily react with residual organic material in the SEM chamber and deposit amorphous carbon on the surface of the CNFs. This is clearly visible from the SEM images in Figure 10.10, which shows the CNF to be apparently thicker after electron beam exposure (Figure 10.10b) compared to before exposure (Figure 10.10a).

Figure 10.11 shows the measured multimode resonance response from this single CNF before any electron beam treatment in the SEM, with four distinct resonance modes detected in the 5–15 MHz range, with $Q$ of 360–760. After irradiation of the CNF with the electron beam, the resonance response is immediately measured again. As shown in Figure 10.11b, the first two major resonance peaks, originally at 5.94 and 6.73 MHz, shift down to lower frequencies at 5.63 and 6.15 MHz, respectively, with noticeable changes in $Q$ factors. In addition, the signal amplitudes are lowered and the two initially higher frequency resonances at 9.70 and 14.10 MHz are no longer present after the electron beam shower. The disappearance of the third and fourth modes can also be explained on the basis of carbon deposition, which makes those peaks less responsive to optical detection.

The above results indicate that our 1D CNF NEMS devices encounter a reduction in resonance frequency due to electron beam irradiation-induced amorphous carbon deposition on the surface. Based on the measured frequency shift of the lowest resonance mode, the added mass can be calculated by using $\delta M/M \approx 2\delta f_0/f_0$, where $\delta M$, $M$, $\delta f_0$, and $f_0$ are the added mass, the total mass of resonator, the added mass-induced frequency shift, and the initial resonance frequency of the device, respectively. Given the measured resonance frequency shift of the fundamental mode in Figure 10.11b, we estimate the deposited mass on the CNF surface to be $\delta M \approx 4.4$ fg, whereas the CNF's own mass is $M \approx 41.5$ fg. Being able to have response to this ~5.2% frequency shift due to ~10.6% loaded mass suggests that the CNFs have exciting prospects for enabling robust adsorption-based mass sensors with a very wide dynamic range.

## 10.3.2 GRAPHENE-BASED NEMS RESONATORS

Just like carbon's other low dimensionality—0D and 1D structures—2D graphene exhibits very interesting properties. Electrons in graphene interact with the honeycomb lattice in a remarkable way which leads to unique electron transport properties. Graphene's charge carriers are referred to as massless Dirac fermions, where

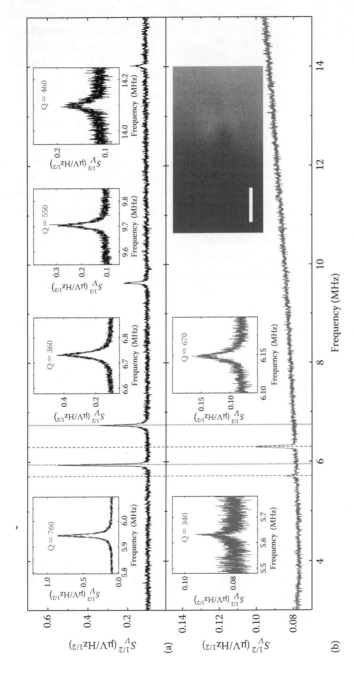

**FIGURE 10.11** Multimode resonance responses measured before and after electron beam radiation. (a) A wide-range resonance response of an individual CNF before electron exposure. (b) Response spectrum after an electron beam shower. Insets show details of individual resonances. Inset image shows a CNF with laser spot. Scale bar: 1 μm.

the electrons move relativistically at speeds of ~$10^6$ ms$^{-1}$ of instead of $3 \times 10^8$ ms$^{-1}$. These unique physical attributes make graphene a remarkable platform in which to also examine interesting low-dimensional physics effects, besides the obvious potential it has for technological applications. The massless carriers and minimal scattering over micron-sized crystallites result in quantum mechanical effects that can survive even at room temperature. Graphene exhibits ballistic room-temperature electron mobility of $2.5 \times 10^5$ cm$^2$/V-s [94], the ability to carry ultrahigh current densities ~$10^9$ A/cm$^2$ [95], an ultrahigh Young's modulus of 1 TPa with a breaking strength of ~40 N/m [96], an excellent elasticity for accommodating strains of up to 20% without breaking, a very high thermal conductivity of ~5000 W/m-K [97,98], and optical absorption of 2.3% [99]. It also appears to be an excellent barrier for gases [100].

In the work conducted here, we have explored the potential of graphene for enabling 2D NEMS resonators. The 2D graphene resonators were fabricated by transferring monolayer CVD-grown graphene to lithographically pre-patterned microtrenches. Although the transfer of graphene from metal growth substrates has become common practice for laboratory device fabrication, more work is necessary to achieve reproducibility and address the issues of defect introduction. As an example, when clean interfaces are achieved between two transferred graphene layers, hybridized properties emerge that open new scientific and technological opportunities. It is also critical to grow graphene directly on high-quality dielectrics such as hexagonal boron nitride (h-BN), which is free of dangling bonds and charge in homogeneities, unlike SiO$_2$. This is crucial in order to maintain graphene's high charge carrier mobility and, in this regard, all future heterostructure architectures with active graphene elements.

Nonetheless, although polycrystalline CVD graphene consists of various defects such as grains boundaries arising from initial growth, and through transfer to other substrates, the mechanical properties of CVD-grown graphene are comparable to those of mechanically exfoliated monocrystalline graphene [101]. In addition to the excellent mechanical properties, polycrystalline CVD graphene can be grown over large areas on copper (Cu) foil, facilitating the fabrication of large suspended structures through transfer techniques to pre-patterned microtrenches. By contrast, achieving such large suspended structures with mechanically exfoliated, monocrystalline graphene has been a significant challenge. CVD growth of graphene is performed by using a 1 inch (1″-diameter) quartz tube furnace. Before growth, Cu foil is carefully cleaned with acetic acid to remove the native Cu oxide and the foil is then annealed at 1000°C for 3 h in hydrogen (H$_2$) to remove any residual oxide; this anneal also allows for the small Cu grains to merge to form larger grains that serve as growth templates for graphene. Figure 10.12a shows an optical microscope image of graphene on Cu foil, showing the smooth surface of merged large Cu grains during the annealing process. Graphene is then synthesized using H$_2$ and methane (CH$_4$) gases at 1000°C for 5 min. During growth, we flow 100 sccm of CH$_4$ with 20 sccm of H$_2$ and at a pressure of 10 torr. After growth, we perform Raman spectroscopy to characterize the CVD-grown graphene. From Raman results, our CVD graphene has very high uniformity and high quality [102]. CVD graphene is then transferred to pre-patterned microtrenches to fabricate graphene resonators by using conventional wet transfer processes. Figure 10.12 demonstrates examples of transferred graphene onto SiO$_2$ substrates and suspended over an array of microtrenches to yield drumhead resonators.

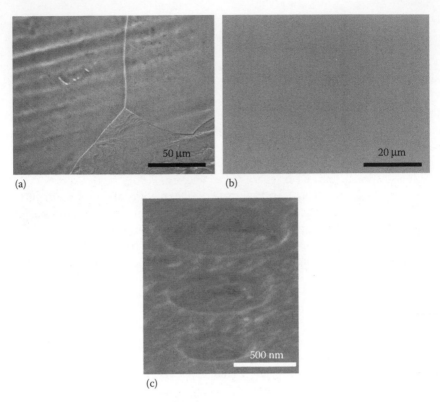

**FIGURE 10.12** CVD graphene on different substrates. (a) An optical microscope image of CVD graphene on Cu foil. Different sizes of Cu grains exhibit different surface morphologies. (b) Transferred CVD graphene on 290 nm-thick SiO$_2$ on Si substrate. Graphene is continuous over entire optical images. (c) An SEM image of suspended graphene on circular microtrenches, demonstrating drumhead graphene resonators.

The resonance characteristics of circular drumhead CVD graphene resonators are measured by using an optical interferometry system, similar to the schematic shown in Figure 10.8a. Figure 10.13 shows an example of the measured thermomechanical noise spectrum from a drumhead graphene resonator. Thermomechanical noise at resonance is given by

$$S_{x,th}^{1/2}(\omega_m) = \left( \frac{4k_BTQ}{\omega_m^3 M_{eff}} \right)^{1/2}$$

where:
   $T$ is the temperature
   $\omega_m$ is the angular resonance frequency
   $k_B$ is Boltzmann's constant
   $Q$ is the $Q$ factor
   $M_{eff}$ is the device effective mass

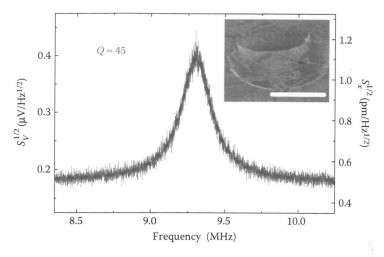

**FIGURE 10.13** Measured thermomechanical resonance response of a monolayer CVD graphene resonator. Inset shows a SEM image. Scale bar: 5 μm.

From this equation, the calculated Brownian motion of the device at room temperature is $S_{x,th}^{1/2}(\omega_m) = 0.98$ pm/Hz$^{1/2}$ and the responsivity and sensitivity of interferometry system with the graphene resonator are $\Re = S_{v,th}^{1/2}/S_{x,th}^{1/2} = 0.358$ μV/pm and $S_{x,sys}^{1/2} = 0.49$ pm/Hz$^{1/2}$, respectively.

The measured resonance characteristics provide precise mechanical properties of graphene. By using the drumhead membrane resonator model, $f_0 = 2.4048(\gamma/\rho t)^{1/2}/\pi d$, where, $d$, $\gamma$, $\rho$, and $t$ are the diameter, tension (N/m), density, and thickness of the resonators, respectively, and the estimated tension in the resonator is $\gamma \sim 0.01$ N/m (strain of $3 \times 10^{-5}$), demonstrating the unique capability of measuring ultrasmall strain levels in a 2D NEMS platform.

## 10.4 SUMMARY AND CONCLUSIONS

In sum, we have successfully shown that dc PECVD-synthesized MWCNTs exhibit ultralow reflectance properties over a wide spectral range from UV to IR for relatively thin (<10 μm) absorber ensembles. The structural characteristics of the MWCNT absorbers were engineered by controlling the bottom-up synthesis parameters during PECVD, which enabled optimization of the optical properties of the nanoabsorbers. Such vertically oriented nanocarbon aerogel structures were also found to be extremely rugged, withstanding temperatures of up to 400°C, without showing signs of any significant degradation in optical properties.

We also demonstrate 1D CNFs and 2D resonant NEMS derived from nanocarbon materials, where the resonance frequency and the quality ($Q$) factor of the devices are measured experimentally using ultrasensitive optical interferometry. The 1D nanocarbon resonators are formed using CNFs which are prototyped into cantilever-shaped 1D resonators of few micrometers long, where the resonance frequency and $Q$s are extracted from measurements of the undriven thermomechanical noise spectrum. The thermomechanical noise measurements yield resonances in the ~3–15 MHz range, with $Q$ of ~200–800. Significant changes in resonance characteristics were observed as a result of electron beam-induced amorphous carbon deposition on the CNFs, which suggests that 1D CNF resonators have strong prospects for ultrasensitive mass detection. We also presented results on NEMS resonators based on 2D graphene nanomembranes, which exhibit robust undriven thermomechanical resonances for the extraction of ultrasmall strain levels.

## ACKNOWLEDGMENTS

JL and PXLF are grateful to Case School of Engineering for support and facilities.

## REFERENCES

1. P. Avouris, Z. Chen, and V. Perebeinos, Carbon-based electronics, *Nat. Nanotechnol.* **2**, 605 (2007).
2. Q. Cao and J. Rogers, Ultrathin Films of Single-Walled Carbon Nanotubes for Electronics and Sensors: A Review of Fundamental and Applied Aspects, *Adv. Mater.* **21**, 29 (2009).
3. A. Bachtold, P. Hadley, T. Nakanishi, and C. Dekker, Logic circuits with carbon nanotube transistors, *Science* **294**, 1317 (2001).
4. Z. Zhang, X. Liang, S. Wang, K. Yao, Y. Hu, Y. Zhu, Q. Chen et al., Doping-Free Fabrication of Carbon Nanotube Based Ballistic CMOS Devices and Circuits, *Nano Lett.* **7**, 3603 (2007).
5. J. Appenzeller, Y. Lin, J. Knoch, Z. Chen, and P. Avouris, Comparing carbon nanotube transistors-the ideal choice: A novel tunneling device design, *IEEE Trans. Electron Dev.* **52**, 2568 (2005).
6. P. R. Bandaru, C. Daraio, S. Jin, and A. M. Rao, Novel electrical switching behaviour and logic in carbon nanotube Y-junctions, *Nat. Mater.* **4**, 663 (2005).
7. J. Li, A. Cassell, H. T. Ng, R. Stevens, J. Han, and M. Meyyappan, Bottom-up approach for carbon nanotube interconnects, *Appl. Phys. Lett.* **82**, 2491 (2003).
8. H. Li, C. Xu, N. Srivastava, and K. Banerjee, Carbon Nanomaterials for Next-Generation Interconnects and Passives: Physics, Status, and Prospects, *IEEE Trans. Electron Dev.* **56**, 1799 (2009).
9. W. B. Choi, D. S. Chung, J. H. Kang, H. Y. Kim, Y. W. Jin, I. T. Han, Y. H. Lee et al., Fully Sealed, High-Brightness Carbon-Nanotube Field- Emission Display, *Appl. Phys. Lett.* **75**, 3129 (1999).
10. F. Lu, L. Gu, M. J. Meziani, X. Wang, P. G. Luo, L. M. Veca, L. Cao, and Y. P. Sun, Advances in bioapplications of carbon nanotubes, *Adv. Mater.* **21**, 139 (2009).
11. C. Yu, S. Saha, J. Zhou, K. Shi, A. M. Cassel, B. A. Cruden, Q. Ngo, and J. Li, Thermal contact resistance and thermal conductivity of a carbon nanofiber, *J. Heat Transfer Trans. ASME* **128**, 234 (2006).
12. P. Wei, W. Bao, Y. Pu, C. N. Lau, and J. Shi, Anomalous thermoelectric transport of Dirac particles in graphene, *Phys. Rev. Lett.* **102**, 166808 (2009).
13. H. Ago, K. Petritsch, M. S. P. Shaffer, A. H. Windle, and R. H. Friend, Composites of Carbon Nanotubes and Conjugated Polymers for Photovoltaic Devices, *Adv. Mater.* **11**, 1281 (1999).
14. Y. Homma, S. Chiashi, and Y. Kobayashi, Suspended single-wall carbon nanotubes: Synthesis and optical properties, *Rep. Prog. Phys.* **72**, 066502 (2009).
15. T. Rueckes, K. Kim, E. Joselevich, G. Y. Tseng, C. L. Cheung, and C. M. Lieber, Carbon Nanotube-Based Nonvolatile Random Access Memory for Molecular Computing, *Science* **289**, 94 (2000).
16. J. E. Jang, S. N. Cha, Y. J. Choi, D. J. Kang, T. P. Butler, D. G. Hasko, J. E. Jung, J. M. Kim, and G. A. J. Amaratunga, Graphene 2.0, *Nat. Nanotechnol.* **3**, 26 (2008).

17. A. B. Kaul, E. W. Wong, L. Epp, and B. D. Hunt, Electromechanical carbon nanotube switches for high-frequency applications, *Nano Lett.* **6**, 942 (2006).
18. M. F. Yu, O. Lourie, M. J. Dyer, K. Moloni, T. F. Kelly, and R. S. Ruoff, Strength and Breaking Mechanism of Multiwalled Carbon Nanotubes Under Tensile Load, *Science* **287**, 637 (2000).
19. M. S. Dresselhaus, G. Dresselhaus, and P. Avouris (Eds.), *Carbon Nanotubes*, Springer, Berlin, Germany (2001).
20. H. W. Kroto, J. R. Heath, S. C. O'Brien, R. F. Curl, and R. E. Smalley, C60: Buckminsterfullerene, *Nature* **318**, 162 (1985).
21. S. Iijima, Helical microtubules of graphitic carbon, *Nature* **354**, 56 (1991).
22. S. Iijima and T. Ichihashi, Single-shell carbon nanotubes of 1-nm diameter, *Nature* **363**, 603 (1993).
23. M. Han, B. Ozyilmaz, Y. Zhang, and P. Kim, Energy Band-Gap Engineering of Graphene Nanoribbons, *Phys. Rev. Lett.* **98**, 206805 (2007).
24. A. K. Geim and K. S. Novoselov, The rise of graphene, *Nat. Mater.* **6**, 183 (2007).
25. L. Dai, *Carbon Nanotechnology: Recent Developments in Chemistry, Physics, Materials Science and Device Applications*, (Ed: L. Dai), Elsevier, Oxford (2006).
26. H. O. Pierson, *Handbook of Carbon, Graphite, Diamond, and Fullerenes: Properties, Processing and Applications*, Noyes Publications, Park Ridge, NJ (1993).
27. A. Krüger, *Carbon Materials and Nanotechnology*, Wiley-VCH, Weinheim, Germany (2010).
28. H. W. Kroto, J. R. Heath, S. C. O'Brien, R. F. Curl, and R. E. Smalley, C60: Buckminsterfullerene, *Nature* **318**, 162 (1985).
29. K. S. Novoselov, A. K. Geim, S. V. Morozov, D. Jiang, Y. Zhang, S. V. Dubonos, I. V. Grigorieva, and A. A. Firsov, Electric Field Effect in Atomically Thin Carbon Films, *Science* **306**, 666 (2004).
30. H. P. Boehm, A. Clauss, G. O. Fischer, and U. Hoffmann, Z., *Thinnest carbon films, Naturforsch. B* **17**, 150 (1962).
31. A. E. Morgan, and G. A. Somarjai, Low energy electron diffraction studies of gas adsorption on the platinum (100) single crystal surface, *Surf. Sci.* **12**, 405 (1968).
32. J. W. May, Platinum surface LEED rings, *Surf. Sci.* **17**, 267 (1969).
33. Y. Gamo, A. Nagashima, M. Wakabayashi, M. Terai, and C. Oshima, Atomic structure of monolayer graphite formed on Ni(111), *Surf. Sci.* **374**, 61 (1997).
34. Y. S. Dedkov, A. M. Shikin, V. K. Adamchuk, S. L. Molodtsov, C. Laubschat, A. Bauer, and G. Kaindl, Intercalation of copper underneath a monolayer of graphite on Ni(111), *Phys. Rev. B* **64**, 035405 (2001).
35. M. Batzill, The surface science of graphene: Metal interfaces, CVD synthesis, nanoribbons, chemical modifications, and defects, *Surf. Sci. Rep.* **67**, 83 (2012).
36. Y. B. Zhang, Y. W. Tan, H. L. Stormer, and P. Kim, Experimental observation of the quantum Hall effect and Berry's phase in graphene, *Nature* **438**, 201 (2005).
37. K. S. Novoselov, A. K. Geim, S. V. Morozov, D. Jiang, M. I. Katsnelson, I. V. Grigorieva, S. V. Dubonos, and A. A. Firsov, Two-dimensional gas of massless Dirac fermions in graphene, *Nature* **438**, 197 (2005).
38. D. Zhan, J. Yan, L. Lai, Z. Ni, L. Liu, and Z. Shen, Engineering the electronic structure of graphene Advanced Materials, *Adv. Mater.* **24**, 4055 (2012).
39. J. W. Evans, P. A. Thiel, and M. C. Bartelt, Morphological Evolution During Epitaxial Thin Film Growth: Formation of 2D Islands and 3D Mounds, *Sur. Sci. Rep.* **61**, 1 (2006).
40. L. D. Landau, Theory of phase changes. I, *Phys. Z. Sowjetunion* **11**, 26 (1937).
41. N. D. Mermin, Crystalline Order in Two Dimensions, *Phys. Rev.* **176**, 250 (1968).
42. P. R. Wallace, The Band Theory of Graphite, *Phys. Rev.* **71**, 622 (1947).
43. K. S. Novoselov, V. I. Fal'ko, L. Colombo, P. R. Gellert, M. G. Schwab, and K. Kim, A roadmap for graphene, *Nature* **490**, 192 (2012).

44. X. Wang, Z. Zhi, and K. Mullen, Transparent, conductive graphene electrodes for dye-sensitized solar cells, *Nano Lett.* **8**, 323 (2009).
45. P. Matyba, H. Yamaguchi, G. Eda, M. Chhowalla, L. Edman, and N. D. Robinson, Graphene and Mobile Ions: The Key to All-Plastic, Solution-Processed Light-Emitting Devices, *ACS Nano* **4**, 637 (2010).
46. X. Miao, S. Tongay, M. K. Petterson, K. Berke, A. G. Rinzler, B. R. Appleton, and A. F. Hebard, High efficiency graphene solar cells by chemical doping, *Nano Lett.* **12**, 2745 (2012).
47. M. D. Stoller, S. Park, Y. Zhu, J. An, and R. S. Ruoff, Graphene-Based Ultracapacitors, *Nano Lett.* **8**, 3498 (2008).
48. F. Schedin, A. K. Geim, S. V. Morozov, E. W. Hill, P. Blake, M. I. Katsnelson, and K. S. Novoselov, Detection of individual gas molecules adsorbed on graphene, *Nat. Mater.* **6**, 652 (2007).
49. J. T. Robinson, F. K. Perkins, E. S. Snow, Z. Wei, and P. E. Sheehan, Reduced graphene oxide molecular sensors, *Nano Lett.* **8**, 3137 (2008).
50. E. W. Hill, A. K. Geim, K. Novoselov, F. Schedin, and P. Blake, Graphene Spin Valve Devices, *IEEE Trans. Mag.* **42**, 2694 (2006).
51. H. B. Heersche, P. Jarillo-Herrero, J. B. Oostinga, L. M. K. Vandersypen, and A. F. Morpurgo, Bipolar Supercurrent in Graphene, *Nature* **446**, 56 (2007).
52. S. Stankovich, R. D. Piner, X. Chen, N. Wu, S. T. Nguyen, and R. S. Ruoff, Synthesis of graphene-based nanosheets via chemical reduction of exfoliated graphite oxide, *J. Mater. Chem.* **16**, 155 (2006).
53. S. Stankovich, D. A. Dikin, G. H. B. Dommett, K. M. Kohlhaas, E. J. Zimney, E. A. Stach, R. D. Piner, S. T. Nguyen, and R. S. Ruoff, *Nature* **442**, 282 (2006).
54. U. Huynh, J. J. Dittmer, and A. P. Alivisatos, Hybrid nanorod-polymer solar cells, *Science* **295** (5564), 2425–2427 (2002).
55. R. R. Nair, P. Blake, A. N. Grigorenko, K. S. Novoselov, T. J. Booth, T. Stauber, N. M. R. Peres, and A. K. Geim, Fine structure constant defines visual transparency of graphene, *Science* **320** (5881), 1308 (2008).
56. D. Derkacs, S. H. Lim, P. Matheu, W. Mar, and E. T. Yu, Improved performance of amorphous silicon solar cells via scattering from surface plasmon polaritons in nearby metallic nanoparticles, *Appl. Phys. Lett.* **89**, 093103 (2006).
57. L. Hu and G. Chen, Building Blocks for Integrated Graphene Circuits, *Nano Lett.* **7** (11), 3249–3252 (2007).
58. L. Tsakalakos, J. Balch, J. Fronheiser, M.-Y. Shih, S. F. LeBoeuf, M. Pietrozykowski, P. J. Codella et al., Strong broadband optical absorption in silicon nanowire films, *J. Nanophoton.* **1**, 013552 (2007).
59. K. Peng, Y. Wu, H. Fang, X. Zhong, Y. Xu, and J. Zhu, Uniform, axial-orientation alignment of one-y dimensional single-crystal silicon nanostructure arrays, *Angew. Chem. Int. Ed.* **44**, 2737–2742 (2005).
60. F. J. Garcia-Vidal, J. M. Pitarke, and J. B. Pendry, Effective medium theory of the optical properties of aligned carbon nanotubes, *Phys. Rev. Lett.* **78**, 4289–4292 (1997).
61. Z.-P. Yang, L. Ci, J. A. Bur, S.-Y. Lin, and P. M. Ajayan, Experimental observation of an extremely dark material made by a low-density nanotube array, *Nano Lett.* **8**, 446 (2008).
62. K. Mizuno, J. Ishii, H. Kishida, Y. Hayamizu, S. Yasuda, D. N. Futaba, M. Yumura, and K. Hata, A black body absorber from vertically aligned single-walled carbon nanotubes, *Proc. Natl. Acad. Sci. U.S.A.* **106**, 6044 (2009).
63. K. Hata, D. N. Futaba. K. Mizuno, T. Namai, M. Yumura, and S. Iijima, Water-assisted highly efficient synthesis of impurity-free single-walled carbon nanotubes, *Science* **306**, 1362 (2004).

64. S. Fan, M. G. Chapline, N. R. Franklin, T. W. Tombler, A. M. Cassell, and H. Dai, Self-oriented regular arrays of carbon nanotubes and their field emission properties, *Science* **283**, 512 (1999).

65. R. Andrews, D. Jacques, A. M. Rao, F. Derbyshire, D. Qian, X. Fan, E. C. Dickey, and J. Chen, Continuous production of aligned carbon nanotubes: A step closer to commercial realization, *Chem. Phys. Lett.* **303**, 467 (1999).

66. P. Eriksson, J. Y. Andersson, and G. Stemme, Interferometric, low thermal mass IR-absorber for thermal infrared detectors, *Phys. Scripta.* **T54**, 165 (1994).

67. J. H. Lehman, R. Deshpande, P. Rice, B. To, and A. C. Dillon, Carbon multi-walled nanotubes grown by HWCVD on a pyroelectric detector, *Infrared Phys. Tech.* **47**, 246 (2006).

68. A. B. Kaul, J. B. Coles, K. G. Megerian, M. Eastwood, R. O Green, and P. R. Bandaru, Ultra-High Optical Absorption Efficiency from the Ultraviolet to the Infrared Using Multi-Walled Carbon Nanotube Ensembles, *Small* **9**, 1058 (2013).

69. C. Nunes, V. Teixeira, M. Collares-Pereira, A. Monteiro, E. Roman, and J. Martin-Gago, Deposition of PVD solar absorber coatings for high-efficiency thermal collectors, *Vacuum* **67**, 623 (2002).

70. Y. Murakami, S. Chiashi, Y. Miyauchi, and S. Maruyama, Direct Synthesis of Single-Walled Carbon Nanotubes on Silicon and Quartz-Based Systems, *Jpn. J. Appl. Phys.* **43**, 1221 (2004).

71. S. Sato, A. Kawabata, D. Kondo, M. Nihei, and Y. Awano, Carbon nanotube growth from titanium–cobalt bimetallic particles as a catalyst, *Chem. Phys. Lett.* **402**, 149 (2005).

72. D. J. Advena, V. T. Bly, and J. T. Cox, Deposition and characterization of far-infrared absorbing gold black films, *Appl. Opt.* **32**, 1136 (1993).

73. C. E. Johnson, Wastewater and Hazardous Solid Waste Disposal in the Aluminum Products Industry, *Metal Finish.* **78**, 21 (1980).

74. S. Kodama, M. Horiuchi, T. Kuni, and K. Kuroda, Ultra-black nickel-phosphorus alloy optical absorber, *IEEE Trans. Inst. Meas.* **39**, 230 (1990).

75. C. Lee, S. Bae, S. Mobasser, and H. Manohara, A Novel Silicon Nanotips Antireflection Surface for the Micro Sun Sensor, *Nano Lett.* **5**, 2438 (2005).

76. J. Zhu, Z. Yu, G. F. Burkhard, C.-M Hsu, S. T. Connor, Y. Xu, Q. Wang, M. McGehee, S. Fan, and Y. Cui, Optical absorption enhancement in amorphous silicon nanowire and nanocone arrays, *Nano Lett.* **9**, 279 (2009).

77. L. Hu and G. Chen, Analysis of Optical Absorption in Silicon Nanowire Arrays for Photovoltaic Applications, *Nano Lett.* **7**, 3249 (2007).

78. P. X.-L. Feng, Nanoelectromechanical switching devices: Scaling toward ultimate energy efficiency and longevity (invited talk), *Proceedings of the 3rd Berkeley Symposium on Energy Efficient Electronic Systems (E3S)*, Berkeley, CA, October 28–29 (2013), pp. 1–2, doi:10.1109/E3S.2013.6705881.

79. A. K. Naik, M. S. Hanay, W. K. Hiebert, X. L. Feng, M. L. Roukes, Towards single-molecule nanomechanical mass spectrometry, *Nat. Nanotechnol.* **4**, 445–450 (2009).

80. D. Rugar, R. Budakian, H. J. Mamin, and B. W. Chui, Single spin detection by magnetic resonance force microscopy, *Nature* **430**, 329–332 (2004).

81. T. Rocheleau, T. Ndukum, C. Macklin, J. B. Hertzberg, A. A. Clerk, and K. C. Schwab, Preparation and detection of a mechanical resonator near the ground state of motion, *Nature* **463**, 72–75 (2010).

82. X. L. Feng, R. He, P. Yang, and M. L. Roukes, Very high frequency silicon nanowire electromechanical resonators, *Nano Lett.* **7**, 1953–1959 (2007).

83. X. M. H. Huang, X. L. Feng, C. A. Zorman, M. Mehregany, and M. L. Roukes, VHF, UHF and microwave frequency nanomechanical resonators, *New J. Phys.* **7**, 247 (2005).

84. J. S. Bunch, A. M. van der Zande, S. S. Verbridge, I. W. Frank, D. M. Tanenbaum, J. M. Parpia, H. G. Craighead, and P. L. McEuen, Electromechanical resonators from graphene sheets, *Science* **315**, 490–493 (2007).

85. T. Rueckes, K. Kim, E. Joselevich, G. Y. Tseng, C.-L. Cheung, and C. M. Lieber, Carbon nanotube-based nonvolatile random access memory for molecular computing, *Science* **289**, 94–97 (2000).

86. J. Moser, J. Güttinger, A. Eichler, M. J. Esplandiu, D. E. Liu, M. I. Dykman, and A. Bachtold, Ultrasensitive force detection with a nanotube mechanical resonator, *Nat. Nanotechnol.* **8**, 493–496 (2013).

87. A. Eriksson, S. Lee, A. Sourab, A. Issacsson, R. Kaunisto, J. M. Kinaret, and E. E. B. Campbell, Direct transmission detection of tunable mechanical resonance in an individual carbon nanofiber relay, *Nano Lett.* **8**, 1224–1228 (2008).

88. A. B. Kaul, E. W. Wong, L. Epp, and B. D. Hunt, Electromechanical carbon nanotube switches for high-frequency applications, *Nano Lett.* **6**, 942–947 (2006).

89. A. B. Kaul, K. G. Megerian, A. T. Jennings, and J. R. Greer, In situ characterization of vertically oriented carbon nanofibers for three-dimensional nano-electro-mechanical device applications, *Nanotechnology* **21**, 315501 (2010).

90. Y. Zhang, A. Chang, J. Cao, Q. Wang, W. Kim, Y. M. Li, N. Morros, E. Yenilmez, J. Kong, and H. J. Dai, Electric-field directed growth of aligned single-walled carbon nanotubes, *Appl. Phys. Lett.* **79**, 3155–3157 (2001).

91. K. B. K. Teo, M. Chhowalla, G. A. Amaratunga, W. I. Milne, G. Pirio, P. Legagneux, F. Wyczisk, J. Olivier, and D. Pribat, Characterization of plasma-enhanced chemical vapor deposition carbon nanotubes by Auger electron spectroscopy, *J. Vac. Sci. Technol. B* **20**, 116–121 (2002).

92. A. V. Melechko, V. I. Merkulov, T. E. McKnight, M. A. Guillorn, K. L. Klein, D. H. Lowndes, and M. L. Simpson, Vertically aligned carbon nanofibers and related structures: Controlled synthesis and directed assembly, *J. Appl. Phys.* **97**, 041301 (2005).

93. J. Lee, P. X.-L. Feng, and A. B. Kaul, Characterization of plasma synthesized vertical carbon nanofibers for nanoelectronics applications, *MRS Proc.* **1451**, 117–122, doi:10.1557/opl.2012.922 (*2012 MRS Spring Meeting, Symposium EE—Nano Carbon Materials & Devices*, San Francisco, CA, April 9–13, 2012).

94. A. S. Mayorov. R. V. Gorbachev, S. V. Morozov, L. Britnell, R. Jalil, L. A. Ponomarenko, P. Blake et al., Micrometer-scale ballistic transport in encapsulated graphene at room temperature, *Nano Lett.* **11**, 2396 (2011).

95. J. Moser, A. Barreiro, and A. Bachtold, Current-induced cleaning of graphene, *Appl. Phys. Lett.* **91**, 163513 (2007).

96. C. Lee, X. D. Wei, J. W. Kysar, and J. Hone, Measurement of the elastic properties and intrinsic strength of monolayer graphene *Science* **321**, 385 (2008).

97. A. A. Balandin, S. Ghosh, W. Z. Bao, I. Calizo, D. Teweldebrhan, F. Miao, and C. N. Lau, Superior Thermal Conductivity of Single-Layer Graphene, *Nano Lett.* **8**, 902 (2008).

98. A. A. Balandin, Thermal properties of graphene and nanostructured carbon materials *Nat. Mater.* **10**, 569 (2011).

99. R. R. Nair, P. Blake, A. N. Grigorenko, K. S. Novoselov, T. J. Booth, T. Stauber, N. M. R. Peres, and A. K. Geim, Fine structure constant defines visual transparency of graphene *Science* **320**, 1308 (2008).

100. J. S. Bunch, S. S. Verbridge, J. S. Alden, A. M. van der Zande, J. M. Parpia, H. G. Craighead, and P. L. McEuen, Impermeable atomic membranes from graphene sheets *Nano Lett.* **8**, 2458 (2008).

101. G.-H. Lee, R. C. Cooper, S. J. An, S. Lee, A. van der Zande, N. Petrone, A.G. Hammerberg et al., High-strength chemical-vapor-deposited graphene and grain boundaries, *Science* **340**, 1073–1076 (2013).

102. J. Lee and P. X.-L. Feng, High frequency graphene nanomechanical resonators and transducers, *Proceedings of the IEEE International Frequency Control Symposium (IFCS 2012)*, Baltimore, MD, May 21–24 (2012), pp. 1–7, doi:10.1109/FCS.2012.6243742.

# 11 Carbon Nanostructures from Biomass Waste for Supercapacitor Applications

*Ankit Tyagi and Raju Kumar Gupta*

## CONTENTS

**ABSTRACT**  This chapter covers the recent advancements for the effective utilization of biomass for the fabrication of carbon nanostructures based electrode materials for energy storage applications in supercapacitors. Biomass is available in large quantity and produced from waste generated from various sources such as agriculture residue, tree leafs, food waste, nutshells, household and industrial activities. They contains celluloses, hemi-celluloses and lignin biopolymers, which can be considered as carbon source and one can get the activated carbon of industrial importance. The basic principles of energy storage devices e.g. supercapacitors, synthesis, characterization and performances of various electrode materials are discussed. The potential of carbon nano particles for energy storage applications has not been fully recognized yet. However, the trends for future developments are very promising.

## 11.1   INTRODUCTION

Energy is an important issue for every country on the earth.[1] At present, we depend majorly on fossil fuel consumption for our energy needs; energy production from fossil fuels will lead to several future impact on world economy and ecology.[2,3] Consumption of fossil fuels gives rise to the $CO_2$ emission, which is rising at an alarming rate due to global economic expansion. $CO_2$ gas is the main contributor to the global warming. An increase in the world population is also forcing us to consume more and more fossil fuel. According to the survey made by "World Energy Council," it has been estimated that by the end of year 2050, world will consume double the energy what we are consuming today.[4] Because of this rapid increase in the energy consumption around the world and the harmful effects of the traditional energy sources on environment, several challenges are imposed to human health, environment, and energy security.[5] We have to move toward the energy production from the renewable sources in order to meet our energy consumption because fossil fuels will not last forever.

Energy in the form of solar energy, wind is available in abundant amount; they are the renewable and readily available sources of energy.[6,7] However, the availability of this energy at the time of need is still an issue because sun is available in day time only and wind does not blow all the time. Another issue associated with renewable energy is the imbalance in its regional distribution due to which there is very less or no renewable energy available in some parts and there is availability of vast amount of renewable energy in other parts of the earth.[8,9] Energy storage devices play an important role in storing renewable energy so that we can utilize it at the time of unavailability of energy source.

This will create new opportunities in the development of energy storage devices and technologies that utilize energy more efficiently. Market for portable electronic devices such as mobiles and the hybrid electric vehicles is also growing rapidly and requires the use of clean energy sources.[10] Li-ion batteries, fuel cells, and supercapacitors as the energy storage systems are receiving most of the attention of scientific community at present. These are important in our daily life and play a vital role in completing the energy requirement of portable electronics such as cell phones, laptops, and notepads. They have also great potentials for future electric vehicles.[2,11] However, most of the batteries are not able to give large power and suffer from slow power delivery, which is required in some applications, for example, at the start of electric vehicles or at the time of braking and accelerating the electric vehicles. In the twenty-first century, it is a challenge for the storage devices to achieve high power density and long-term cycling life.[11,12] Supercapacitors can overcome this problem as they can store and release energy with high-rate capability and they are high power density devices.

## 11.2 COMPARISON OF SUPERCAPACITORS WITH CONVENTIONAL CAPACITORS, BATTERIES, AND FUEL CELLS

Comparison of all these energy storage devices can be made on the basis of Ragone plot, which gives relationship between the specific energy (Wh/kg) versus specific power (W/kg) (Figure 11.1).

A Ragone plot showed that batteries and low-temperature fuel cells are typical high energy devices having energy density in between 10 and 1000 Wh/kg, although they store large amount of energy, but they are not able to deliver this energy at a faster rate because of their low power density. Conventional capacitors such as metal electrode capacitors or electrolytic capacitors have very high power density on the order of $10^6$ W/kg, but the low energy density is a serious problem with this type of energy storage system, and they are not able to meet the requirement of long duration. Therefore, we are not able to meet all the needs of today's market in both the cases.[14] Electrochemical capacitors are also called supercapacitors that fill the gap between batteries and conventional capacitors. They may store 100 or 1000 times more energy compared to conventional capacitors and can release high charge in less time compared to batteries. Hence, they may improve the performance of conventional capacitors in terms of energy density and battery performance in terms of power density. They are also expected to have much longer cycle life than batteries (Table 11.1).

### 11.2.1 SUPERCAPACITORS

Capacitors that store the energy within the electrochemical double layer at the electrode/electrolyte interface are known under various names which are trademarks

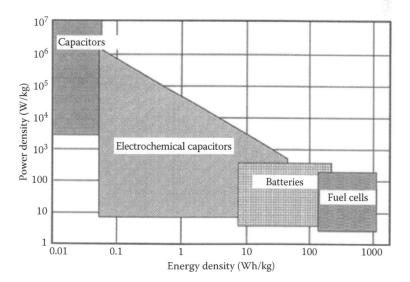

**FIGURE 11.1** A Ragone comparison of power and energy densities for electrochemical capacitors, storage batteries, and fuel cells. (Choudhury, N.A. et al., *Energ. Environ. Sci.*, 2, 55–67, Copyright 2009. Reproduced by permission of the Royal Society of Chemistry.)

**TABLE 11.1**

**Comparison of the Properties of Battery, Electrostatic Capacitor, and Supercapacitor**

| Basis | Battery | Electrostatic Capacitor | Supercapacitor |
|---|---|---|---|
| Discharging time | 0.3–3 h | $10^{-3}$–$10^{-6}$ s | 0.3–30 s |
| Charging time | 1–5 h | $10^{-3}$–$10^{-6}$ s | 0.3–30 s |
| Energy density | 10–100 Wh/kg | <0.1 Wh/kg | 1–10 Wh/kg |
| Specific power | 50–200 W/kg | >10,000 W/kg | ~1000 W/kg |
| Charge–discharge efficiency | 0.7–0.85 | ~1 | 0.85–0.98 |
| Number of cycles | 500–2000 | >500,000 | >100,000 |

*Source:* Zhang, Y. et al., *Int. J. Hydrogen Energ.*, 34, 4889–4899, 2009.

or established colloquial names such as *double-layer capacitors*, *supercapacitors*, *ultracapacitors*, *power capacitors*, *gold capacitors*, or *power cache*. *Electrochemical double-layer capacitor*" is the name that describes the fundamental charge storage principle of such capacitors.[14]

### 11.2.2 TYPES OF SUPERCAPACITORS

Supercapacitors can be classified into three categories depending on the charge storage mechanism at the electrode/electrolyte interface: electric double-layer capacitors, pseudocapacitors, and hybrid electrochemical capacitors. They are described in Sections 11.2.2.1 through 11.2.2.3.

### 11.2.2.1 Electric Double-Layer Capacitors

In the case of electric double-layer capacitors, double layers of charges are formed at the electrode/electrolyte interface due to electrostatically reversible adsorption of electrolyte ions on the electrode surface. Electrodes are generally high surface area carbon electrodes having mesoporosity. There is no chemical reaction between the electrode and the electrolyte in this type of charge storage mechanism. Current will flow due to the formation and deformation of double layer; hence, the process is highly reversible and millions of charge/discharge cycles are possible resulting in long lifetime of the electric double-layer capacitors.[6,15,16]

### 11.2.2.2 Pseudocapacitors

Pseudocapacitors store electric energy due to faradaic charge transfer between the electrode and the electrolyte. Electrode materials for pseudocapacitors include transition metal oxides such as $RuO_2$, $MnO_2$, $Co_3O_4$, or NiO, and conducting polymers such as polyaniline (PANI) or polypyrrole (PPy). These materials store energy through oxidation–reduction, electrosorption, or doping–dedoping process. Specific capacitance of pseudocapacitors is 10–100 times higher than that of electric double-layer capacitors. It has also been seen that pseudocapacitors show higher working voltage than electric double-layer capacitors. In pseudocapacitors, electrochemical process occurs near the surface, so that only a thin layer of electrode material is faradaically active and takes part in the charge storage mechanism. However, pseudocapacitors usually have lower power densities and cycling life than electric double-layer capacitors.[6,7,15-18]

### 11.2.2.3   Hybrid Electrochemical Capacitors

As the name suggests, hybrid electrochemical capacitor is a hybrid of electric double-layer capacitor and pseudocapacitor. In the last few years, there are several electrochemical capacitors developed by researchers with asymmetrical configuration, which include a carbon electrode having double-layer capacitance and a pseudocapacitive material having faradic charge storage at the other electrode. In general, pseudocapacitive materials have been used as cathode in most of the hybrid electrochemical capacitors developed till date. Pseudocapacitive electrodes store the charge through faradic process, which helps in improving the overall specific capacitance, and extend the working voltage of the hybrid electrochemical capacitors. The energy density of hybrid electrochemical capacitors is higher than that of electric double-layer capacitors.[6,15,16,18]

### 11.2.3   Charge Storage Mechanism in Electric Double-Layer Capacitors

The electric double-layer capacitors are usually made of carbon electrodes having high surface area. Electric double-layer capacitance is developed at the electrode/electrolyte interface where electric charges are distributed on the electrode surface and oppositely charged ions are arranged in the electrolyte. The charge storage mechanism depends on the surface dissociation as well as ion adsorption from both the electrolyte and crystal lattice defects. Figure 11.2 shows that electric double-layer capacitance developed at the interface between the electrode and the electrolyte. Electric charges arrange on the electrode surfaces in order to maintain the condition of electroneutrality; electrolyte ions of opposite charge counterbalance the amount of charge that builds up on the

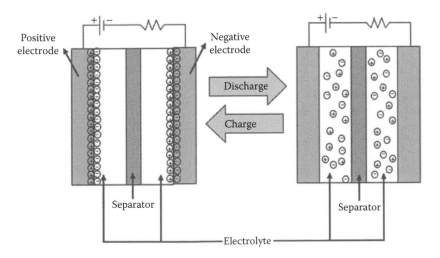

**FIGURE 11.2**  Mechanism of charge/discharge process for electric double-layer capacitor. (Reproduced with permission from Manaf, N.S.A. et al., *ECS J. Solid State Sci. Technol.*, 2, M3101–M3119, Copyright 2013, The Electrochemical Society.)

electrolyte side. During the process of charging, the electrons travel from the positive electrode to the negative electrode through an external load. Within the electrolyte, cations move toward the negative electrode and anions move toward the positive electrode. In this way, energy is stored at the double-layer interface. During discharge, the reverse processes take place. Because there is no charge transfer across the electrolyte/electrode interface, there are no chemical or composition changes in this process (i.e., nonfaradaic processes). Consequently, charge storage in electric double-layer capacitors is highly reversible, leading to very high cycling stabilities (up to $10^6$ cycles). It is generally said that the higher the surface area the higher the specific capacitance. However, it is not always true as many researchers showed that high surface area carbon may have low specific capacitance than lower surface area carbon. Porosity of the surface plays a major role in the charge storage mechanism. All the pores are not electrochemically active. Only the pores having size more than 5Å are electrochemically active and the optimal filling of poresis is achieved for pores having size close to 7–8Å in aqueous and organic electrolytes. Generally, high surface area carbon having most of the mesopores (2–50 nm in size) is considered as a good electrode material for the electric double-layer capacitor. Apart from the above factors, specific capacitance also depends on the surface groups attached on the surface of carbon and the size of electrolyte ion taking part in the electrochemical process. Higher oxygen containing or acidic surface functional groups on the surface of carbon may result in higher specific capacitance.[18–25]

## 11.2.4   MODELS FOR CHARGE STORAGE IN ELECTRIC DOUBLE-LAYER CAPACITORS

To understand the electrical processes occurring at the interface between a solid conductor and an electrolyte in detail, various models have been developed gradually over the years.

### 11.2.4.1   Helmholtz's Double-Layer Model

Helmholtz was the first who introduced the concept of double layer at the electrode/electrolyte interface in 1853. After that he compared the metal/metal interface with the metal/aqueous electrolyte interface in 1879. Metals are the good conductors of charge so that there is no electric field inside the metallic electrode at equilibrium. Consequently, he proposed that all the charges in the metallic conductors stay at its surface and the counter charge in the solution also resides at the surface to maintain the electroneutrality. Thus, two layers of charges are formed at the electrode/electrolyte interface having opposite polarity and are separated by a short distance (order of few angstroms). A capacitor is formed at the interface. Differential capacitance of this capacitor is given by the following equation:

$$C_H = \frac{\partial \sigma}{\partial \psi} = \frac{\varepsilon \varepsilon_0}{d}, \text{ where } \sigma = \frac{\varepsilon \varepsilon_0}{d} \psi$$

where:
   $C_H$ is the differential capacitance per unit area at the interface
   $\sigma$ is the surface charge density

$\psi$ is the voltage drop across the double layer
$\varepsilon$ is the dielectric constant of the medium
$\varepsilon_0$ is the permittivity of the free space
$d$ is the interlayer spacing

This model failed as it predicts the constant differential capacitance at the interface. But for real systems it is not constant, but it varies with potential, and hence considering $d$ constant is not correct. Thus, we need a more complicated model to explain the interfacial phenomena.[26-28]

### 11.2.4.2  Gouy–Chapman Model

Gouy and Chapman both independently proposed the idea of diffuse layer so this model was named as Gouy–Chapman model. In the early 1900s, Gouy found that capacitance in the double layer is not constant and it depends upon the applied potential across the electrodes and the concentration of the electrolyte. He revealed that the charge on the metallic electrode is confined to the surface, but for electrolyte it is not always true, especially for the low concentration of electrolyte solutions. Considerable amount of thickness of the electrolyte solution is required in order to balance the charges on the electrode surface. According to this model, electric forces are attractive or repulsive depending on the polarity of the electrolyte ions acting between the electrode and the electrolyte ions, together with the tendency of thermal motion of the ions which tries to randomize the electrolyte ions. Thus, the diffuse layer of charges is formed on the electrolytic side of the interface, and the constant $d$ in the Helmholtz model should be replaced by the average separation of ions in the diffuse layer. From this point, they concluded that the capacitance depends on the applied potential on the electrode because the diffuse layer becomes compact as we increase the applied potential. Similarly, if we increase the concentration of the electrolyte solution, the compact layer becomes compact and the differential capacitance at the electrode/electrolyte interface increases. Mathematical formulation of differential capacitance by this model uses Poisson's equation and the Boltzmann equation. Poisson's equation explains how the charge per unit volume depends on the variation of the electrode potential with the distance from the electrode surface, and the Boltzmann equation gives the distribution of the electrolyte ions because of their thermal motion. According to Gouy–Chapman model, differential capacitance at the interface is given by the following equation:

$$C_G = \frac{d\sigma^M}{d\psi_0} = \left(\frac{2z^2e^2\varepsilon\varepsilon_0 n^0}{kT}\right)^{1/2} \mathrm{Cosh}\left(\frac{ze\Phi_0}{2kT}\right) = \kappa\varepsilon\varepsilon_0 \mathrm{Cosh}\left(\frac{ze\Phi_0}{2kT}\right)$$

where:

$$\kappa = \frac{1}{\text{Diffusion layer thickness}} = \left(\frac{2n^0 z^2 e^2}{\varepsilon\varepsilon_0 kT}\right)^{1/2}$$

$\sigma^M$ is the surface charge density on the metallic side
$\psi_0$ is the potential at the electrode surface

$z$ is the magnitude of charge on the ion
$e$ is the electronic charge
$n^0$ is the concentration of each ion in the solution
$k$ is the Boltzmann constant
$T$ is the absolute temperature

The weakness of this model is that they considered ions as point charge. Because of this, ions can approach the surface arbitrarily closer and the capacitance can increase infinitely with the increase in potential according to this model, which is not possible for the real systems.[26–28]

### 11.2.4.3 Stern and Grahame Model

In 1924, Stern modified the Gouy–Chapman model by including both compact layer and diffuse layer. In this model, he considered that ions are having a finite size and can approach the surface closer till their ionic radius. The closest approach of the ion toward the electrode surface also depends on the solvation of ion in the solvent because of which the radius of the ion increases. The plane formed by the locus of the centers of the specifically adsorbed ions is called the *inner Helmholtz plane* (IHP). The solvated ions can approach the surface of the electrode only up to the IHP. The plane formed by the centers of the solvated ions is called the *outer Helmholtz plane* (OHP). Solvated ions can interact only with the electrode through long-range electrostatic forces and does not depend on the chemical properties of the ions; these ions are called nonspecifically adsorbed ions. This model is the combination of the Helmholtz and the Gouy–Chapman model, and the differential capacitance by this model is given by the following equation:

$$\frac{1}{C_S} = \frac{d}{\varepsilon\varepsilon_0} + \frac{1}{\left(2\varepsilon\varepsilon_0 z^2 e^2 n^0 / kT\right)^{1/2} \cosh\left(ze\Phi_2/2kT\right)}$$

where:
  $d$ is the distance between the electrode surface and the OHP
  $\Phi_2$ is the potential of the OHP

This equation can also be written as the combination of two capacitors in series:

$$\frac{1}{C_S} = \frac{1}{C_H} + \frac{1}{C_G}$$

This equation gives the result close to the experimental values in most of the situations. When the concentration of the electrolyte is high or the applied potential is high, a very compact layer is formed and Helmholtz layer capacitance is dominated over diffusion layer thickness; for very dilute electrolyte solutions or small electrode potentials, diffusion layer capacitance is higher compared to Helmholtz capacitance because a very loose compact layer is formed in this case (Figure 11.3).[26–29]

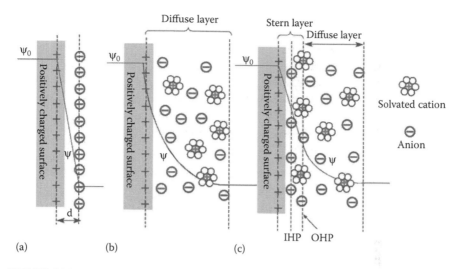

**FIGURE 11.3**  Models of the electrical double-layer at a positively charged surface: (a) the Helmholtz model; (b) the Gouy–Chapman model; and (c) the Stern model, showing the IHP and OHP. The IHP refers to the closest approach of specifically adsorbed ions (generally anions) and the OHP refers to that of the nonspecifically adsorbed ions. The OHP is also the plane where diffuse layer begins. $d$ is the double-layer distance described by the Helmholtz model. $\psi_0$ and $\psi$ are the potentials at the electrode surface and the electrode/electrode interface, respectively. (Zhang, L.L. and Zhao, X.S., *Chem. Soc. Rev.*, 38, 2520–2531, Copyright 2009. Reproduced by permission of the Royal Society of Chemistry.)

#### 11.2.4.4  Bockris, Devanathan, and Müller Model

J. O'M. Bockris, M. A. V. Devanathan, and K. Müller (BDM) proposed a model for electric double-layer in 1963 and they included the effect of solvent during electric double-layer formation. The earlier models neglected the intermolecular forces among the dipoles, which is an important parameter in deciding the rate of change of dipole orientation with applied electrode potential. They suggested that the first layer of water is present at the surface of electrode within the IHP having dielectric constant 6. The dipoles of water molecules in this layer are aligned in a fixed direction due to the charge on the electrode surface. Some water molecules may be displaced by specifically adsorbed ions. In the second layer of water molecules, the dipoles of the water molecules are not fixed in direction compared to the first layer and have dielectric constant 32 (Figure 11.4).[27,29,30]

### 11.2.5  Charge Storage Mechanism for Pseudocapacitors

The major problem with the carbon-based electric double-layer capacitors is their low specific energy. Most of the commercially available products have specific energy up to 10 Wh/kg, whereas the most lead acid batteries have specific energy up to 35–40 Wh/kg. The specific energy storage further increases up to 150 Wh/kg for the Li-ion batteries. The most active transition metal oxide for pseudocapacitors is hydrous $RuO_2$ because of its high-specific capacitance and high electrochemical stability. Being a noble metal oxide, $RuO_2$ is very costly and has some harmful

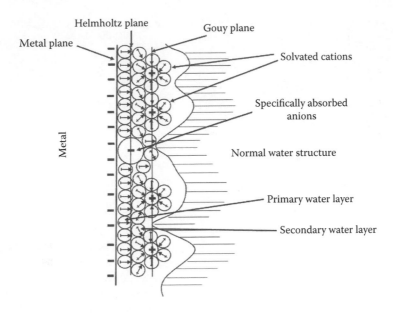

**FIGURE 11.4**  BDM double-layer model including layers of solvent.

environmental effects. Therefore, the scientists are trying to replace it with an environment-friendly and low-cost material. The U.S. military is using $RuO_2$-based supercapacitors for missile and aerospace applications, because cost is not an issue for these applications. $MnO_2$ is the most promising material after $RuO_2$ for pseudo-capacitors due to its high theoretical specific capacitance of ~1370 F/g, low-cost and environmental benign. Many researchers tried to study the charge storage mechanism of the $MnO_2$; two mechanisms of them got attention. In the first mechanism, it is considered that the electrolyte metal ions ($C^+ = Li^+$, $K^+$, $Na^+$, etc.) or $H^+$ ions intercalate and deintercalate into the bulk $MnO_2$. During intercalation, reduction and oxidation occur during the deintercalation process.

$$MnO_2 + H^+ + e^- \rightleftharpoons MnOOH$$

or

$$MnO_2 + C^+ + e^- \rightleftharpoons MnOOC$$

The second mechanism considered the adsorption of electrolyte metal cations at the surface of $MnO_2$ electrode.

$$MnO_{2\,surface} + C^+ + e^- \rightleftharpoons \left(MnO_2^- C^+\right)_{surface}$$

In both mechanisms, the oxidation state of Mn changes from +3 to +4 during oxidation and +4 to +3 during reduction process. It is also seen that there is no change in the oxidation state for the bulk material so that only a thin layer of $MnO_2$ electrode is electrochemically active. Although the uptake of cations during the intercalation process is still not very clear, $H^+$ can directly intercalate into $MnO_2$ structure but not the cations.[6,7,17,31,32] Electrochemical activity of $MnO_2$ is known to be dependent on the water content in the $MnO_2$ and also on the thermodynamic stability of $MnO_2$ phase.[31,33] In addition, manganese dissolution and oxygen evolution reaction are two factors that strongly influence the cycling properties of the $MnO_2$ electrode. It has been observed that most of the $MnO_2$-based pseudocapacitors show ~20% reduction in specific capacitance after ~1000 charge/discharge cycles, suggesting the difficulty in achieving long-term cycling stability for a device (Figure 11.5).[6,7,31,32]

## 11.2.6    Components of Supercapacitors

Supercapacitors mainly consist of two electrodes, a separator, and an electrolyte. The electrodes are made up of a metallic collector, and of an active material, which is a high surface area carbon, a metal oxide, or a conducting polymer. These two electrodes are separated by a membrane called separator. Separators allow the movement of ions through it but inhibit the conduction of electrons in order to prevent the short circuiting. Space between the electrodes is impregnated with an electrolyte. The electrolyte may be of solid-state electrolyte, organic electrolyte, ionic electrolyte, or aqueous type depending on the application. The working voltage of supercapacitor is determined by the decomposition voltage of the electrolyte and depends mainly on the environmental temperature, current intensity, and the required lifetime (Figure 11.6).[34]

### 11.2.6.1  Electrolytes

The type of electrolyte used in supercapacitors has a considerable effect on its performance, that is, the amount of energy stored, and how quickly this stored energy can be released. The relationship between the amount of energy stored ($E$), the operating

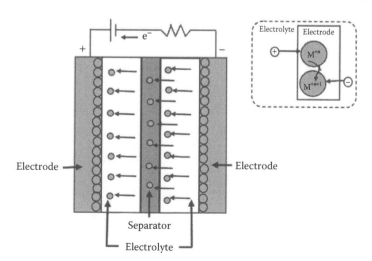

**FIGURE 11.5**    Energy storage mechanism for pseudocapacitors. (Reproduced with permission from Manaf, N.S.A. et al., *ECS J. Solid State Sci. Technol.*, 2, M3101–M3119, Copyright 2013, The Electrochemical Society.)

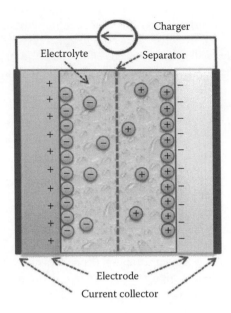

**FIGURE 11.6** Schematic representation of a supercapacitor cell. (Bose, S. et al., *J. Mater. Chem.*, 22, 767–784, Copyright 2012. Reproduced by permission of the Royal Society of Chemistry.)

voltage ($V$), and the specific capacitance ($C$, i.e., $E = \frac{1}{2}CV^2$) is important for the determination of energy density, so that the decomposition voltage of the electrolyte determines the operating voltage of the electrochemical capacitors. The electrolyte may be of the solid-state, organic, or aqueous type. Organic electrolytes are produced by dissolving quaternary salts in organic solvents. Their dissociation voltage may be greater than 2.7 V.[34] Aqueous electrolytes are typically KOH or $H_2SO_4$, presenting a dissociation voltage of only 1.23 V. Ionic liquids are generally used these days for attaining higher voltage window up to maximum 4.2 V. Electrolyte conductivity has a marked effect on the equivalent series resistance, which determines the power output. Aqueous electrolytes due to their high conductivity and low viscosity tend to produce faster rates of charge/discharge. Relatively higher concentration of electrolytes is required to minimize the equivalent series resistance and hence maximize the power density. However, strong acids and strong bases determine the materials of construction for the cell. The use of concentrated electrolytes also increases the rate of self-discharge displayed by the capacitor.[3,14]

### 11.2.6.2 Separators

Most of the separators available in the market are designed for batteries only. Hence there is a need to develop the separator that is specific for the electrochemical capacitors in order to achieve the exceptional performance of the electrochemical capacitors. If organic electrolytes are used, polymer (typically polypropylene) or paper separators are applied. With aqueous electrolytes, glass fiber separators as well as ceramic separators are generally used.[34]

## 11.2.7    BIOMASS WASTE USED FOR SUPERCAPACITOR APPLICATIONS

Puthusseri et al. have prepared high surface area carbon derived from waste paper via the hydrothermal method followed by KOH activation in the inert environment and used it for supercapacitor applications. The Brunauer–Emmett–Teller surface area was reported to be 2341 m²/g. The as-synthesized material had both mesopores and micropores with 70% pore volume due to the micropores and rest due to mesopores, having 2–3 nm size. A solid-state supercapacitor device was fabricated with ionic liquid polymer gel electrolyte having maximum power density of 19,000 W/kg and energy density of 31 Wh/kg. The Li-ion electrochemical capacitor was also constructed using waste paper carbon as cathode which showed an excellent energy density of 61 Wh/kg (Figure 11.7).[36]

Karthikeyan et al. have used pine cone to produce high surface area carbon. They selected pine cone or coulter pines and their seeds because it is a common biomass, which has existed for over 200 million years, low cost, and present in abundant amount. They showed that raw pine cone has highly organized crystalline cellulose fibers, lignin, carbohydrates, and uronic acid in it, and microfibers have an external diameter of 45 μm. They produced the maximum surface area (~3950 m²/g with an average pore size of 2.9 nm) for the 1:5 ratio of KOH to pine cone carbon when heated under Ar atmosphere, which is much higher than that reported for carbonaceous materials derived from various other biomass precursors. They showed the

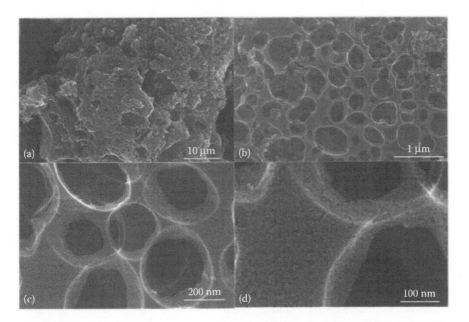

**FIGURE 11.7**    (a–d) Field emission scanning electron microscope images of carbon from waste paper at different magnifications. (From Puthusseri, D., Aravindan, V., Anothumakkool, B., Kurungot, S., Madhavi, S., and Ogale, S. From Waste Paper Basket to Solid State and Li-HEC Ultracapacitor Electrodes: A Value Added Journey for Shredded Office Paper, *Small*. 2014. 10. 4395–4402. Copyright Wiley-VCH. Reproduced with permission.)

maximum specific capacitance of 198 F/g or energy density of 61 Wh/kg for 0–3 V voltage window at a current density of 0.25 A/g for activated carbon with 5 wt.% KOH, using symmetrical supercapacitor cell configuration. They also reported that small oxygen-containing groups present in activated carbon surface are crucial to ensure long cyclability and maintained ~90% retention in specific capacitance after 20,000 cycles.[37]

Hao et al. have used bagasse as a raw material and produced hierarchical porous carbon aerogel for the supercapacitor applications. Bagasse is an industrial waste from sugar industry, a leftover after the extraction of sucrose from sugarcane plant. Demand of sugar is growing in the world so that more and more bagasse will produced as industrial waste, present in abundant amount and considered as a renewable source of biomass. It contains high proportion of cellulose, which was extracted through alkaline hydrolysis. Bagasse carbon aerogel was obtained through freeze drying of cellulose solution, which was further activated with KOH at different temperatures in order to get activated carbon aerogel. It was used for supercapacitor applications and a specific capacitance of 142.1 F/g at a current density of 0.5 A/g was achieved. As-prepared electrode showed 93.9% retention in specific capacitance after 5000 cycles.[38]

Jin et al. have used big bluestem, also called *Andropogon gerardii* or prairie tallgrass, which is a type of grass species found in the Midwest United states. Originally, these are used to produce biofuels via thermochemical processes, which produce biochar as a waste. They reported that biochar wastes generated from biofuel production have 30% of initial biomass and can be utilized further for the supercapacitor applications. They used three different activation agents ($NaHCO_3$, NaOH, and KOH) to activate the big bluestem biochar through carbonization under $N_2$ flow. The maximum specific surface area obtained was 552, 1616, and 2490 m²/g when $NaHCO_3$, NaOH, and KOH were used as activating agents, respectively. The carbon produced after activation has multilayer graphene structure as confirmed through Raman spectroscopy. Symmetrical two-electrode configuration was used for electrochemical characterization. The maximum specific capacitance was 283 F/g for KOH-activated sample in 6 M KOH electrolyte at the current density of 0.1 A/g.[39]

Peng et al. have used pomelo peel as a biomass source to generate carbon for the supercapacitor applications. Pomelo is a fruit found in Southeast Asia and the south of the Yangtze River in China; it belongs to the citrus family. Pomelo peel was carbonized under Ar atmosphere, followed by activation with KOH in order to get high surface area carbon (P-AC, 2191 m²/g).This activated carbon was again heated under Ar atmosphere to get low oxygen content activated carbon (H-AC, 2057 m²/g). The specific capacitance of P-AC was found to be 342 F/g, whereas it was 286 F/g for H-AC under similar conditions. Symmetric supercapacitor with H-AC sample showed that it can work up to 1.7 V voltage window with 1 M $NaNO_3$ aqueous electrolyte and can retain 90% specific capacitance even after 1100 cycles at the current density of 2 A/g during charging/discharging.[40]

Xu et al. have prepared two-dimensional porous carbon with high surface area (1069 m²/g) and high micropore volume ratio (83%) through the KOH activation of carbonized pistachio nutshell. They have reported that pistachio nutshell having 2D lamellar structure remains as it is after carbonization. Thickness of each slice in carbonized shell is ~200 nm. Because of this structure, the ion transport length decreases and helps in fast charge transport in supercapacitors. After KOH activation, more pores are generated having an average pore size of 0.76 nm; this size is slightly higher than that of TEA$^+$ ion (0.68 nm) in tetraethylammonium tetrafluoroborate organic electrolyte. These desolvated TEA$^+$ ions may approach the electrode surface more closely so that thickness of the double layer decreases, resulting in higher capacitance (20.1 μF/cm$^2$). The electrochemical activity of carbon was tested in aqueous as well as organic electrolytes. High energy density of 10/39 WH/kg at a power of 52/286 kW/kg was reported in 6 M KOH aqueous electrolyte and 1 M TEA BF$_4$ in ethylene carbonate-diethyl carbonate (1:1) organic electrolyte, respectively (Figure 11.8).[41]

Jin et al. have prepared activated carbon fibers at different activation time from waste wood shavings generated from fir (*Cunninghamia lanceolata*), waste from a furniture factory in China. First, they prepared fibers through a self-made spinning apparatus and activated them with the help of steam–nitrogen mixture gas. Maximum specific and micropore surface areas were 3223 and 2300 m$^2$/g, respectively, and the average pore size of 3–4 nm for 180 min activated sample. The electrochemical characterization showed a specific capacitance of 280 F/g at 0.5 A/g current density and retained 81.8% at 10 A/g in 1 M H$_2$SO$_4$. As-prepared carbon fibers showed high capacitance retention of 99.3% after 2000 charge/discharge cycles. Electrochemical impedance spectroscopy confirmed that as-prepared activated carbon fibers were having small equivalent series resistance (0.72–0.78 Ω) and charge transfer resistance (0.21–0.35 Ω).[42]

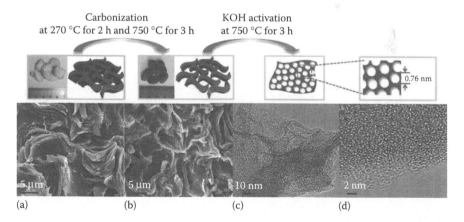

(a)          (b)          (c)          (d)

**FIGURE 11.8** The simulation model as well as the scanning electron microscopic (SEM) and high-resolution transmission electron microscopic (HRTEM) images of four characteristic states of the carbon representing the reaction process. (a) SEM image of the pistachio nutshell, which has a continuous, uniform, and lamellar natural structure. (b) SEM image of the carbonized nutshell, which possesses a curved 2D lamellar structure with certain shrinkage compared with that of the natural structure. The resultant porous carbon preserves a wealth of natural plant features of the pistachio nutshell. (c and d) HRTEM images with different magnification times highlight the porous structure of AC-SPN-3. The diameter of the pore is about 0.5–1 nm. (Reproduced with permission from Xu, J. et al., *Sci. Rep.*, 4, Copyright 2014, Nature Publishing Group.)

Wang et al. have used kenaf stems to produce hierarchical porous carbon with the help of Ni etch process using hydrochloric acid. Different sized pores were produced via carbonizing carbon surface using a different concentration of Ni (34 ± 10 nm). The maximum specific surface area was found to be 1408 $m^2$/g with a pore size of 3.92 nm for the lowest concentration of Ni salt. It was also reported that electric conductivity of the carbon is lowest for this sample (1.52 s/cm). Etched and pyrolyzed carbon showed poor conductivity than only pyrolyzed carbon due to its more fluffy structure. Fourier transform infrared spectroscopy (FTIR) and X-ray photoelectron spectroscopy (XPS) results confirmed that as-prepared carbon was having oxygen-containing groups attached to its surface. As-prepared sample showed highest specific capacitance of 327 F/g at 2 mV/s scan rate and retained 84.7% of the specific capacitance at 100 mV/s scan rate. Charge/discharge characterization showed 95.6% specific capacitance after 5000 cycles at a current density of 1 A/g (Figure 11.9).[43]

Bhattacharjya et al. have prepared activated carbon from cow dung via KOH activation and used it for supercapacitor applications. Cow dung is present in abundant amount on the earth, currently used as fertilizer and cooking fuel in most of the developing countries. They partially carbonized the cow dung, mixed with KOH in

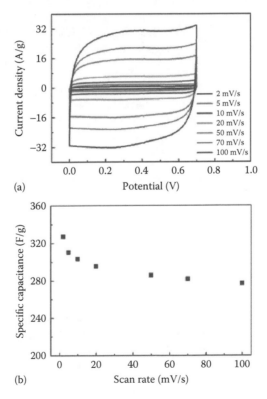

**FIGURE 11.9**  (a) Typical C–V curves of PKSC$_{0.016}$ electrode obtained at different scan rates of 2, 5, 10, 20, 50, 70, and 100 mV/s. (b) The plot of specific capacitance of PKSC$_{0.016}$ electrode as a function of scan rate. (Wang, L. et al., *RSC Adv.*, 4, 51072–51079, Copyright 2014. Reproduced by permission of the Royal Society of Chemistry.)

different mass ratio, and then carbonized the mixture in $N_2$ atmosphere, resulting in highly porous carbon with irregular surface morphology. Interconnected pores were obtained after carbonization for the optimized amount of KOH. The carbon as-prepared had Brunauer–Emmett–Teller surface area of 1984 $m^2$/g and 68% micropore volume ratio. Electrochemical characterizations were carried out using full-cell symmetric supercapacitor configuration in organic electrolytes at 0–2.5 V. The maximum specific capacitance was 124 F/g at a current density of 0.1 A/g for 2:1 ratio of KOH and precarbonized char (ACDC-2), and remained at 117 F/g at an increased current density of 1 A/g. The energy density of cell was found to be 28 Wh/kg at 1 A/g current density. Charging/discharging was done at 1 A/g current density and showed 85% retention of specific capacitance after 1000 cycles. Electrochemical impedance spectroscopy results showed that the internal resistance is minimum for ACDC-2 sample (6.35 $\Omega$), whereas equivalent series resistance remains almost similar for all the samples.[44]

Li et al. have used walnut shell to prepare activated carbon/ZnO composite for the supercapacitor applications. They used hydrothermal technique followed by heat treatment at different temperatures to get activated carbon/ZnO composite. Irregular morphology was obtained for carbon and flowerlike ZnO clusters were present on the surface of activated carbon. Raman and FTIR showed the graphitic nature of carbon and the presence of oxygen-containing groups on the surface of activated carbon, respectively. Precursor $ZnCl_2$ played multiple important roles together; it helped in increasing the surface area, improved the surface oxygen groups, and also transformed into ZnO, resulting in improving the pseudocapacitance of the supercapacitor. The specific surface area was obtained up to 818.9 $m^2$/g for activated carbon/ZnO composite and 1072.7 $m^2$/g for activated carbon at 800°C after removing ZnO. The maximum specific capacitance of 117.4 F/g was obtained for activated carbon/ZnO composite at a current density of 0.5 A/g during charge/discharge experiment in 6 M KOH aqueous electrolyte. There was no obvious decay in specific capacitance after 1000 charge/discharge cycles at 0.5 A/g current density.[45]

Huang et al. have prepared phosphorus- and oxygen-rich activated carbon through phosphoric acid activation of fruit stones (a mixture of apricot and peach stones), a lignocellulosic waste. They reported that the specific surface area decreased as the carbonizing temperature increased (up to 800°C) due to pore shrinkage of carbon structure or the blockage of the pores by surface functionalities and increased slightly at temperature of 900°C and 1000°C due to the decomposition and volatilization of phosphocarbonaceous compound. XPS showed that carbon, oxygen, and phosphorus were present in all the samples, but maximum phosphorus (2.58%) was present in 900°C sample (APP900). Electrochemical characterization for two-electrode configuration in 1 M $H_2SO_4$ showed that oxygen functionalities contributed in pseudocapacitance at low current densities (0.05 A/g), but also increased the equivalent series resistance and resulted in deterring the specific capacitance of supercapacitors. APP900 sample showed the maximum specific capacitance of 165 F/g at 0.1 A/g current density for an operating voltage window of 1.5 V in 1 M $H_2SO_4$ aqueous electrolyte. The charging/discharging was done up to 20,000 cycles and 99% capacitance retention was achieved for APP900 sample.[46]

Jin et al. have used distillers dried grains with soluble (DDGS) generated biochar waste for the supercapacitor applications. DDGS is the by-product of the corn-based ethanol industry; it contained 31.8% crude protein as its main constituent, natural fiber fat, and moisture. They converted it into the activated carbon through KOH activation at two different temperatures (950°C and 1050°C). The samples were half graphitic in nature as confirmed by Raman spectroscopy. The maximum surface area was found to be 2959 $m^2/g$ with a pore diameter variation from mesoporous (2–5 nm) to microporous (<2 nm). The electrochemical characterizations were performed in 1 M tetraethylammonium tetrafluoroborate in acetonitrile organic electrolyte. The maximum specific capacitance was found to be 150 F/g for 950°C sample at a current density of 0.5 A/g. The specific capacitance degraded 12% after 1000 cycles.[47]

Sun et al. have prepared porous graphene-like nanosheets (PGNSs) with a large surface area of 1874 $m^2/g$ via low-cost simultaneous activation–graphitization route from coconut shell, a renewable biomass waste. PGNSs exhibited a high specific capacitance of 268 F/g, which is much higher than that of activated carbon fabricated by only activation (210 F/g) at 1 A/g. These also had excellent cycling properties and retained 99.5% of Coulombic efficiency after 5000 cycles in KOH electrolyte. They also tested the performance of PGNSs in organic electrolytes and found an outstanding capacitance of 196 F/g at 1 A/g having an energy density of 54.7 Wh/kg at a high power density of 10 kW/kg.[48]

Peng et al. have collected five different types of tea leaves from China and prepared activated carbon by carbonizing them at high temperature followed by KOH activation. They found high specific surface area ranging from 2245 to 2841 $m^2/g$ for such activated carbon. As-prepared material was tested electrochemically in a three-electrode system in KOH electrolyte and showed a maximum specific capacitance of 330 F/g at 1 A/g. All materials showed 92% retention in specific capacitance after 2000 cycles.[49] He et al. have prepared high surface area carbon derived from rice husk. They impregnated the rice husk with $ZnCl_2$ and carbonized at 1123 K for 1 h and achieved a specific surface area of more than 1442 $m^2/g$ for a different mass ratio of $ZnCl_2$ to rice husk. Three-electrode system was used to characterize this material electrochemically in KOH electrolyte. As-prepared material retained a specific capacitance of 243 F/g after 1000 charge–discharge cycles at a current density of 0.05 A/g.[50]

Li et al. have prepared supercapacitor electrode material by carbonizing the chicken egg shell membrane, a common livestock biowaste. They achieved a three-dimensional structure by carbonization of egg shell membrane, which was composed of interwoven connected carbon fibers containing 10% by weight oxygen and 8% by weight nitrogen. As-prepared material was tested electrochemically in three-electrode system in acidic ($H_2SO_4$) and basic (KOH) electrolytes. They reported specific capacitances of 297 and 284 F/g in basic and acidic electrolytes, respectively, and observed only 3% decay in specific capacitance after 10,000 cycles at a current density of 4 A/g.[51] Cuna et al. have prepared biocarbons and activated BCs from *Eucalyptus grandis* wood dust. They reported different activation methods. The activating agent solution-to-wood weight ratio was 7:1 for LiOH, 5:1 for $H_3PO_4$, 6:1 for KOH, and 1:1 for $ZnCl_2$. The electrical conductivity of the obtained carbons varies rapidly from $10^{-8}$ to 1 S/cm in the temperature range of 500°C–700°C. The maximum specific capacitance of 203 F/g was reported for chemical activation with $ZnCl_2$ in acidic electrolyte (2 M aqueous $H_2SO_4$) at 1 mV/s scan rate for 0–1 V voltage window.[52]

## 11.3   CONCLUSION

Biomass is produced in such a large quantity from various industrial institutions and household activities that it can be considered as a renewable source. The disposal of the biomass is an important issue these days, and this problem can be resolved by utilizing generated biomass to produce various carbon nanostructures for the energy storage applications. The carbons thus produced have different morphologies, sizes, porous structures, surface functionalities, and crystallinities, and such properties of carbon nanomaterials have a great significance for their practical applications in supercapacitors. The properties change as the source of the carbon changes, and a great amount of research is going on to control the properties of the carbon nanostructures produced from biomass waste. In sum, the potential of carbon nanostructures from various biomass wastes for energy storage applications has not been fully recognized yet. This chapter summarizes few reports on utilizing carbon nanostructures obtained from biomass waste for supercapacitor applications.

## ACKNOWLEDGMENTS

The authors thank their parental institute for providing the necessary facilities to accomplish this work.

## REFERENCES

1. S. Shi, C. Xu, C. Yang, J. Li, H. Du, B. Li, and F. Kang, *Particuology*, 2013, 11, 371–377.
2. M. Winter and R. J. Brodd, *Chemical Reviews*, 2004, 104, 4245–4270.
3. P. J. Hall, M. Mirzaeian, S. I. Fletcher, F. B. Sillars, A. J. R. Rennie, G. O. Shitta-Bey, G. Wilson, A. Cruden, and R. Carter, *Energy & Environmental Science*, 2010, 3, 1238–1251.
4. www.worldenergy.org.
5. T. Chen and L. Dai, *Materials Today*, 2013, 16, 272–280.
6. G. Yu, X. Xie, L. Pan, Z. Bao, and Y. Cui, *Nano Energy*, 2013, 2, 213–234.
7. D. P. Dubal, J. G. Kim, Y. Kim, R. Holze, C. D. Lokhande, and W. B. Kim, *Energy Technology*, 2014, 2, 325–341.
8. P. Simon and Y. Gogotsi, *Nature Materials*, 2008, 7, 845–854.
9. X. Lu, M. Yu, G. Wang, Y. Tong, and Y. Li, *Energy & Environmental Science*, 2014, 7, 2160–2181.
10. A. L. M. Reddy, S. R. Gowda, M. M. Shaijumon, and P. M. Ajayan, *Advanced Materials*, 2012, 24, 5045–5064.
11. J. Jiang, Y. Li, J. Liu, X. Huang, C. Yuan, and X. W. Lou, *Advanced Materials*, 2012, 24, 5166–5180.
12. J. N. Tiwari, R. N. Tiwari, and K. S. Kim, *Progress in Materials Science*, 2012, 57, 724–803.
13. N. A. Choudhury, S. Sampath, and A. K. Shukla, *Energy & Environmental Science*, 2009, 2, 55–67.

14. R. Kötz and M. Carlen, *Electrochimica Acta*, 2000, 45, 2483–2498.
15. Y. Zhang, H. Feng, X. Wu, L. Wang, A. Zhang, T. Xia, H. Dong, X. Li, and L. Zhang, *International Journal of Hydrogen Energy*, 2009, 34, 4889–4899.
16. B. E. Conway, *Journal of the Electrochemical Society*, 1991, 138, 1539–1548.
17. L. L. Zhang and X. S. Zhao, *Chemical Society Reviews*, 2009, 38, 2520–2531.
18. N. S. A. Manaf, M. S. A. Bistamam, and M. A. Azam, *ECS Journal of Solid State Science and Technology*, 2013, 2, M3101–M3119.
19. M. Jayalakshmi and K. Balasubramanian, *International Journal of Electrochemical Science*, 2008, 3, 1196–1217.
20. M. Nakamura, M. Nakanishi, and K. Yamamoto, *Journal of Power Sources*, 1996, 60, 225–231.
21. A. G. Pandolfo and A. F. Hollenkamp, *Journal of Power Sources*, 2006, 157, 11–27.
22. D. Qu and H. Shi, *Journal of Power Sources*, 1998, 74, 99–107.
23. E. Raymundo-Piñero, K. Kierzek, J. Machnikowski, and F. Béguin, *Carbon*, 2006, 44, 2498–2507.
24. H. Liu and G. Zhu, *Journal of Power Sources*, 2007, 171, 1054–1061.
25. T. A. Centeno and F. Stoeckli, *Journal of Power Sources*, 2006, 154, 314–320.
26. H. Gerischer, *Berichte der Bunsengesellschaft für physikalische Chemie*, 1973, 77, 658–658.
27. B. E. Conway, *Electrochemical Supercapacitors: Scientific Fundamentals and Technological Applications*, Springer, New York, 2013.
28. A. J. Bard and L. R. Faulkner, *Electrochemical Methods: Fundamentals and Applications*, Wiley, New York, 1980.
29. M. Matsumoto, *Surfactant Science Series*, 1998, 76, 87–100.
30. J. O. M. Bockris, M. A. V. Devanathan, and K. Muller, *On the Structure of Charged Interfaces*, Proceedings A The Royal Society, London, 1963.
31. C. D. Lokhande, D. P. Dubal, and O.-S. Joo, *Current Applied Physics*, 2011, 11, 255–270.
32. M. Toupin, T. Brousse, and D. Bélanger, *Chemistry of Materials*, 2004, 16, 3184–3190.
33. B. D. Desai, J. B. Fernandes, and V. N. K. Dalal, *Journal of Power Sources*, 1985, 16, 1–43.
34. A. Schneuwly and R. Gallay, Properties and applications of supercapacitors: From the state-of-the-art to future trends, Proc PCIM, 2000.
35. S. Bose, T. Kuila, A. K. Mishra, R. Rajasekar, N. H. Kim, and J. H. Lee, *Journal of Materials Chemistry*, 2012, 22, 767–784.
36. D. Puthusseri, V. Aravindan, B. Anothumakkool, S. Kurungot, S. Madhavi, and S. Ogale, From Waste Paper Basket to Solid State and Li-HEC Ultracapacitor Electrodes: A Value Added Journey for Shredded Office Paper, *Small*, 2014, 10, 4395–4402.
37. K. Karthikeyan, S. Amaresh, S. N. Lee, X. Sun, V. Aravindan, Y.-G. Lee, and Y. S. Lee, *ChemSusChem*, 2014, 7, 1435–1442.
38. P. Hao, Z. Zhao, J. Tian, H. Li, Y. Sang, G. Yu, H. Cai, H. Liu, C. P. Wong, and A. Umar, *Nanoscale*, 2014, 6, 12120–12129.
39. H. Jin, X. Wang, Z. Gu, J. D. Hoefelmeyer, K. Muthukumarappan, and J. Julson, *RSC Advances*, 2014, 4, 14136–14142.
40. C. Peng, J. Lang, S. Xu, and X. Wang, *RSC Advances*, 2014, 4, 54662–54667.
41. J. Xu, Q. Gao, Y. Zhang, Y. Tan, W. Tian, L. Zhu, and L. Jiang, *Scientific Reports*, 2014, 4, 5545–5550.
42. Z. Jin, X. Yan, Y. Yu, and G. Zhao, *Journal of Materials Chemistry A*, 2014, 2, 11706–11715.
43. L. Wang, Y. Zheng, Q. Zhang, L. Zuo, S. Chen, S. Chen, H. Hou, and Y. Song, *RSC Advances*, 2014, 4, 51072–51079.
44. D. Bhattacharjya and J.-S. Yu, *Journal of Power Sources*, 2014, 262, 224–231.
45. Y. Li and X. Liu, *Materials Chemistry and Physics*, 2014, 148, 380–386.

46. C. Huang, A. M. Puziy, T. Sun, O. I. Poddubnaya, F. Suárez-García, J. M. D. Tascón, and D. Hulicova-Jurcakova, *Electrochimica Acta*, 2014, 137, 219–227.
47. H. Jin, X. Wang, Z. Gu, G. Anderson, and K. Muthukumarappan, *Journal of Environmental Chemical Engineering*, 2014, 2, 1404–1409.
48. L. Sun, C. Tian, M. Li, X. Meng, L. Wang, R. Wang, J. Yin, and H. Fu, *Journal of Materials Chemistry A*, 2013, 1, 6462–6470.
49. C. Peng, X.-B. Yan, R.-T. Wang, J.-W. Lang, Y.-J. Ou, and Q.-J. Xue, *Electrochimica Acta*, 2013, 87, 401–408.
50. X. He, P. Ling, M. Yu, X. Wang, X. Zhang, and M. Zheng, *Electrochimica Acta*, 2013, 105, 635–641.
51. Z. Li, L. Zhang, B. S. Amirkhiz, X. Tan, Z. Xu, H. Wang, B. C. Olsen, C. M. B. Holt, and D. Mitlin, *Advanced Energy Materials*, 2012, 2, 431–437.
52. A. Cuña, N. Tancredi, J. Bussi, A. Deiana, M. Sardella, V. Barranco, and J. Rojo, *Waste Biomass Valor*, 2014, 5, 305–313.

# Index

**Note**: Locator followed by '*f*' and '*t*' denotes figure and table in the text

Printed and bound by CPI Group (UK) Ltd, Croydon, CR0 4YY

24/10/2024

01778281-0006